MATERIALS
IN
INDUSTRY

Hot cells and remote control manipulators for handling radioactive materials at the Whiteshell Nuclear Research Establishment, Pinawa, Canada. The chemical apparatus on the left analyzes the gaseous products of nuclear fission.

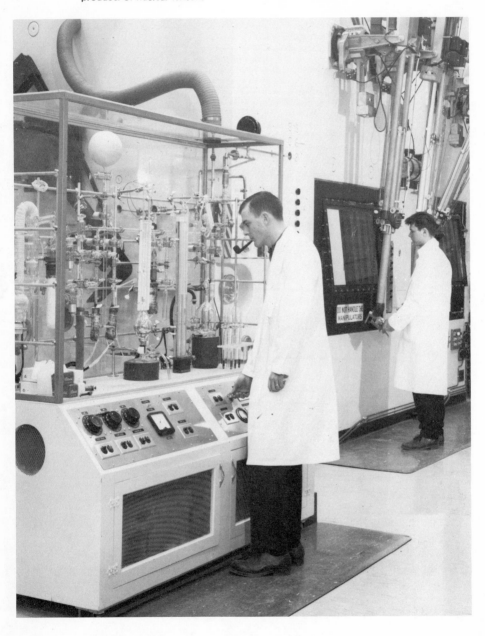

Prentice-Hall, Inc.    *Englewood Cliffs, New Jersey*

**W.J. PATTON**

*Department of Mechanical Technology*
*Red River Community College*

# MATERIALS IN INDUSTRY

**second edition**

*Library of Congress Cataloging in Publication Data*

PATTON, W J
  Materials in industry.

  1. Materials   I. Title.
TA403.P32  1975      620.1'1      75–1235
ISBN  0–13–560722–1

10  9  8  7  6  5  4  3  2  1

Printed in the United States of America

PRENTICE-HALL INTERNATIONAL, INC., London
PRENTICE-HALL OF AUSTRALIA, PTY LTD., Sidney
PRENTICE-HALL OF CANADA, LTD., Toronto
PRENTICE-HALL OF INDIA PRIVATE LIMITED, New Delhi
PRENTICE-HALL OF JAPAN, INC., Tokyo
PRENTICE-HALL OF SOUTHEAST ASIA (PTE.) LTD., Singapore

# CONTENTS

**3**   **PHYSICAL PROPERTIES OF MATERIALS**   **60**

*Part* **2**   **NATURAL MATERIALS**

**4**   **WATER AND INDUSTRIAL GASES**   **85**

This book is a revised edition of *Materials in Industry*. While the author is gratified by the reception received by the original book, in the imaginative technical area of materials it is always possible to do better. Hopefully, this version is better. The range of topics has been narrowed somewhat and the discussion of more critical topics has been expanded.

This revision also has a difference in emphasis. The objective is to show how to use materials imaginatively in many applications. The range of illustrations and of problems has been greatly extended so that the reader can obtain a more powerful feeling for applications. The problems set up for the reader to try call for selecting materials for such diverse commodities as plastic drafting boards, hockey sticks, artificial arteries, and beehives. These problems should be found interesting to attempt. They are all founded in practical experience with materials.

The presentation follows the historical development of materials. First the ceramics, then the metals, and lastly the organics, all in order of ascending complexity. The last type of materials discussed is the biomaterials. This group serves as a final review of earlier sections of the book, as well as to introduce to the reader this most fascinating group of materials. Some conservatives will say that the biomaterials have negligible importance in engi-

*AUTHOR'S*
*REMARKS*

neering: I have attempted to counter that argument, and I think that younger practitioners, with an eye to their own futures, will understand the significance of biomaterials.

I have enjoyed my own professional successes in the engineering area of materials, which I consider almost a form of engineering art, and I confess to having had much pleasure in setting up this book and its suggested problems. But the book will fail if it does not convey to the reader the power and scope of the techniques described and the sheer pleasure of working with materials and selecting them. Let the reader read what is said here and then obtain samples of the many materials so that he may test and try them. I have said that materials are art in engineering: art is only for people who will practice in the media of the art, not for scholars. The information here is presented to be *used*, not just to be learned.

W. J. P.

# MATERIALS
# SCIENCE

*Part* **1**

Sometime within the next thirty or forty million years, the Earth, the planet that is mankind's present habitation, will celebrate its four billionth birthday. The method of determining the age of the earth (see Chapter 2) is an outstanding example of the techniques of Materials Technology. But the method was discovered late in the age of the earth, actually in the twentieth century.

The earth was created with a bountiful supply and range of raw materials to serve the needs of the human race that was to inhabit it so late in its history. Until about a hundred years ago, man used the earth's materials as he found them: either rock or clay dug from the earth's crust, or wood products from the earth's forested regions, or simply the water and air around him. A few more fortunate civilizations used a restricted range of metals in small quantities. But within the last hundred years, men have begun to synthesize and formulate secondary materials based on the primary resources of the earth. These synthetic materials grow rapidly more numerous as our materials technology and our solid state physics advance in knowledge and techniques, and include such indispensable materials as:

beryllium copper for remarkable fatigue resistance
stainless steels for vacuum chambers

# MATERIALS FOR A TECHNICAL CIVILIZATION

1

polyurethane rubber for press-forming of metals and for the tires of fork-lift trucks
synthetic fire-resistant oils for the hydraulic fluid power systems of warships and hydraulic
 presses
silicon carbide for grinding wheels
hydrazine for rocket propellants
plastics for packaging
semiconductor materials for transistors
epoxies for high-strength cements and repairs to concrete floors
zirconium alloys for nuclear fuel rods
leaded steels for fast machining
very low carbon steels for porcelain enamelling

The materials in this list may appear to be special materials for special purposes. This is true, but our civilization has largely abandoned general-purpose materials and replaced them with special-purpose materials formulated exactly for the purpose intended. This statement is true even for supposedly general-purpose materials such as wood, portland cement, concrete, and structural steel—there are, for example, not less than 250 alloys of structural steel for buildings and structural work. Not even water is exempt from exact formulation. The use of raw water is a rarity: water is treated with chemicals, or deoxidized, or deionized, or desalted.

The rapid increase in the number and variety of industrial materials is explained in part by the increasing demands made upon these materials.

Fig. 1-1   A projection microscope for the microscopic examination and photography of materials. Material specimens are placed on the platform and their bottom surfaces examined in the viewer on the right-hand side.

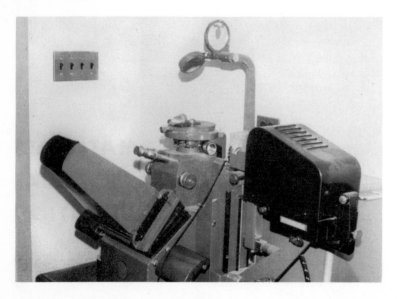

A hundred years ago, metals were required to have only the properties of reasonable tensile strength, reasonable ductility, and sufficient softness for such machining operations as punching and drilling. Nowadays, if a new metal is offered to the market, a wider range of requirements is imposed, including such qualities as weldability, machinability, corrosion resistance, controlled grain size, close tolerance on dimensions, fatigue resistance, hot strength, suitability for low temperature operation, plasticity, vibrational damping capacity, suitable electrical and magnetic characteristics, low neutron absorption for nuclear work, low vapor pressure for aerospace applications, impact and wear resistance, and many more.

The starting point for the synthesis of present-day materials is the supply of materials in the earth's crust, plus water and the earth's atmosphere. The following table gives the percentages of the principal elements in the earth's crust.

## ELEMENTS OF THE EARTH'S CRUST

| | | | |
|---|---|---|---|
| oxygen | 46.5% | rubidium | 0.03% |
| silicon | 27.6% | strontium | 0.03% |
| aluminum | 8.13% | chromium | 0.02% |
| iron | 5.00% | carbon | 0.009% |
| calcium | 3.64% | nickel | 0.008% |
| sodium | 2.83% | copper | 0.007% |
| potassium | 2.59% | tin | 0.004% |
| magnesium | 2.09% | lead | 0.0016% |
| titanium | 0.44% | uranium | 0.0004% |
| hydrogen | 0.14% | silver | 0.00001% |
| manganese | 0.10% | | |

It is interesting to note that the first metals used by mankind were the rare metals—copper, tin, and lead. Iron, the second most abundant metal, on which our steel civilization is based, did not assume its dominant position until about 100 years ago. Aluminum, which is one-twelfth of every shovelful of clay, is really a twentieth-century metal. Magnesium, also a common metal, again is a metal of this century, and titanium, one two-hundredth of the earth's crust, is not yet twenty years old in its industrial applications. By and large, the most abundant metals were the last ones used, and this unusual twist of fate is explained by the difficulties of extracting these metals from their ores and refining them.

Special attention must be given to silicon (27.6 percent) and carbon (0.009 percent). These two nonmetallic elements lie in the same column of the Periodic Table and therefore have similarities in their chemical behavior. Both elements may be used to synthesize very large molecules. Silicon is the basic element in such ceramic materials as quartz, mineral ores, glass, portland cement, silicon carbide, firebrick, building brick, and sand. Carbon is the basic element for a million organic materials, including automobile tires, antifreeze for engine radiators, fuels (including solid rocket fuels), anesthetics, drugs, dyes, soaps and detergents, paints and finishes, solvents, refrigerants, plastics, nylon, textiles, and graphite electrodes for welding and melting of metals.

The next table presents the elements in the Earth's atmosphere, expressed as volume percentages. Water vapor is omitted, since the amount of water vapor varies.

### ELEMENTS IN THE EARTH'S ATMOSPHERE

| nitrogen | 78.03% | neon | 0.00123% |
|---|---|---|---|
| oxygen | 20.99% | helium | 0.0004% |
| argon | 0.94% | krypton | 0.00005% |
| carbon dioxide | 0.03% | xenon | 0.000006% |
| hydrogen | 0.01% | | |

On a weight basis, air analyzes 75.5 percent nitrogen, 23.2 percent oxygen, 1.33 percent argon. All the gases in the table have a variety of interesting and critically important uses in industry, especially in the fields of metal melting, welding, refrigeration, and gas-filled discharge tubes. These uses are discussed in Chapter 4.

Finally, it must be remarked that the salt water of the oceans is a prolific source of minerals and metals, as yet unexploited to any considerable degree. The major solid constituents of salt water are listed in the following table. Minor constituents, such as nickel and gold, are not given.

### MAJOR SOLID CONSTITUENTS OF SEA WATER

| $NaCl$ | 2.71% | $MgSO_4$ | 0.12% |
|---|---|---|---|
| $MgCl_2$ | 0.54% | $CaSO_4$ | 0.08% |
| $KCl$ | 0.04% | $CaCO_3$ | 0.01% |
| $MgBr_2$ | 0.01% | | |

Sodium chloride is occasionally extracted from sea water, though underground brines are a richer source of supply. The most important industrial products extracted from sea water, aside from animal products, are magnesium oxide for firebrick and the metal magnesium. A great many other materials will in the future be drawn from the vast storehouse represented by the oceanic areas of the earth.

Each of the 92 elements of the earth has its special characteristics, and all these characteristics have a use in industry. Tungsten is valued for its unusually high melting point, while bismuth is equally useful for its low melting point. Beryllium offers the advantage of low density, while lead is used for its high density. Hydrogen gas is a useful coolant because of its remarkably high specific heat. Oxygen is a magnetic gas. Nickel is magnetostrictive, that is, its dimensions change if its magnetic field strength is changed. Cesium has one electron in its outer electron shell and is therefore a suitable material for photocell cathodes. Argon does not dissolve in metals, an unusual characteristic for a gas. Whatever the special properties of an element or compound, our technical civilization can find a use for them.

But in the brilliant catalogue of successes produced by the men and women whose vocation is solid state physics and materials technology, there are of course a few failures to record—two important failures in particular. After much recent effort and cost, we find ourselves unable to produce a ductile ceramic material, and this is a major defeat. Perhaps an even more serious frustration is our inability to produce materials capable of extended operation at temperatures above 2000°F. This is one of our greatest technical handicaps. We can make materials with sufficiently high melting points— hafnium carbide, for example, melts at 7500°F—but for one reason or another, such as oxidation or loss of strength, none of these materials can be used at temperatures near their melting points. There is at the moment no prospect of overcoming these two technical deficiencies in the near future.

## 1.1 a convenient classification
## of solid materials

It is impossible to be conversant with the thousands of industrial materials unless one can be guided by some broad and simple generalizations. Such general rules will be set out in this book as they arise. Some of them will have no scientific validity, such as the division of all gases into ideal and imperfect gases in Chapter 4, where an ideal gas is one that obeys a simple formula while an imperfect gas is one that does not. All such general rules, too, will have their exceptions. Nothing more than convenience is argued for such generalizations.

Most of the industrial materials are solids. Here we can set out the first of our broad and somewhat unscientific generalizations: almost all solids are either *metals*, *ceramics*, or *organics*.

A *metal* is an elemental substance (not a chemical compound) which readily conducts both heat and electric current. All the 80 or so metals share these two characteristics, but almost no other characteristics, though most metals are hard, lustrous, and have a high modulus of elasticity (stiffness). Most metals also are ductile. Among the solid materials, metals predominate because of their useful characteristics of hardness, strength, rigidity, formability, machinability, conductivity, and dimensional stability.

*Ceramic materials* are rock or clay mineral materials. Familiar examples of ceramic materials are sand, glass, brick, cement, concrete, minerals, clay, grinding wheels, plaster, fiberglass insulation, spark plug bodies. Most ceramics occur naturally and are simply dug out of the ground with or without further processing. Such natural ceramic materials include the ores from which we extract all our metals. Besides these natural ceramics, an increasing number of synthetic ceramics are in use, such as silicon carbide, tungsten carbide, boron nitride. In general, the ceramic materials are silicates, oxides, carbides, nitrides, or borides. Though an interesting group of materials, the ordinary person views them as devoid of the glamor that surrounds the metals and the plastics. Their dominant characteristics are these:

1. they are rock-like in appearance
2. they are brittle
3. they are hard materials, and most may be used as abrasives
4. they are resistant to high temperatures
5. they do not conduct electric currents
6. they are not easily corroded
7. they are opaque to light

The *organic* materials are in general a development of the twentieth century. These are the countless synthetic or manufactured materials based chemically on carbon; they include the following:

| | |
|---|---|
| plastics | textiles |
| rubbers | lubricants |
| paper | detergents |
| refrigerants | paints and finishes |
| fuels | foods, vitamins, and medicines |
| adhesives | explosives |

The structural organic materials, which include the rubbers, plastics, and wood, have the following characteristics:

1. light weight: specific gravity approximately 1.0
2. poor conductors of heat and electric current
3. combustible
4. soft and ductile
5. not dimensionally stable
6. poor temperature resistance

Archeologists designate the first ages of Man as Stone Ages, that is, ceramic ages. In these early periods ceramic materials were used for tools. The first use of metals, thousands of years before Christ, initiated the Metal Ages, which carry us to the present day. A technical man would probably incline to the view that the major event in materials technology occurred in the decade 1855–1865; in this decade the development of the Bessemer converter and the open-hearth furnace made structural steel cheap and its use widespread. Before this time, construction materials and many consumer articles were largely of wood and ceramics, with a limited use of wrought iron and cast iron, especially for bridges. The Steel Age in which we live is thus not much more than 100 years old. Steel is the dominant material in manufactured products, being consumed on this continent at the rate of about 150,000,000 tons per year.

But a study of trends in materials usage and of price trends of materials, together with an assessment of developments in that most versatile technology, plastics science, suggests some remarkable changes in materials usage in the near future. It appears to be clearly indicated that in or before the year 2000 the tonnage of plastics consumed will exceed that of steels. Indeed, it is already true that the number of cubic inches of plastics annually consumed exceeds that of steels. The Steel Age therefore will represent a comparatively brief historical interval of not more than 150 years so that this outstanding technical age is close to its end. It would seem that plastics and rubbers will assume a dominant position in the materials of construction and manufacture. At the time of writing, plastics have fallen in price and wood prices have increased, with the inevitable result that furniture, fence posts, shelving, house siding, moldings, and other products customarily made of wood are increasingly being made of plastics materials.

No one, however, should read these trends as indicating that plastics will make other materials such as wood wholly obsolete. Every material excels in certain characteristics and therefore has its peculiar applications. The author can think of only one material that has become obsolete because of advances in technology—the metal radium, used as a radioactive isotope until the end of World War II but now wholly replaced by the much superior artificial isotopes iridium-192 and cobalt-60. Wrought iron is no longer produced on this continent but it is still in use in Europe.

## 1.2   selection of materials

The selection of a material for a specific application is invariably a thorough, lengthy, and expensive investigation. Almost always more than one material is suited to the application, and the final selection is a compromise that weighs the relative advantages and disadvantages. The varied requirements demanded of any material may however be reduced to three broad demands:

1. service requirements
2. fabrication requirements
3. economic requirements

The service requirements of course are paramount. The material *must* stand up to service demands. Such demands commonly include dimensional

Fig. 1-2 The terminal operation in materials technology: the disposal of wastes. In order to design this incinerator furnace, the physical properties of the refuse must be determined.

stability, corrosion resistance, adequate strength, hardness, toughness, heat resistance. In addition to any such basic requirements, other properties may be required, such as low electrical resistance, high or low heat conductivity, fatigue resistance, or others.

Fabrication requirements have tended to grow in importance. It must be possible to shape the material and to join it to other materials. Primary shaping methods include casting or sintering: those materials which cannot be cast, such as tungsten, must be sintered to shape from powders. The cast or sintered shapes are then usually formed into semifinished shapes such as bars or sheets by processes such as rolling, extruding, or wire-drawing. Secondary industries, often called fabricators or converters, then modify these semifinished shapes by either of two methods: machining or forming. In machining, metal is removed by cutting it out as chips on such machines as machine tools and grinders. Since this is an inefficient method of producing the final shape, whenever possible, machining is replaced by forming methods such as pressing, forging, swaging, or hot and cold heading. The final operation is joining: this may be done by fasteners, such as bolts, studs, and pins, or by welding, brazing, soldering, or cementing methods. Extra fabrication operations such as heat treatment may also be required. By and large, then, the assessment of fabrication requirements concerns questions of machinability, hardenability, heat-treatability, ductility, castability, and weldability, qualities which are sometimes quite difficult to assess.

Finally, there are the economic requirements. The standard of living of any nation can only be improved by producing goods at lower cost, and whether a country's economy be capitalist, socialist, or communist, cost factors are microscopically scrutinized. The objective is the *minimum overall cost* of the component to be made, and this objective is sometimes attained only by increasing one or more of the cost components. For example, a more expensive free-machining metal may be substituted for a standard metal, since the savings in machining cost may overweigh the increased cost of the more expensive metal. Again, the substitution of a light metal, such as aluminum, for a heavier metal such as steel increases the raw material cost but may result in substantial freight savings to the customer.

In a capitalist economy, a less tangible economic requirement is that the product must sell under competitive market conditions. To improve the sales of the product, expensive stainless steel may be substituted for cheaper painted steel, or vinyl-coated steel may be substituted for plain steel—changes made to improve the market for the product, and which at first thought may appear to be economically unwarranted. This may be a false conclusion, however. The most important variable affecting costs is not a technical one at all, but the marketing variable of volume. Anything that cannot be sold in volume simply cannot be economically produced, and the materials man who feels that marketing problems are not his concern is certainly unwise, knowledgeable though he may be in his field.

## PROBLEMS

1  Other than cost factors, list all the characteristics required of the following products. Consider both service and manufacturing requirements and, where necessary, physical appearance.
   (a) an ideal drafting board. Include the characteristic of self-healing when indented by a compass point. Does the standard basswood drafting board meet all your requirements?
   (b) an ideal hockey stick. Does the standard wood hockey stick meet all your requirements?
   (c) sheet material for writing, drawing, and documentation. Does paper meet all your requirements?
   (d) a plumb bob
   (e) insulation on low-voltage electric wire
   (f) a rain gutter for a house. Do not overlook resistance to ultra-violet deterioration.
   (g) the lining of an oven for a domestic kitchen

(h) a garage door

(i) chain for a chain drive

(j) a bandsaw blade

(k) the surface material for a piano key

(l) the material for an automobile tire

2  Why is wood rarely used for building small boats at the present time?

3  What types of adhesives are presently in use? What were the common adhesives thirty years ago?

4  If there were no plastic materials, what materials might be used to make telephones?

5  If leather were unobtainable, could any of the plastics or rubbers serve for footwear and as well provide a sufficiently attractive appearance? Would you use the same material for both uppers and lowers?

6  What material might be substituted for steel in horseshoes? (This question is answered later in this book.)

7  Can you suggest a material that might substitute for a horse blanket (sheet material under the saddle)? Requirements: must not abrade the horse's skin, must cushion shock from the rider, must absorb perspiration, must be cleanable.

8  What materials are currently in use for house siding?

9  Wood is a cellular material. Compare plastic foams with wood. Are plastic foams superior to wood in general properties or general usefulness?

10  What are the general advantages of concrete as a construction material?

11  Magnesium metal is highly combustible under certain circumstances—it was used for fire bombs in World War II. Find out whether or not this metal can be safely welded with the oxyacetylene torch.

12  Materials have to be joined to the same material or to different material. List the joining methods that you know.

13  Can you find any applications of adhesives in your own automobile?

14  Why are automobile tires black?

15  List the many types of ceramic materials used as construction materials. What ceramic material is used in interior plasterboard?

16  There is much discussion of the possibility of exhausting the earth's resources. What metals are unlikely ever to be exhausted even at high rates of consumption?

17  Can you select a material without regard to its manufacturing capabilities?

18  In your opinion, who would make the better material technologist: a person who had never made a mistake in materials selection or a person who had?

## 2.1 the structure of the atom

The thirty or more fundamental particles of physics that explain matter and energy are distinguished by two types of electric charge: positive and negative. In addition, a number of particles, such as the neutron and the photon, have no electric charge associated with them. The fundamental unit of electric charge is the charge associated with an electron (or a proton). This small unit of charge may be explained in macroscopic terms by noting that the passage of $6.28 \times 10^{18}$ electrons per second through an electric circuit would register exactly 1 ampere on an ammeter.

An atom of any element has the structure of a small nucleus of closely bound heavy particles, collectively called "nucleons," surrounded by a "cloud" of electrons. The nucleons are a mixture of uncharged neutrons and positively charged protons. The total mass of the nucleons is substantially the mass of the atom, since the electrons have little mass. However, the volume occupied by the atom is determined by the volume of the cloud of electrons surrounding the nucleus. The radius of a nucleus is of the order of $10^{-14}$ meters; the radius of an atom is about $1-2 \times 10^{-10}$ m.

# THE SCIENCE
# OF MATERIALS

# 2

In the nuclei of the lighter elements, the number of neutrons is generally equal to the number of protons. But in the heavier elements, lead, uranium, or thorium, for instance, this equal proportion of nuclear particles is not possible if the nucleus is to be stable: the presence of too many protons in the nucleus would produce repulsive forces due to like charges, such that the binding energy holding the nucleus together would be overcome. Therefore in the larger nuclei the number of neutral neutrons exceeds the number of protons, perhaps by as much as 50 percent.

The structure or arrangement of electrons in any atom is a highly complex subject which cannot be properly explained except in terms of the mathematical concepts of quantum mechanics. Since the purpose of this book is the broad subject of the selection and service requirements of industrial materials, we can apply certain simplified concepts to this matter of electronic structure. Such concepts will not be scientifically exact, but we can take comfort in the thought that not even research physicists have worked out entirely satisfactory theories.

The energy quantities applicable to atomic energy levels are of course extremely minute when compared to the standard industrial energy units of Btu, horsepower, or watt. A more suitable energy unit for atomic purposes is the *electron volt*, a unit with the additional virtue of being much easier to understand and use in computation than the larger units. The electron volt, abbreviated "ev," is the energy acquired by an electron when accelerated across a 1-v drop. Thus if 300 v is applied between the cathode and the anode of a vacuum tube, an electron leaving the cathode will arrive at the positive anode with an energy of 300 ev. Similarly, an electron traversing an 18-v welding arc will acquire 18 ev of energy in doing so. Halfway across the arc its energy will be 9 ev. Note that this energy in the two examples cited is kinetic energy. A voltage is a force and therefore will accelerate the electron throughout the voltage drop. The electron volt is related to the watt through the following conversion factors:

$$1 \text{ electron volt} = 1.602 \times 10^{-19} \text{ joules}$$

$$1 \text{ watt} \qquad = 1 \text{ joule per second}$$

The electron volt is of course not restricted to kinetic energy. The electrostatic attraction between an electron and an atomic nucleus is measured as potential energy in electron volts.

## 2.2  electron energy levels

The electrons of an atom are in orbit about the nucleus. For the case of the earth's orbiting path about the sun, it is known that equilibrium arises because the earth traverses a path such that the gravitational attraction of the sun for the earth is balanced by the opposing centrifugal force of the earth's motion. Suppose now that this concept is applied to the case of a negatively charged electron orbiting about its nucleus. The attraction of the positively charged protons in the nucleus for the electron would be balanced by the centrifugal force arising from the circular motion of the electron. This is not the end of the matter, however. It can be shown from simple mechanics and Newton's law of motion that the electron, like the earth, is accelerating simply because its motion is circular instead of straight-line, the centrifugal force implying an acceleration (force = mass × acceleration). But an electron is electrically charged, and an accelerating electric charge must radiate energy. For example, the acceleration of electric charges in a transmitting antenna causes the radiation of radio waves to the atmosphere. The orbiting electron would thus continuously radiate energy as it circled the nucleus. This continuous loss of energy would result in the collapse of the electron into the nucleus in a remarkably short period of time.

But except for nuclear radioactivity, atoms are normally stable. The reasoning given above is wrong because it assumed that the energy of the atom was continuously variable. It is not. The energy state of any atom has only a certain set of permitted values, which will be different for different elements. This quantized set of energy values is infinite in number. In order to make a somewhat difficult subject simple, we will deal with this matter of atomic energy levels by discussing what is called "the Bohr model of the hydrogen atom," which is a somewhat simplified description of the simplest of the atoms.

The hydrogen atom has a single electron, and according to the simpler concepts of quantum mechanics, this electron will be found only in certain permitted orbits in relation to the nucleus, and in no other positions. In these permitted positions the electron does not radiate or otherwise lose energy. Figure 2.1 sketches these orbits. Rather than refer to these positions as orbits, it is preferable to describe them as energy levels, since there will be a definite value of potential energy associated with each orbit. There are an infinite number of these distinct energy levels. The infinitieth level is that level farthest from the atom. In this level, mutual electrostatic attraction of the negative electron for the positive nucleus is zero, and this level is therefore assigned an energy of zero electron volts. In this condition the nucleus has of course lost its electron, and it is in the ionized condition. In any energy level

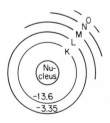

Fig. 2-1 Energy levels of the hydrogen atom (Bohr model of the hydrogen atom).

closer to the nucleus than the ionized condition, the electron energy is assigned a minus value.

In the energy level closest to the nucleus, the K level, where the electron would normally be found, the hydrogen electron has an energy of $-13.6$ ev. If then the electron is bound to the nucleus by an energy of 13.6 ev in the K level, then a minimum energy of 13.6 ev will be required to ionize the atom by pulling the electron off the nucleus. For a practical case, suppose single atoms of hydrogen (i.e., not hydrogen molecules, $H_2$) were used as shielding gas in a welding arc. Then this gas would ionize if the arc voltage reached 13.6. It should be noted that these statements do not apply to the diatomic molecule of hydrogen, but only to single atoms. In the case of the molecule there are two nuclei to attract the electron, an entirely different configuration.

The K level of $-13.6$ ev has a nominal diameter of 1.06 A. An Angstrom, abbreviated "A," is $10^{-8}$ centimeters, or $10^{-10}$ meters. The electron moves around this orbit at about 1 percent of the speed of light.

The electron may be moved out of the K level to a more remote level by providing energy to the atom from heat, a voltage, a collision with another particle, or by other means. The energy of the electron in any energy level of the hydrogen atom is given by the simple relationship

$$E = -\frac{13.6}{n^2}$$

where $E$ = energy in electron volts

$n$ = number of the energy level: K = 1, L = 2, M = 3, etc.

Thus in the K level, $E = -13.6/1^2 = -13.6$ ev

in the L level, $E = -13.6/2^2 = -3.35$ ev

in the M level, $E = -13.6/3^2 = -1.5$ ev

The energy differences between the outer electron levels become very small. The minimum energy that will be required to move the electron out of the K shell is (13.6 − 3.35) or 10.25 ev, this being the difference in energy between the K and L levels. An energy less than this amount cannot change the energy level. If the electron should be displaced to a more remote energy level, it would be expected to drop back to the K level at a later time. To do so, it must release energy. Such energy is carried away as a photon (particle or wave) of electromagnetic radiation. The characteristics of such radiation will be dealt with presently.

To be more exact in our concepts, we should note that it is not possible to locate an electron precisely in space and time. Any experiment which attempts to fix the position of the electron must necessarily disturb the electron and ruin the precision of the experiment. Physicists therefore speak of a *probability distribution* for the electron in its energy level. The probability of finding the electron at the nucleus is of course zero. The maximum probability for the location of the electron is at a radius of 0.53 Angstroms, and there is less probability of finding the electron closer or farther from the nucleus. This position of maximum probability is made the position of the K level.

The hydrogen atom is of course the simplest atom, and it presents the simplest possible case. However, the basic concepts set out here hold for any atom:

1. The total energy of the atom electron system is quantized in a restricted range of permitted values.
2. The precise positions of the electrons at any instant cannot be determined, and only the probability of finding an electron at any position can be specified. The position of highest probability sets the size (and the shape, which is not discussed here) of the atom.

## 2.3  more complex atoms
## than hydrogen

The fundamental particles in the atom are specified by two numbers. One number is the atomic number $Z$. This is the number of protons in the nucleus and also the number of electrons if the atom is not ionized. The other number is the mass number $A$, which is the number of nucleons (protons plus neutrons). The mass and density of the atom are determined largely by $A$: most of the chemical properties of the atom are determined by $Z$. *The element is identified exclusively by the number of protons.* Thus any atom containing 53 protons is iodine and only iodine. Any atom containing 8 protons is oxygen and only oxygen. The number of electrons may vary because of ionization; the number of neutrons in the nucleus may also vary. But if the number of protons should vary, then the element is no longer the original element. To cite an example, let us take the case of the most frequently used radioactive element in industry and medicine: cobalt. Natural cobalt has a nucleus of 27 protons and a total of 59 nucleons (32 neutrons). This atom is made radioactive by adding 1 more neutron to the nucleus, for a total of 60 nucleons. The radioactive atom is still a cobalt atom, since the number of protons is unchanged. Radioactive cobalt serves its functions, such as whole-body radiation of a patient in a hospital or photography of a thick weld, by means of certain nuclear transformations which will be discussed later in this chapter. As a result of these transformations, the end condition of the atom is 28 protons and 32 neutrons. The change in the proton count to 28 indicates that the atom is now a nickel atom.

Atoms of any element with varying numbers of neutrons are termed *isotopes*. Oxygen, for example, has three principal isotopes, $_8O^{16}$, $_8O^{17}$, $_8O^{18}$. Hydrogen has 3 isotopes also, $_1H^1$, $_1H^2$ (called deuterium and sometimes given as $_1D^2$), and $_1H^3$ (called tritium, used in dials of watches which can be read in the dark). The number of protons $Z$ is the subscript, and mass number or atomic weight $A$ is the superscript. Frequently in industry only the atomic weight is given, thus Iridium-192, or Cobalt-60.

An atom of atomic number $Z$ has a nuclear charge of $+Ze$, where $e$ is the fundamental electric charge, and has $Z$ electrons. The electron energy levels are quantized, but determination of these energy levels is complicated by the repulsion of the orbiting electrons from each other.

The Periodic Table of the Elements is an arrangement of the 102 or more known elements in the order of their atomic numbers $Z$, beginning with hydrogen, $Z = 1$, so that elements with similar properties fall in vertical groups in the table. Figure 2.2 is one arrangement of the Periodic Table, with several subgroups between Groups II and III.

The famous inert gases, helium, neon, argon, etc., form a group with certain industrially important characteristics:

1. All are gases at ambient temperatures.
2. All are chemically inert.
3. They do not dissolve in metals.

Although they share a number of common properties, the inert gases have their individual characteristics, too, such as differences in their ionization potential (helium supplies a hotter welding arc than argon because of its higher ionization potential), and differences in their radiation characteristics (an electric discharge through a neon-filled tube produces a reddish light, while xenon gives a white light). All these elements, except helium, contain 8 electrons in their outermost filled electron energy level. Helium has only 2 electrons, both in the K level.

The maximum number of electrons that can be contained in any energy level of an atom follows this simple principle:

$$\text{The K level can hold a maximum of } 2 \times 1^2 = 2 \text{ electrons}$$

| | | | |
|---|---|---|---|
| L | // | $2 \times 2^2 =$ | 8 electrons |
| M | // | $2 \times 3^2 =$ | 18 electrons |
| N | // | $2 \times 4^2 =$ | 32 electrons, etc. |

The inert gases, having a complete population of 8 electrons in the outermost filled level, do not normally engage in chemical activity, such activity being occasioned by incompletely filled shells of fewer than 8 electrons. Again, actual electron configurations are more complex than this simplified presentation would suggest: every level above the K level is subdivided into two or more sublevels.

Fig. 2-2 Periodic table of the elements.

**Periodic Table of the Elements**

Metalloids and Non-metals — METALS — Transition Metals

| I | II | III | IV | V | VI | VII | VIII |
|---|---|---|---|---|---|---|---|
| 1 **H** Hydrogen | | | | | | | 2 **He** Helium |
| 3 **Li** Lithium | 4 **Be** Beryllium | 5 **B** Boron | 6 **C** Carbon | 7 **N** Nitrogen | 8 **O** Oxygen | 9 **F** Fluorine | 10 **Ne** Neon |
| 11 **Na** Sodium | 12 **Mg** Magnesium | 13 **Al** Aluminum | 14 **Si** Silicon | 15 **P** Phosphorus | 16 **S** Sulfur | 17 **Cl** Chlorine | 18 **Ar** Argon |
| 19 **K** Potassium | 20 **Ca** Calcium | ... | 32 **Ge** Germanium | 33 **As** Arsenic | 34 **Se** Selenium | 35 **Br** Bromine | 36 **Kr** Krypton |
| 37 **Rb** Rubidium | 38 **Sr** Strontium | ... | 50 **Sn** Tin | 51 **Sb** Antimony | 52 **Te** Tellurium | 53 **I** Iodine | 54 **Xe** Xenon |
| 55 **Cs** Cesium | 56 **Ba** Barium | ... | 82 **Pb** Lead | 83 **Bi** Bismuth | 84 **Po** Polonium | 85 **At** Astatine | 86 **Rn** Radon |
| 87 **Fr** Francium | 88 **Ra** Radium | 89 **Ac** Actinium | | | | | |

**Transition Metals**

| 21 **Sc** Scandium | 22 **Ti** Titanium | 23 **V** Vanadium | 24 **Cr** Chromium | 25 **Mn** Manganese | 26 **Fe** Iron | 27 **Co** Cobalt | 28 **Ni** Nickel | 29 **Cu** Copper | 30 **Zn** Zinc |
|---|---|---|---|---|---|---|---|---|---|
| 39 **Y** Yttrium | 40 **Zr** Zirconium | 41 **Nb** Niobium | 42 **Mo** Molybdenum | 43 **Tc** Technetium | 44 **Ru** Ruthenium | 45 **Rh** Rhodium | 46 **Pd** Palladium | 47 **Ag** Silver | 48 **Cd** Cadmium |
| 57 **La** Lanthanum | 72 **Hf** Hafnium | 73 **Ta** Tantalum | 74 **W** Tungsten | 75 **Re** Rhenium | 76 **Os** Osmium | 77 **Ir** Iridium | 78 **Pt** Platinum | 79 **Au** Gold | 80 **Hg** Mercury |

Also: 31 **Ga** Gallium, 49 **In** Indium, 81 **Tl** Thallium (Group III)

**Lanthanides (Rare Earth Metals)**

| 58 **Ce** Cerium | 59 **Pr** Praseodymium | 60 **Nd** Neodymium | 61 **Pm** Promethium | 62 **Sm** Samarium | 63 **Eu** Europium | 64 **Gd** Gadolinium | 65 **Tb** Terbium | 66 **Dy** Dysprosium | 67 **Ho** Holmium | 68 **Er** Erbium | 69 **Tm** Thulium | 70 **Yb** Ytterbium | 71 **Lu** Lutetium |
|---|---|---|---|---|---|---|---|---|---|---|---|---|---|

**Actinides**

| 90 **Th** Thorium | 91 **Pa** Protoactinium | 92 **U** Uranium | 93 **Np** Neptunium | 94 **Pu** Plutonium | 95 **Am** Americium | 96 **Cm** Curium | 97 **Bk** Berkelium | 98 **Cf** Californium | 99 **Es** Einsteinium | 100 **Fm** Fermium | 101 **Md** Mendelevium | 102 **No** Nobelium | 103 **Lw** Lawrencium |
|---|---|---|---|---|---|---|---|---|---|---|---|---|---|

Hydrogen has 1 electron. Helium has 2 electrons. The second electron of helium completely fills the K level, giving the quality of inertness to helium. The next element in the Periodic Table, the metal lithium, has 3 electrons, the third being in the L level. Lithium is in Group I of the table. This group comprises the alkali metals, all of which have 1 electron in the outermost populated level. This single electron is the distinguishing characteristic of the group, just as was 8 electrons for the inert gases. The single outermost electron of the alkali metal group is not closely bound to the nucleus because of the masking effect of electrons in levels closer to the nucleus. As a result the ionization potential for these elements is only 4–5 electron volts. Such low levels of energy are obtainable from ultraviolet radiation and account for the use of these metals in the cathodes of photoelectric cells.

Group II elements are also metals, with 2 electrons in the outermost populated level. Beryllium and magnesium crystallize in a hexagonal crystal pattern of atoms: so do the metals of the subgroup titanium, zirconium, and hafnium, a subgroup also characterized by two outermost electrons.

The subgroup vanadium, columbium (also called niobium by scientists and nuclear industry personnel), tantalum and the subgroup chromium, molybdenum, tungsten are important metals in metallurgical practice. These are the "refractory" or high melting-point metals, with the following common characteristics:

1. Their crystal structure is body-centered cubic.
2. All are used as alloying elements in steels.
3. All readily form carbides, especially within the internal structure of steels (the preceding subgroup, titanium, zirconium, and hafnium, have an even stronger tendency to form carbides).

The subgroup copper, silver, and gold is again an industrially significant group, their common characteristics being these:

1. Their conductivity for both heat and electric current is outstanding.
2. They are exceedingly ductile and can readily be beaten out to very fine foils.
3. Their crystal structure is face-centered cubic.

Group III metals have 3 electrons in the outer populated level. Note that the group number in the Periodic Table is the number of outer electrons. There are no metals to the right of Group III in the table, so that a metal might be defined as an element with no more than 3 electrons in the outer populated level.

Group IV contains the semiconductor materials silicon and germanium, used in transistors and semiconductor rectifiers. Silicon is also the base element for the cements used in the construction industry. The group also includes the important element carbon. The presence of 4 outer electrons gives silicon and carbon their remarkable chemical versatility.

We need not address any general remarks to Groups V and VI. These include elements with a variety of special uses in materials technology, such as the formulation of low melting-point alloys. Group VII is the "halogen" group. Since the halogens have 7 electrons in the outer level, they combine readily with the alkali metals of Group I to produce a closed configuration of 8 electrons.

## 2.4 aspects of relativity theory

The theory of relativity, like quantum mechanics and solid state physics, is one of the powerful scientific developments of the twentieth century which has greatly extended the applications of materials technology. The theory of relativity is concerned with the speed of light. Here we are concerned only with that aspect of the theory relating energy and mass.

Mass can be converted into energy and energy can be converted into mass. To convert from mass to energy, the famous equation of Einstein is used:

$$E = mc^2$$

where $E$ = the energy produced by the disappearance of mass

$m$ = mass

$c$ = speed of light

Because $c^2$ is an enormous figure, the equation indicates that a small amount of mass will produce a large quantity of energy.

The Einstein equation may also be written

$$E = mc^2 = \frac{m_0 c^2}{\sqrt{1 - (v/c)^2}} = m_0 c^2 + \tfrac{1}{2} mv^2$$

where $v$ = velocity of the body

$m_0$ = the rest mass of the body, i.e., its mass when at rest.

This form of the equation shows that the total energy of a moving body is the sum of its mass energy and its kinetic energy. Suitable units for numerical work with the Einstein equation are:

$$E = \text{joules}$$
$$m = \text{kilograms}$$
$$c = \text{meters per second}$$

As an example, we will calculate the amount of energy in joules produced by the conversion of 1 atomic mass unit to energy. This is almost equivalent to converting 1 neutron (1.00898 atomic mass units) or 1 proton to equivalent energy.

$$1 \text{ amu} = 1.660 \times 10^{-27} \text{ kilograms}$$
$$E = 1.660 \times 10^{-27} \times (3 \times 10^8)^2 = 1.494 \times 10^{-10} \text{ joules}$$

In terms of electron volts, this is 931 Mev (million electron volts) or almost 1 billion ev. The conversion from joules to electron volts is made by way of the conversion factor 1 ev $= 1.602 \times 10^{-19}$ joules.

The reverse process of converting energy to mass occurs in certain cosmic ray processes, in nuclear operations, and in linear accelerator operations. The linear accelerator is a modification of the X-ray tube which is used to photograph heavy thicknesses of steel. Typically in such processes, a negative and a positive electron are created from a gamma ray. The process is called *pair production*. To understand pair production, consider that the mass of an electron is the equivalent of 0.51 Mev. Now a photon with an energy of 0.51 Mev cannot create an electron, even though it has the required amount of energy, for the photon has no charge, and a negative charge is required for an electron. But a photon with an energy of at least 2 × 0.51 Mev, or 1.02 Mev, can be converted into mass in the following process, which is pair production. Two electrons will be produced by annihilation of the photon, one being a negative electron and the other a positive electron or positron. The sum of 1 negative charge and 1 positive charge equals the charge on the photon, which is zero. However, the positron very quickly disappears by combining with a negative electron, the two particles being annihilated to produce two photons of 0.51 million electronvolts. The probability of pair production is approximately proportional to $(E - 1.02)$, where $E$ is given in Mev. Thus pair production becomes more probable at energy levels of 5 Mev or higher.

The mass of a body, measured at zero velocity, is termed its *rest mass*. If the body is in motion, then its mass becomes greater:

$$m = \frac{m_0}{\sqrt{1 - (v/c)^2}}$$

The faster the body, the greater its mass, but the increase is negligible until the velocity approaches about one-sixth the speed of light, or about 30,000 miles per second. Such velocities are found in industry, principally in high-voltage equipment. Thus an electron accelerated across a 100,000-volt X-ray tube or electron-beam welding machine will acquire a terminal velocity of about one-half the speed of light. The above equation suggests that if a body can be accelerated to the speed of light it would have infinite mass, but it can be shown that no mass can be accelerated to this velocity. The only particles that can, and do, travel at the speed of light are those that have no rest mass, such as the photon, now to be discussed.

Any wave energy which can be radiated through space as a voltage wave accompanied by a magnetic field is termed electromagnetic radiation. The radiation of radio and television signals from a transmitting antenna, and received by a receiving antenna, is perhaps the most familiar example to cite.

Figure 2.3 is a composite diagram of information on all the types of electromagnetic radiation. Actually, all the various types—radio, infrared, visible light, ultraviolet, and gamma ray—are identical in nature and differ

Fig. 2-3   Electromagnetic radiation.

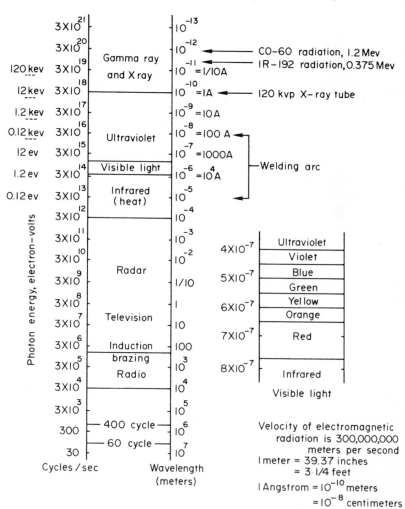

only in frequency. There is no clear-cut distinction between any two types. But since certain frequency ranges of radiation are transmitted or received by special devices, such ranges are given special names. Thus if the radiation can be detected by the human eye, it is called *visible light*. If it comes from an X-ray tube, it is called *X-radiation*. If it has a heating effect, it is called *infrared* or *radiant heat*. It is interesting to note that the prominent natural types of radiation in the world around us are infrared to ultraviolet, especially from the sun, and that the human eye is sensitive to a narrow band of radiation in the middle of this broad range of "natural" radiation.

All electromagnetic radiation travels at the velocity of light through vacuum space. Such radiation includes both an alternating electric field and an alternating magnetic field normal to the electric field, as in Fig. 2.4.

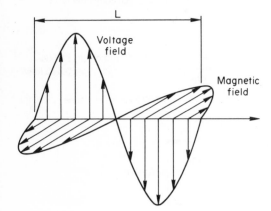

Fig. 2-4 Relationship of electric and magnetic fields in radiation.

One positive and one negative half-wave of the alternating wave are together termed one wavelength, as for 60-cycle alternating current. The frequency is the number of wavelengths that pass any fixed point in one second. As with most types of wave motion, the product of the frequency times the wavelength is equal to the velocity $c$:

$$fL = c$$

Thus to determine the wavelength of 60-cycle alternating current,

$$60 \times L = 186,000 \text{ mps}$$

$$L = 3100 \text{ miles}$$

Each a-c wave of radiation may also be considered as an energetic particle of zero rest mass, called a *photon*. Indeed, for the very high-energy radiation in the gamma ray range, it is not usual to use the concept of a wave, this being less useful than the notion of a particle. The photon, being the wave, contains all the energy of the wave. It necessarily travels at the speed of light; indeed it exists only when traveling at this velocity. On colliding with another body, the photon transfers all its energy to the body. Being then without either mass or energy, it ceases to exist. The energy which is the photon varies directly with the frequency:

$$E = hf$$

where $h$ is some constant.

In industrial uses of high-frequency radiation beyond the radio range, a variant of this formula is used, based on the wavelength of the energy in Angstroms ($1 \text{ A} = 10^{-8}$ cm). If the photon energy is proportional to the frequency, then it is proportional also to the reciprocal of the wavelength, since $fL = c$. The formula used in industry for determining photon energy is the following:

$$\text{photon energy in ev} = \frac{12430}{\text{wavelength in Angstroms}}$$

The constant 12430 is often made 12345 to make it easier to remember. The error is about 1 percent, which for industrial purposes is unimportant.

**Example.** Determine the wavelength of X-radiation if the X-ray tube uses an impressed voltage of 300,000.

$$300,000 \text{ ev} = 12345/L_A \quad \text{and} \quad L_A = 0.04 \text{ A}$$

Figure 2.3 shows the photon energy ranges of the high-frequency types of radiation.

## 2.6 biological and photographic effects of electromagnetic radiation

A practical understanding of photon energy, its implications, and its uses is best gained in a brief review of its biological and photographic effects.

Radio, television, and radar waves are of very low energy as measured in electron volts because of their low frequency. The human body is ordinarily insensitive to them. Infrared radiation is more powerful, with energies ranging up to a maximum of about 1.3 ev. This radiation we can sense as radiant heat. Still farther up the scale of radiation is visible light, with maximum energy of 3 ev in the violet range. Such radiation is sufficiently powerful to make the retina of the eye respond.

Damaging radiation begins at ultraviolet energy levels, 3 ev or more. The ultraviolet components of the sun's spectrum can produce sunburn. The more powerful ultraviolet radiation from a welding arc can produce a variety of reactions in the human skin and eye. Both the retina, the light-receiving portion of the eye, and the cornea, the lens of the eye, are especially sensitive to the high energy of ultraviolet radiation, and if exposed, the result can be retinitis, conjunctivitis (the conjunctiva is the inner lining of the eyelid), or keratitis (thickening of the cornea). Serious reactions of the skin to radiation of the welding arc are not common, but cases of dermatitis and even carcinoma have been found or suspected.

Ultraviolet radiation is also powerful enough to depolymerize the long molecules of such organic materials as polyethylenes and rubbers. Carbon black is added to polyethylene pipe and automobile tires to absorb such radiation and thus protect the material.

Radiation energies above approximately 10 Kev (10,000 electronvolts) are termed X-ray and gamma-ray radiation. Gamma and X-radiation are actually identical except for their source: X-radiation is produced by vacuum tubes and gamma radiation from radioisotopes; otherwise there is no difference whatever between them. This radiation is powerful enough to penetrate solid matter, and its penetrating power increases with its photon energy. Gamma-ray or X-ray radiation above 1 Mev can penetrate 6 in. of steel without excessive attenuation: even at 100 Kev it is possible to radiograph steel 0.375 in. thick. Physiological damage to the human body results from the ionization produced by high-energy photons. This damage is more severe in the heavier and therefore more absorbing types of tissue, such as bone.

All types of electromagnetic radiation which are above the general radio range are used in photography. That is, it requires a minimum energy of about 1 electron volt, which is infrared, to develop photographic emulsions, although special film is required for infrared photography. Visible light, 1.5 to 3 ev, is used for ordinary camera film, which therefore will also be sensitive to the shorter wavelengths of ultraviolet and X-radiation. Because X-radiation actually penetrates the film, X-ray film carries an emulsion on both sides.

## 2.7 electron-photon interaction

The hydrogen electron can be displaced to an outer energy level from the K level by acquiring energy. It can jump back to a lower energy level also (if there is an available hole or electron position for it to fill), but it must lose energy to do so. This energy is removed by a photon. A large jump will release a photon in the ultraviolet range, while a small jump will release a photon of radio or infrared energy.

Suppose this hydrogen atom to be part of an acetylene molecule which is being oxidized in an oxyacetylene flame. The heat of the flame causes the electron to jump from the K shell to an outer shell, let us say the M shell of energy 1.5 ev. As the acetylene molecule travels out of the flame and cools, it loses energy, causing the electron to fall back to the L or K shell from the M shell. Presumably the hydrogen atom is attached to an $H_2O$ molecule after combustion, but we are ignoring any effect on energy levels due to the molecular configuration. Let us suppose the electron falls back to the L shell. To do so, it must lose 1.85 electron volts, the energy difference between M and L. This lost energy of 1.85 ev becomes a single photon of radiation. The wavelength of the radiation is 6700 Angstroms. This is red visible light. If the electron had jumped back to the K shell, the photon would have an energy of 12.1 ev, which is ultraviolet.

Figure 2.3 indicates that about 10 Kev is the minimum photon energy for X-radiation. Since the maximum electron jump in the hydrogen atom is 13.6, it is not possible to produce X rays by means of electron jumps in hydrogen atoms.

Figure 2.5 shows the approximate K, L, and M energy levels of sodium and molybdenum. Sodium has 2 electrons in the K level, 8 in the L level, and 1 in the M level. The maximum electron jump in this atom is 1041 ev, from the K shell to the infinitieth level of ionization. Such a jump would produce a photon of ultraviolet. Molybdenum, which has 42 electrons, can produce a maximum photon energy of 20,000 ev, which is in the longwave X-ray range.

A photoelectric cell is a simple application of these quantum concepts. This device is a vacuum tube diode; a diode is a two-electrode device, one electrode being an electron emitter and the other an electron collector and

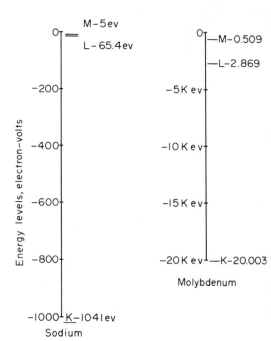

Fig. 2-5 Energy levels in sodium and molybdenum atoms.

therefore having a voltage positive with respect to the electron emitter. The photocell can create a small electric current of a few microamps if exposed to radiation with a certain minimum photon energy, which is in the range of visible light to ultraviolet. The current arises from a flow of electrons from a radiation-sensitive cathode, not a heated cathode in the case of a photocell, to the collecting anode. The cathode material is some metal which is easily ionized by the loss of one electron. Suitable metals are Group I metals such as cesium and rubidium, with a single electron in the outermost populated level. Such electrons can be separated from their atoms by photon energies of a few electron volts. The released electrons are attracted to the anode by a positive voltage, thus creating a small current.

## 2.8   the laser and fluorescence

Electromagnetic radiation is energy. Every small boy has tried using a magnifying glass to concentrate the heat of the sun in order to char a piece of paper. Using the concepts of quantum mechanics, the laser is a more sophisticated substitute for the small boy's magnifying glass.

Before discussing the laser, we ought first to discuss the case of a heating element in an electric stove. When current is switched to the heating element, it radiates a certain amount of infrared. As the temperature of the element rises, greater electron jumps become possible, and the element radiates a spectrum of infrared and visible light. Photons of red light are produced when an electron jumps back toward the nucleus through an energy difference of about 1 electron volt. Those electrons which return toward the nucleus through smaller energy differences emit photons of infrared radiation. The radiation occurs over a very broad band of wavelengths, and it is not very intense.

The laser radiates in the same frequency range as the electric element, but with greater intensity and with closer control over the characteristics of the radiation. A great many solids, liquids, and gases will "lase"; here we will consider only the solid-state laser.

Suppose that in the laser crystal there are two characteristic energy levels which the electrons typically inhabit: a higher energy level when the electrons receive photon energy, and a lower energy level when they release photon energy. Three processes are possible:

1. When radiation interacts with electrons in the lower energy level, a photon is absorbed and the electron is raised to the upper level.
2. An electron can spontaneously emit a photon by dropping back to the lower level.
3. Radiation can interact with an electron in the upper level, causing it to emit a photon of radiation. In this case the photon emitted will travel

in the same direction and be in phase (reinforce) with the stimulating photon. The radiation will then be twice as powerful, that is, it will be amplified. This third case is the laser process.

The laser solves the problem of producing huge amounts of power from radiation by two characteristics: the electron jumps all produce the same photon energy, and all the electrons make the jump at the same instant. Despite the publicity it has received, the high-power, solid-state laser is hardly an impressive piece of equipment. High-power lasers consist of a ruby (aluminum oxide) crystal which has been doped with a few billion chromium atoms. Another type of laser crystal is made of certain types of glass doped with the rare earth metal, neodymium. The back end of the crystal is heavily silvered to reflect any internal radiation: the front end face is partially silvered. The intense beam of radiation is emitted from this front end. The crystal end faces must be ground smooth and flat to about a millionth of an inch. A xenon-filled flash tube is used to excite the electrons of the laser crystal. Figure 2.6 is a diagram of such a laser.

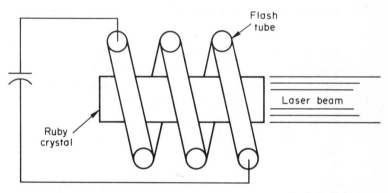

Fig. 2-6   Ruby laser.

The laser is fired by discharging the capacitor through the xenon tube, which then puts out a powerful pulse of white light. The chromium atoms in the ruby crystal absorb the yellow and green fraction of this visible radiation. This absorption pumps vast numbers of electrons from a lower energy level 1 to a higher energy level 3. The more powerful the flash, the greater the number of electrons excited to the higher level. From level 3 the electrons rapidly fall back to an intermediate level 2. This is a small energy drop, which appears as heat in the crystal. A few electrons fall back to energy level 1, producing photons of red light. This light is reflected back and forth between the ends of the crystal, stimulating the electrons in level 2 to fall in unison back to level 1, thus producing a pulse of red light of remarkable power.

Laser action is an example of *fluorescence*, which means the emission of radiation of longer wavelength during exposure to radiation of shorter wavelength. When the laser crystal is irradiated with shorter-wave green and yellow radiation, it delivers longer-wave red radiation. Virtually all materials can be made to fluoresce, not necessarily in the visible range, though they may require powerful radiation such as X rays to do so. If the emission of radiation from the substance persists after exposure to the primary radiation is discontinued, this is called *phosphorescence.*

*Phosphors* are materials which emit visible light when irradiated at high photon energies, such as ultraviolet. In parts of our country phosphorescent wood can be found which glows green in the dark. Such wood has been caused to phosphoresce by ultraviolet radiation from the sun, the phosphor material being certain salts in the wood. Fluorescent light bulbs have a phosphor painted on the glass bulb which fluoresces in the visible light range when irradiated by the ultraviolet discharge through the mercury vapor of the lamp. Similarly, cathode-ray tubes, such as television picture tubes, are coated with a phosphor surface.

## 2.9   energy bands in solids

The preceding sections make it obvious that there are many more allowable energy levels in an atom than there are electrons to fill them. So far, however, the discussion has been concerned with individual atoms. The more usual case of an aggregation of atoms such as might be found in a crystal must next be considered.

In the case of an isolated atom, the energy levels are determined by the forces within the atom, such as the attraction of the nucleus for the electron, and the repulsion between electrons. If we next concern ourselves with a crystal of iron, zirconium, or other element, any electron will be under the influence of the many nuclei and electrons that surround it on all sides. The influence of these neighboring atoms will change the energy levels within the atom. Taking the case of a system of two atoms only, in close proximity to each other—the usual interatomic distance in metal crystals is about 2–5 A—the energy levels shift slightly, and each energy level of the single atom is split into two levels separated by an extremely small energy interval of very much less than 1 electron volt.

In any solid, enormous numbers of atoms are packed into a small volume, so that interaction between the atoms causes the splitting of energy levels. Each energy level now becomes a multitude of very closely spaced energy levels. These closely packed ranges of energy levels, which in the single atom were one energy level, are now called *energy* bands, because the difference between energy levels within any permitted band may be of the order of only $10^{-19}$ ev, which is a negligible energy difference. The energy band appears to be continuous over its range, although the actual number of energy levels within the energy band is actually finite, and not infinite. We no longer have permitted and forbidden energy levels, but permitted and forbidden energy bands. Inner energy bands such as the K band corresponding to the K shell of a single atom are narrow-range energy bands with only a narrow interval of energy across the band. Higher energy levels, such as the M band, are broadened into wide energy bands.

The single hydrogen atom could radiate energy only in a relatively few permitted monochromatic frequencies, corresponding to the energy differences between levels K to L, K to M, M to N, etc. Because of the multitude of energy levels within an energy band, solids can absorb and emit radiation over a continuous range of frequencies, and thus give a continuous spectrum of radiation, as in the case of a hot plate of an electric stove. All solids show essentially the same spectrum when heated, and there is no characteristic spectrum by which the solid may be identified. All solids, metal, ceramic, or other, begin to produce visible radiation when heated above 1100°F, the beginning of the red-hot range of temperatures.

In all solids the very lowest energy bands are completely filled by electrons. The highest normally filled energy band is called the *valence band* or *filled band*. The next higher band is called the *conduction band* or the *empty band*. At any temperature above absolute zero there are some electrons excited from the valence (filled) band to the lowest normally empty (conduction) band, and therefore some holes in the valence band. For silicon at room temperature the energy gap between the empty conduction band and the highest normally filled band is 1.09 ev, and for germanium it is 0.72 ev.

Fig. 2-7   Energy bands in solids.

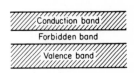

No electric current can be carried by electrons in a filled band. If a solid has a partially filled band, then it is a good conductor of electricity. That is, a partially filled band (conduction band) conducts. Now it is a familiar fact that all good electrical conductors (chiefly metals and carbides) are opaque to heat and visible light. Such radiation has low photon energy. When such radiation falls on a conductor, the photon energy is absorbed as excitation energy to lift electrons in the conduction band to other energy levels within the same band. The energy interval between levels in the conduction band is small, corresponding to the small photon energy. Hence a good conductor should be opaque to light and heat.

Most nonmetals are insulators. In the case of an insulator the highest occupied energy band is completely filled. Although it might seem that such a case would be rare, it is actually the most common. In the case of an insulator, not only is the highest occupied band filled but also the unoccupied forbidden energy band just above the filled band is wide, that is, several electron volts, and thermal excitation does not provide enough energy to jump electrons from the filled band to the next higher permitted band. The electrons then are more or less restricted to the occupied band, and the material is an insulator for both heat and electricity.

In addition to electrical conductors and insulators, there is a third type of material, the semiconductors, with a conductivity between that of conductors and insulators. This class includes silicon and germanium. A semiconductor has a narrow forbidden band between the filled and the conduction band. Heat energy at room temperature is sufficient to raise some electrons into the higher conduction band where they can carry electric current. The number of electrons raised to the conduction band will increase with temperature.

Many good insulators are transparent to heat and light, glass and certain plastics, for example. Since the low-energy heat and light photons cannot excite the electrons in a filled band to the conduction band, the light must pass through unabsorbed in such materials. It has been found that as the wavelength of the radiation shortens to the ultraviolet region, these solids begin to absorb the radiation, as is true for glass. This would be expected for those insulators in which the energy gap of the forbidden region corresponds to the energies of ultraviolet photons. Semiconductors, having narrow forbidden zones of about 1 ev, are opaque to visible light but transparent in the infrared.

To illustrate, we may compare the behavior of three elements, each containing 4 electrons in the outermost electron occupied level. These are diamond (carbon), germanium, and lead. Diamond is an insulator, germanium is a semiconductor, and lead is a conductor. In the case of diamond, the filled and the conduction band are widely separated by a large forbidden band, so that a considerable amount of energy is required to excite an electron into the conduction band. Hence the insulating characteristic of diamond. The two energy bands in germanium are close enough, 0.72 ev, for a considerable number of electrons to make the jump into the conduction band. Finally, in the case of lead, the two bands actually overlap one another, and the valence electrons are free to move in the conduction band as free electrons in a conductor.

## 2.10   the nucleus

The energy released by electronic rearrangements within the atom is of the order of a relatively few electron volts, although in atoms of higher atomic number, several thousand electron volts are possible. The electron is an active and volatile particle because it can be removed from its atom by the expenditure of a relatively few electron volts of work. By means of such small energy expenditures we can produce electric currents, operate electronic tubes and transistors, and produce such chemical reactions as the combustion of carbon to carbon dioxide. Previous sections of this chapter have centered interest on the electron, leaving by implication a picture of the nucleus as a tight, dense, and inert cluster of protons and neutrons serving only as a datum for electron energy levels. It is now well known that such a picture is false, for even a stable and quiescent nucleus is a sleeping giant and a Promethean source of energy. The great difference between nuclear and electronic sources of energy is indicated in the following table of the particle energies employed by the human race:

## PARTICLE ENERGIES IN ELECTRON VOLTS

| | | |
|---|---:|---|
| 1 water molecule falling over Niagara Falls | 0.00015 | ev |
| 1 visible photon of red light | 1 | ev |
| 1 carbon atom burned to $CO_2$ | 4 | ev |
| fission of 1 uranium nucleus | 200,000,000 | ev |

(1 ev = $1.602 \times 10^{-19}$ joules; 1 watt = 1 joule per second)

The great disparity between nuclear and electronic energy releases is explained by their relative binding energies. For example, the ionization potential binding the hydrogen electron to the nucleus is 13.6 ev, but the binding energy per nucleon in the nucleus of deuterium is close to 1 Mev. For most nuclei, particularly heavier elements, the binding energy per nucleon holding the nucleus together is in the range of 6 to 9 million electron volts, or about one hundred times the maximum energy levels that hold the electron to the nucleus.

For the release of nuclear energy, the nucleus must be made unstable, that is, the binding energy holding the nucleus together must be disrupted. Figure 2.8 is a stability diagram of the known isotopes of the light elements. Proton number (i.e., atomic number) $Z$ is the abscissa and neutron number is the ordinate. Stable isotopes are designated by circles: unstable or radioactive isotopes are designated by solid circles. It is apparent from this figure that only isotopes with a ratio (neutrons)/(protons) = 1, approximately, are stable, for the case of the light nuclei. A line of stability corresponding to the stable neutron-proton ratio is drawn through the figure. With increasing atomic number the line of stability leans increasingly toward a higher count of neutrons, that is, $N/Z > 1$. The greater fraction of neutrons in the heavier nuclei is explained by the greater electrostatic repulsion forces as the number of protons increase: the protons must be increasingly "diluted" with neutrons in order to hold the nucleus together. The heaviest

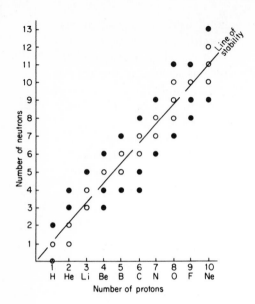

Fig. 2-8 Stability diagram for isotopes of the light elements. Those isotopes are stable which have a neutron-to-proton ratio of approximately one.

stable nucleus is $_{83}Bi^{209}$ with $N/Z = 126/83$, or about 1.5. All nuclei heavier than this are radioactive.

The farther that the $N/Z$ ratio of a nucleus is from the line of stability, the more unstable the nucleus will be. Some isotopes, lying above the line of stability, have a high count of neutrons. Others, below the line, have a high count of protons. The radioactive decay process, for both conditions, will be such as to bring the $N/Z$ ratio toward the line of stability. Isotopes lying above the line of stability will emit negative beta particles, actually electrons, thus changing neutrons into protons. The neutron thus involved will decay into a proton, an electron, and a neutrino (uncharged). Note that no electrons or neutrinos reside in the nucleus, but are created during radioactive decay in a fashion somewhat analogous to the creation of a photon. On the other hand, those isotopes lying below the line of stability have an excess of protons. This condition is adjusted either by release of a positron (positively charged electron) or by acquiring a neutralizing electron from the surrounding electron cloud about the nucleus. This second method of reducing the proton count in effect converts a proton to a neutron. The emission of a gamma photon accompanies all such types of radioactivity. In all these radioactive processes, neutrons are changed into protons or protons into neutrons, but since the proton count is altered, the original element, after radioactivity, becomes a different element.

A number of radioactive isotopes, or radioisotopes as they are usually termed, are found in nature, such as Carbon-14, tritium or Hydrogen-3, uranium, radium, thorium, polonium, radon, and others. A few of these were used in industry before the invention of the nuclear reactor, but their use now is in general not common. Instead, industry uses artificial radio-isotopes manufactured in nuclear reactors. The most important by far of the industrial and medical radioisotopes is Cobalt-60; the next most important is Iridium-192. The powerful gamma radiation of Cobalt-60, with photon energies exceeding 1 Mev, is used in the radiography of steel thicker than about $2\frac{1}{2}$ in. Iridium-192, with gamma radiation in the range of about 0.5 Mev, is used to radiograph thinner sections of steel and the light metals such as aluminum and titanium.

The production of artificial radioisotopes is simple in principle and will be explained in terms of Co-60. For the production of Co-60, natural cobalt, which is Co-59, is encased in an aluminum capsule. The capsule is pushed into one of the irradiation channels in a nuclear reactor where it is bombarded with neutrons. The capsule must remain in the reactor for a considerable period of time. With the passage of time, more and more Cobalt-59 nuclei capture an extra neutron, thus converting to Co-60. The quantity of Co-59 cannot be left in the reactor until every atom of cobalt captures an extra neutron, for two reasons: First, to convert every atom of cobalt would require an infinite time. Second, while waiting for the last million nuclei to acquire an extra neutron, the first million to be converted would be already radioactive and would be decaying to a nonradioactive stability, so that the net result would be that nothing would be gained. Note that all artificial radioisotopes are made unstable by an increase in neutron count, thus placing the nucleus above the line of stability.

For the gamma-ray photography of welds and castings, the isotope is enclosed in a container made of lead or spent uranium, called a *gamma camera*. Such a gamma camera is shown in Fig. 2.9. To photograph a weld,

Fig. 2-9 Gamma camera. By means of the remote control crank shown, the radioactive source can be cranked out of the shielded container into the source tube at the bottom of the illustration.

a sheet of X-ray film is placed against one side of the weld and the weld is irradiated from the other side by pushing the isotope out of its shielding container into the source tube attached to the camera.

## 2.11  half-life

The radioisotopes decay by emission of alpha and beta particles and gamma rays to an ultimately stable condition, in the case of the artificial isotopes, by beta and gamma emission. It is not possible to predict when any specific nucleus will disintegrate. But for any large number of nuclei, the *half-life* is the time period in which half the radioactive nuclei will decay.

The mathematical equation for radioactive decay is one of the most important in engineering science and characterizes a great many processes, such as the loss of voltage or charge on a capacitor.

Let $N_0$ = the number of unstable nuclei at time $t = 0$

$N$ = the number of unstable nuclei at time $t$. $N$ is equal to or less than $N_0$.

Let  $k$ = the decay constant and

$e = 2.718$ = base of the natural logarithms.

It has been determined that the rate at which nuclei decay is proportional to the number of remaining undecayed nuclei. Mathematically this is expressed as

$$\frac{dN}{dt} = -kN$$

By separation of variables $N$ and $t$ and integration, the mathematical law of decay is obtained:

$$N = N_0 e^{-kt}$$

In a time interval $1/k$ the number of unstable nuclei always decreases by a factor of $1/e$, where $1/e = 0.368$. In electrical work, the time constant of an induction coil or capacitor circuit is the time interval required for the voltage to fall to 36.8 percent of the initial voltage. Similarly, the time constant for an isotope would be the time interval required to reduce the number of unstable nuclei to 36.8 percent of their original number. However, the time constant concept is not employed in radioactivity: the half-life is substituted. This is the time interval in which the number of unstable nuclei is reduced by one-half. The half-life $= 0.693/k$.

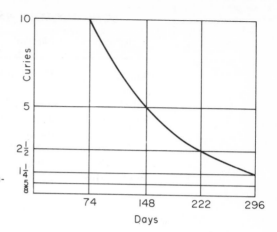

Fig. 2-10  Decay curve for Iridium-192.

The half-life of isotopes may vary from microseconds to years. For Co-60 the half-life is 5.3 years, and for Ir-192 it is 74 days. Thus if one were to purchase 100 active atoms of Ir-192, in 74 days only 50 would still be active, in 148 days only 25 would be active, and in 222 days only 12 would be active (see Fig. 2.10).

The only useful characteristic of an isotope is its nuclear activity. This radioactivity is measured in *curies*, abbreviated "c." The curie is therefore the unit in which radioactive sources are purchased, instead of pounds, ounces, or grams. A curie is that quantity of isotope in which $3.7 \times 10^{10}$ atoms are decaying every second. The reason for the use of this unusual unit of activity is explained by the fact that prior to the invention of the nuclear reactor radium was the radioactive material used by industry and medicine, and in 1 gram of radium there are $3.7 \times 10^{10}$ atomic disintegrations every second. Since radium has a half-life of 1520 years, its activity is almost constant. But for a short-lived isotope such as Ir-192, the 1 curie initially purchased will have decayed to 0.5 curie in 74 days.

The number of curies used for industrial processes depends on the use to which the isotope is put. For the radiography of welds, castings, and forgings, from 10 to 100 curies would be employed, but for such purposes as leak testing, a few microcuries or millicuries might be sufficient.

## 2.12  radioactive isotopes

The availability and use of radioisotopes since 1950 has brought remarkable benefits to science, industry, medicine, and agriculture. The utility of radioisotopes is constantly being expanded, and these materials must be considered standard factors in industrial and scientific activity.

## SOME INDUSTRIAL RADIOISOTOPES

| Isotope | Radiation, Mev | Half Life |
|---|---|---|
| Radium (no longer used) | 12 rays, 0.24–2.20 | 1620 years |
| Cobalt-60 | 1.33 and 1.17 | 5.3 years |
| Iridium-192 | 0.61, 0.58, 0.468, 0.316, 0.308 | 74 days |
| Thulium-170 | 0.084 | 124 days |
| Xenon-113 | 0.081 | 5.3 days |
| Ytterbium-169 | 0.052 | 32 days |

The use of radioisotopes began with the natural one, radium. This had limited use for the gamma ray inspection of welds and castings and as a somewhat dubious treatment for terminal cancer in the medical field. It was also used, with great carelessness, for luminous watch dials. The invention of the nuclear reactor, with its unlimited supply of neutrons, opened up the possibility of producing artificial isotopes uniquely suited to specific requirements.

There is no interest in a general-purpose radioisotope. For any particular purpose, two properties are of especial importance to the user of isotopes: (1) the energy or wavelength of the radiation, and (2) the half-life. For flaw detection in a thick steel casting, high-energy radiation such as that provided by Co-60 is desirable; for similar work on the light metals, such radiation is too powerful to be sufficiently absorbed by the metal, and an isotope with less potent radiation is used, such as Ir-192. In the matter of half-life, there is again some divergence in requirements. An isotope used for gauging the thickness of a paper or metal sheet as it comes off the rolls must have a long half-life, so that its emission strength will be substantially constant with time, but for the leak testing of a pipeline a short half-life is better because safety precautions need not be enforced for an inconveniently long period.

Thulium-170 and Ytterbium-169, whose constants are given in the table above, are used for radiographing thin sections of light materials such as are processed in the aerospace industries. Note that Ytterbium-169 has characteristic radiation approximately the same in wavelength as a 50 kv X-ray tube, which is the smallest standard voltage available in industrial X-ray tubes. Xenon-113 is used for leak testing. The method of leak testing is to inject a few millicuries of Xe-113 into the pressure vessel to be tested and to search the exterior of the vessel with a radiation-monitoring instrument such as a Geiger-Müller counter.

Hydrogen has three isotopes: H-1, H-2 or deuterium, and H-3 or tritium. Deuterium is used as a neutron moderator, but it is not radioactive.

Tritium is radioactive. Its most familiar use is in luminous watch dials, where it is used to activate a phosphor. It is not an especially dangerous isotope, since most of the emission is beta radiation which is rapidly absorbed by air or by the back metal panel of the watch casing. Radium too is occasionally used for such watches, however, and is more dangerous than tritium because it emits penetrating gamma radiation of sufficiently high energy to penetrate the metal of the watch. Although luminous watches must not be considered especially dangerous, if several brands of such watches are monitored with a radiation detector, a surprising variation in radiation will be noted between brands. The European Nuclear Energy Agency has proposed the following maximum quantities of radioisotopes for watches. They are an indication of the relative hazards of these materials.

| | |
|---|---|
| radium | 0.15 microcuries |
| tritium | 7.5 millicuries |
| prometheum | 150 microcuries |

The Radiation Protection Division of the Canadian Federal Department of National Health and Welfare found in a recent survey that perhaps 5 percent of watches have amounts of radium exceeding these values. Watches of Swiss manufacture will in the future conform to the limits given here.

A great many isotope installations are used for measurement and control in industrial processes, such as the gauging and automatic control of sheet thicknesses. For this purpose a small isotope source is positioned on one side of the sheet, with a radiation detector such as a cadmium sulfide cell on the opposite side. The thickness of the sheet can be monitored by measuring the relative absorption of gamma rays by the sheet. Alternately the detector may be mounted on the same side of the sheet as the source, in which position it will measure the backscattered secondary radiation, which is almost linear with sheet thickness.

## 2.13 radiation hazard

Much ignorance and misunderstanding surround the matter of radiation hazard. Most of the more melodramatic aspects of radiation hazard, such as malformed births and sterility, are sheer fiction. Indeed, an intelligent person, after a comprehensive survey of the facts of radiation hazard, might well conclude that there is little danger from radiation short of a lethal dose. This would not be quite true, however. Facing a future of increased radiation for the human race, the medical profession realizes the consequences of wrong decisions and is wisely reluctant to make definitive statements in the matter of radiation hazard. Radiation hazard may well be much less than we once suspected, but we are not yet certain of this, and a quarter of a century is an insufficient experimental period. Certainly radiation can be dangerous if the exposure dose is high enough, and there is always the possibility of accident caused by inadvertence, unawareness, and carelessness and of a history of radiation disease caused by such errors in the past.

The curie, the unit of radioactivity, was defined in Sec 2.11. The number of curies is of course halved in every half-life. The curie, however, does not measure the amount of radiation received. This is measured by the *roentgen*, abbreviated r. To understand the roentgen, consider Fig. 2.11, which is a sketch of a simple apparatus that may be used to demonstrate the effects of radiation. A dry cell is connected through a current meter to a sheet metal cylinder. A concentric wire within the cylinder is connected to the positive terminal of the dry cell. Since the circuit is definitely an open circuit, no current will flow.

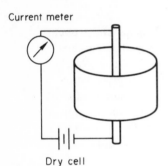

Current meter

Fig. 2-11   A simple radiation meter.

Dry cell

Suppose however that the air between cylinder and wire is exposed to gamma radiation. A weak current will be indicated on the meter.

This apparatus illustrates perhaps the most basic characteristic of radiation: *radiation ionizes atoms*, whether the atoms are in air or the human body or elsewhere. Radiation passing through the insulating air of this little apparatus frees electrons from their energy levels within the atom. These ionized electrons and the positive air ions are attracted to the wire and the cylinder respectively, and thus a current is produced. This apparatus can be considered a crude radiation meter, for increased radiation dosage of the air produces a larger current through the meter. Ionization of the air can be produced by any type of radiation of sufficient energy to remove electrons: alpha, beta, gamma, neutrons, protons, or cosmic ray products.

The amount of radiation received is measured in roentgens. If 2080 million ion pairs, that is, electron plus positive ion, are produced in a cubic centimeter of air measured at 0°C and 14.696 psi as a result of radiation, then the radiation dose is 1 roentgen. Ionization requires the expenditure of energy, and 1 roentgen represents 87 ergs in air. This is a minute amount of energy: 1 joule $= 10^7$ ergs, and 1 joule per second $= 1$ watt.

From the point of view of health hazard we are more concerned with damage caused by a radiation dose of 1 roentgen to the human body than to air. The human body is composed of somewhat denser material than air and hence has a higher absorption for radiation than does air. Exposure of the body to 1 roentgen results in an absorbed dose of 95 ergs per gram of soft tissue. More damage therefore is done to the human body than to air.

Individuals vary in their resistance to specific diseases and environmental conditions, and this is true for radiation. A lethal dose of radiation would be several hundred roentgens if the whole body were exposed to this dose for a short period such as 1 hour. An exposure to 1 roentgen for 1 hour presumably would produce no medically discernible effect. Some typical exposures are these:

| | |
|---|---|
| X-ray chest plate | 0.02–0.5 r |
| luminous watch dial | perhaps 40 mr/year |
| atmospheric fallout | perhaps 3 mr/year |
| uranium in the ground | perhaps 40 mr/year |
| cosmic rays | perhaps 20 mr/year |

where 1 mr = 1 milliroentgen. Regulations allow radiation workers an exposure of approximately 1 roentgen total per quarter year from X-ray and gamma-ray apparatus in addition to the background radiation received by the whole population.

## 2.14 absorption, transmission, and reflection

A highly important part of the information needed for the selection of materials concerns the effect of the surface or body of the material upon incident radiant and acoustic energy. Consider the following examples:

1. Photographic darkrooms must be painted with nonreflective paints.
2. Ceiling tile must absorb acoustic energy without reflecting an excessive fraction of that energy.
3. Beryllium "windows" are used on X-ray tubes because this light metal has a very low absorption for X-radiation.
4. Lead is used as a shield against X- and gamma radiation because of its high absorptivity for such high-energy radiation.
5. Polished aluminum surfaces are used in automobile sealed-beam headlights.

6. Photographic filters are used to absorb certain colors (frequencies) in photography.
7. The dark glass used in arc welding helmets transmits less than one ten-thousandth of the incident light of the arc.
8. Punched tape is commonly used in computer work and in numerically controlled machine tools. If the holes in the tape are to be read by photocells, then the paper tape must have a sufficiently high absorptivity for light, so that light can reach the photocell only through the holes in the tape.

Suppose that an infrared or a visible light wave falls upon a sheet of window glass. The glass will produce three effects on the incident radiation:

1. Most of the radiated energy will be passed or *transmitted* by the glass. We say that glass is transparent to heat and light. But if higher-energy radiation, such as ultraviolet or X-radiation, is incident on the glass, such energy is powerful enough to raise electrons to a higher energy level; such radiation is absorbed by the glass.
2. A lesser fraction of the incident radiation is *reflected* by the surface of the glass.
3. A fraction of the radiation will be *absorbed* by the glass. The absorbed fraction of the radiant energy will raise the temperature of the glass window.

Now the following simple relationship must hold (Fig. 2.12):

The energy incident on the glass
= transmitted energy + reflected energy + absorbed energy.

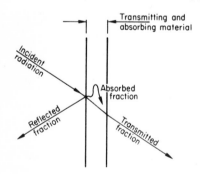

Fig. 2-12 Transmission, absorption, and reflection of incident radiation.

This of course is simply a special statement of the most fundamental of scientific theories, the law of conservation of energy. In other words, the sum of the fractions transmitted, reflected, and absorbed must equal 1. The decimal fraction transmitted is termed the *transmittance*, the decimal fraction reflected is termed the *reflectance*, and the fraction absorbed is termed the *absorptivity*.

$$T + R + A = 1$$

Although the sum of these three fractions is always equal to 1 for any wavelength, the fractions themselves are dependent on wavelength. Thus the transmittance of beryllium metal for visible light is zero, but for X-radiation it is close to unity. Note also that while reflectivity is a surface characteristic, transmittance and absorptivity will depend on the thickness of the material. Thus a single sheet of white typewriter paper transmits 40 percent of visible light incident on it, but a pad of 100 sheets transmits no measurable amount of light.

Suppose a surface to have a high absorptivity for radiation at a certain wavelength. The absorbed energy of radiation will raise the temperature of the absorbing body and therefore cause it to radiate heat. Rough-surfaced bodies absorb and emit heat well, but smooth-surfaced bodies are less efficient in these characteristics, since they are reflective.

A rather interesting aspect of absorptivity concerns the color of metals. Most metals have distinctive colors, which experienced metals men use for identification. Stainless steels have a grayish cast, titanium is faintly blue, etc. Let us consider the familiar case of copper and silver. Figure 2.13 shows that the absorptivity of silver is quite low over the whole range of visible light. Silver therefore efficiently reflects all visible wavelengths to produce a white, silvery color. Aluminum has the same low absorption in the same frequency range, and thus produces a similar color. But copper absorbs the shorter wavelengths above red: the reflection of red frequencies accounts for its color.

Fig. 2-13 Absorptivity of copper, silver, and aluminum. The curves are typical only, since actual values are markedly affected by the condition of the metal surface, and would be less for polished surfaces.

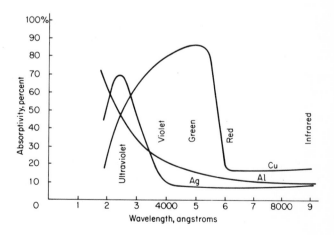

The following table presents some typical absorptivities. The values are for broad-spectrum radiation, not for monochromatic radiation.

TYPICAL ABSORPTIVITIES (at room temperature)

| | |
|---|---|
| aluminum sheet | 0.03–0.055 |
| aluminum paint | 0.50 |
| copper, polished | 0.02 |
| firebrick | 0.70 |
| concrete | 0.9 |
| paper | 0.93 |
| steel, polished | 0.15–0.35 |
| steel, oxidized | 0.7 |
| water | 0.95 |

## 2.15 absorption coefficients for

## transmitted radiation

The word "transparent" should not be understood merely in terms of visible light. Thus all substances are transparent to X-radiation (gamma radiation). Absorption of radiation by transparent materials depends on the thickness of the material. The effect of thickness may be introduced by means of a simple example. It was noted earlier that a sheet of typewriter paper transmits 0.4 of any visible light incident on one side of the paper. If then two sheets of this paper are used as a transmitting medium, the second sheet will transmit 0.4 of the light passed by the first sheet, or $0.4 \times 0.4$, which is 0.16 of the light incident on the first sheet. See Fig. 2.14. Three sheets of paper will transmit $0.4 \times 0.4 \times 0.4$ or 0.064 of the light incident on the first sheet. And so on. Mathematically, the transmitted radiation follows the same equation as was used for radioactive decay:

$$I = I_0 e^{-\mu t}$$

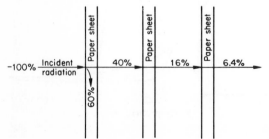

Fig. 2-14 Transmission of visible radiation by paper sheets. The thickness of the sheet is equivalent to a 40-percent value layer, if non-reflective conditions.

where $I$ = intensity of radiation after transmission

$I_0$ = intensity before transmission through the medium

$t$ = thickness of the medium, in inches or centimeters

$\mu$ = linear absorption coefficient, or fractional decrease in intensity per unit thickness of material.

The equation assumes monochromatic (single wavelength) radiation.

In the case of X-radiation, the concept of the "half-value layer" is used, just as the half-life was used for radioactive decay. A *half-value layer* of any material is the thickness of that material required to reduce the intensity of the radiation to one-half. The half-value layer (HVL) = $0.693/\mu$. The more powerful the energy of the radiation in electron volts, the larger the half-value layer will be: Co-60 (about 1.2 Mev) requires a HVL in steel of 0.87 inches, while Ir-192 (about 0.5 Mev) has a HVL of 0.44 inches of steel.

A "filter" must be understood to be a material with a controlled absorption characteristic. The colored glass filters used with visible light cameras reduce the range of wavelengths reaching the photographic film; such filters in effect make the radiation somewhat more monochromatic. Filters are used for the same purpose in X-ray photography, though the materials used for X-ray filters are metals such as copper, aluminum, and lead.

## 2.16 crystal structure

Most solid materials are aggregations of crystals, such as the metallic solid of Fig. 2.15, the individual crystals usually being referred to as *grains*. Each such crystal or grain is actually a single molecule with its atoms arranged in a regular repeating pattern called a space lattice. Although a total of fourteen space lattices is possible, only five will be described here. These

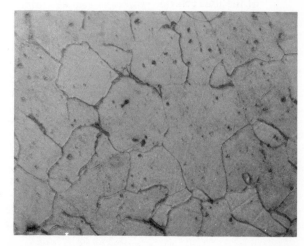

Fig. 2-15 Micrograph of commercial iron.

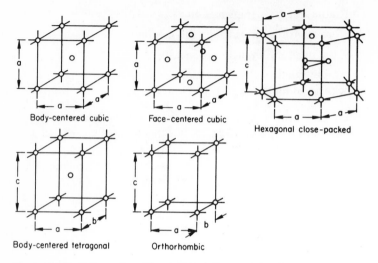

Body-centered cubic    Face-centered cubic

Hexagonal close-packed

Body-centered tetragonal    Orthorhombic

Fig. 2-16   Crystal structures. The small circles representing the positions of atoms are not drawn to the full scale of the atom. Interatomic distances are in the range of 2-5 Angstroms for metals.

space lattices form three-dimensional coordinate systems, the unit distances of the coordinate system being the locations of the atoms in the space lattice. Figure 2.16 shows unit cells of the space lattices to which most of the metals conform. The unit distances between atoms are designated $a$, $b$, and $c$ in three-dimensional space.

*Body-centered cubic* (bcc)—a cubic cell, with all three sides of length $a$, and all angles right angles. One atom in the center of the cell.
*Face-centered cubic* (fcc)—a cubic cell, with all three sides of length $a$, and all angles right angles. One atom in the center of each of the six faces of the unit cell.
*Hexagonal close-packed* (hcp)—three coplanar axes at 120° apart, all at right angles to the fourth axis. $a \neq c$.
*Body-centered tetragonal* (bct)—similar to bcc, except that $a = b \neq c$.
*Orthorhombic*—all angles right angles, but $a \neq b \neq c$.

Neglecting a few exceptions, most metals at room temperature adopt the bcc, fcc, or hcp crystal structure. The distance between atom centers in metal crystals is between 2 and 5 Angstroms. The following is a list of typical interatomic distances at room temperature, in Angstroms:

| | |
|---|---|
| aluminum | 4.0490 A |
| chromium | 2.8845 |
| copper | 3.6153 |
| iron | 2.8664 |
| nickel | 3.5238 |

## CRYSTAL STRUCTURES AT ROOM TEMPERATURE

| bcc | fcc | hcp |
|-----|-----|-----|
| iron | some stainless steels | titanium |
| most steels | aluminum | zinc |
| chromium | copper | zirconium |
| columbium | gold | hafnium |
| molybdenum | lead | rhenium |
| tantalum | nickel | magnesium |
| tungsten | platinum | beryllium |
| vanadium | silver | cadmium |

The bcc group of metals includes only metals with high melting points. The fcc group is distinguished by the great ductility of its metals; all commercially available metal foils are made from metals of this group. The fcc group also does not become brittle at low temperatures, which explains the use of these metals for refrigeration and cryogenic equipment. (Cryogenics is the technology of low temperatures below −250°F.)

## 2.17   phases

A *phase* is any recognizable state or condition of a substance. Most materials have at least three phases: solid, liquid, and gas. No material has fewer than two phases. Many materials, especially the radioactive metals plutonium and uranium, may have as many as six solid phases. The possession of more than one solid phase implies more than one crystal structure.

Carbon has five phases: gas, liquid, amorphous carbon, graphite, diamond. Iron has three solid phases, termed alpha (bcc), gamma (fcc), and delta (bcc) (there is no beta phase in iron). The reason there are two bcc phases in iron is simply that the two are separated by the fcc phase—alpha exists at room temperature, gamma at red heat, and delta close to the melting point. Each phase of any material exists within a definite temperature range. Thus the ice phase of water cannot exist above 32°F.

Now if heat energy is supplied to a material, one of two possible changes will occur:

1. The temperature of the material will be raised.
2. The temperature will remain fixed, but the phase will change.

But if the temperature is changing, the phase will not change, and conversely. This is familiar from the fact that the phase change from water to steam is isothermal, like all other phase changes. Heat may change the temperature or the phase, but not both at the same time. If the temperature is changing, the heat supplied is termed "sensible heat," since the heat effect can be sensed by means of a thermometer. If the phase is changing, the heat is termed "latent heat" ("latent" meaning hidden, since a thermometer is of no help).

Suppose a crystalline material with a single solid phase, such as beryllium, is to be heated. The heat energy introduced into the material will have several effects. One effect will be to expand the interatomic distance between atoms, resulting in an increase in dimensions of the body (the delta phase of plutonium is an exception; it contracts when heated to a higher temperature). Another effect may be changes in electron energy levels, which will be ignored here. A third effect is to set up increased vibrational motion of the atoms about their regular position in the space lattice. It is these vibrational motions of the atoms that give a metal its electrical resistance: if a metal is heated, the amplitude of vibration increases, making more difficult the passage of the electrons of the electric current and thus increasing the resistance.

If the heat energy supplied to the material is continually increased, the vibrational amplitude will increase to the point where some of the interatomic bonding of the space lattice is broken. When this occurs, some of the atoms are freed from the lattice and acquire a degree of mobility. We then say that melting has begun. The phase is being changed to the liquid phase.

A liquid phase however is not completely disordered in its atomic groupings. Small groups of atoms continually form into a space lattice, break up again, and reform. These groupings are short both in space and time, and it is common to refer to such groupings as *crystallites*.

There are a number of materials, such as glass and polyethylene, which, though apparently solid at room temperature, possess only the short-order crystallite groups instead of the long-range space lattice. This noncrystalline condition in a presumably solid phase is termed the *amorphous* condition. Such materials, when heated, do not exhibit a sharp and definite melting point, but instead grow continually less viscous as the temperature is raised. Since such materials behave like liquids or have liquid-like structures, the question arises: Are they actually liquids? The answer is yes. Glass must be considered a stiff liquid. So is firebrick, though it is harder than steel. These and other amorphous ceramic materials are often referred to as *glasses*.

## 2.18 solid solutions
## and equilibrium diagrams

An *alloy* is a mixture of two or more metals, or metal and nonmetallic elements. The most important by far of the alloy groups is the carbon-iron alloy group called "steels," which includes perhaps 25,000 different alloy mixtures. The use of alloys is more usual than pure or nearly pure metals.

The pure metals do not provide highest strengths, but they are employed when the highest possible electrical and thermal conductivity, or the highest resistance to corrosion, is required. Alloying is employed when strength is the paramount requirement, but always at a sacrifice in corrosion resistance. The steels, being alloys, have little corrosion resistance, while pure iron is considerably better in this respect.

The pure metals, being crystalline, have definite melting points. But alloys of two or more metals, although they also are invariably crystalline, do not usually have definite melting points, but melt over a range of temperatures.

The melting and phase characteristics of metal alloys and of ceramic mixtures are displayed by means of *equilibrium diagrams*, also called constitution diagrams, of which Fig. 2.17 is an example. This is an equilibrium diagram for copper-nickel alloys. On such a diagram, the alloy composition is the abscissa, and temperature is the ordinate.

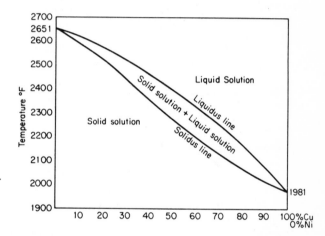

Fig. 2-17 Copper-nickel equilibrium diagram.

For pure nickel, the melting point is 2651°F (1455°C), and the melting point of pure copper is 1981°F (1083°C). These two melting temperatures are the end points of the curves on the diagram. Connecting these two points across all possible alloys of the two metals are two lines. The upper line is called the *liquidus* line. Above the liquidus line the alloy mixture is in the liquid phase. The lower line is called the *solidus* line. Below the solidus line the alloy mixture is in the solid phase. Between the solidus and the liquidus any alloy is partially solid and partially liquid, that is, the alloy mixture freezes or melts in the range of temperatures between the solidus and liquidus lines. For copper-nickel alloys, this melting range is about 100°F (36°C). The solidus line, the liquidus line, and any other such line appearing on an equilibrium diagram, are boundary lines separating phases.

Copper and nickel dissolve each other in any and all proportions, producing an infinite range of solid solutions. Since both metals are fcc, it is reasonable to expect that a solid solution of the two metals will likewise crystallize as face-centered cubic. The two types of metal atoms will be interspersed in the fcc lattice at random. The interatomic distances are altered from those of the pure metals, producing a degree of distortion in the lattice structure. It is this lattice distortion that creates the additional strength of an alloy.

A different type of equilibrium diagram is shown in Fig. 2.18. This is the diagram for the familiar lead-tin solders. In addition to liquidus and solidus lines, there are two other boundary lines, a result of the presence of two alloy phases labelled alpha and beta.

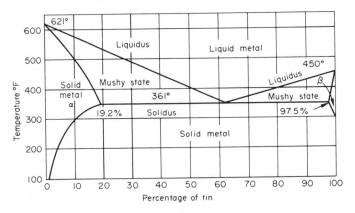

Fig. 2-18   Lead-tin equilibrium diagram.

Consider an alloy of 10 percent Sn—90 percent Pb, which happens to be a low-cost solder with a high liquidus temperature and with too much lead to "wet out" well—lead does not readily wet other materials, hence the addition of tin to the solder. Suppose this alloy to be deposited in the liquid phase in a soldered joint at 650°F (343°C). When it cools to 570°F (299°C), the liquidus line is reached, and the alloy beings to freeze. Freezing continues as the temperature falls, until at 515°F (268°C) solidification is complete. The freezing range is thus 55°F (30°C). The frozen alloy is in the alpha solid phase. But when the alloy is cooled to 300°F (149°C) it crosses another phase line, indicating a phase transformation. Below 300°F (149°C) the alloy will be not a solid solution but a mixture of two phases, alpha plus beta.

For a mixture of 61.9 percent Sn—38.1 percent Pb, notice that the solidus and liquidus temperatures are the same. This alloy has a definite melting point, like pure lead and pure tin. Its composition is called the *eutectic* composition of the alloy. The eutectic mixture is always the alloy mixture with the lowest melting temperature and also the one with a definite melting point. Referring to the copper-nickel equilibrium diagram (Fig. 2.17), no eutectic is indicated for mixtures of these two metals. This is always

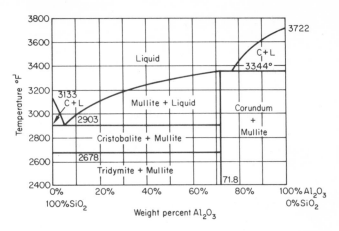

Fig. 2-19  Equilibrium diagram of silica-alumina. Tridymite and cristobalite are phases of silica; mullite is $3Al_2O_3 \cdot 2SiO_2$.

true for any pair of metals which are soluble in each other in all proportions and at all temperatures.

Equilibrium diagrams are as useful in ceramic engineering as in metallurgy. Figure 2.19 is an equilibrium diagram for mixtures of silica and alumina. The eutectic composition is about 5 percent $SiO_2$—95 percent $Al_2O_3$. To show the practical application of this equilibrium diagram, we may consider the case of a standard firebrick, which has a composition of about 40 percent alumina and 60 percent silica. Such a firebrick cannot resist furnace temperatures higher than the solidus temperature of 2903°F (1595°C). By increasing the alumina content we can produce a firebrick with improved resistance to the corrosion of molten slags, but notice that the increase in alumina does not alter the solidus temperature until the alumina content reaches 72 percent. Above 72 percent alumina a new phase appears, called mullite, with a solidus temperature of 3344°F (1840°C). Mullite brick then must be used for the linings of furnaces where firebrick requirements are exceedingly severe. (There are a rather large number of silica-high alumina phases, not shown here, including—besides mullite—andalusite, kyanite, sillimanite, and corundum. Not all are used in firebrick.)

The ceramic oxides have very high melting points, with the exception of a few such as iron oxide and borax. A high melting point or solidus temperature is an advantage in a firebrick, but in other circumstances a high melting point in an oxide may produce serious difficulties. Consider the problem of welding aluminum. Aluminum has a melting point of about 1200°F (649°C), but the metal always has a very thin surface of aluminum oxide. If the aluminum metal is to be melted during welding, this cannot be done without first melting the surface oxide. Unfortunately aluminum oxide melts at 3722°F (2050°C). If a welding temperature sufficient to melt the oxide is employed, an enormous hole will be burned in the aluminum metal.

The difficulty is circumvented by the practical employment of the eutectic concept. An oxide "flux" is added to the surface of the aluminum. The flux mixes with the alumina to produce a low-melting eutectic composition which makes the welding of the low-melting metal possible. Typical fluxes for reducing the softening points of oxides and silicates are borax, iron oxide, and others. Unfortunately the word "flux" has many meanings, but in general it means a ceramic additive, usually for the purpose of reducing a softening or melting point. Fluxes are employed in the melting and welding of most metals and in the manufacture of portland cement.

## 2.19 defects and interstitials
## in crystal lattices

The case of the substitutional solid solution, typified by copper-nickel solid solutions, was briefly discussed in the previous section. In such substitutional solid solutions, one metal atom can substitute for another metal atom in the space lattice. The effect is to produce some distortion in the space lattice, leading to variations of interatomic bonding, and the macroscopic effect being a stronger metal. If a metal is strong, this statement simply means that the interatomic bonding of the space lattice is strong.

Sometimes alloying results in vacancies in the atomic positions in the space lattice. Where this occurs, a *defect structure* is produced, again with lattice distortion in the region of the defect. Vacancy defects can also be produced by high-energy radiation, as when a high-velocity particle removes an atom from the space lattice.

A more important type of space lattice variation is the *interstitial solid solution*. The most important practical uses of the interstitial solid solution occur in the case of the space lattice of iron, which therefore will be used as an example of this effect.

Iron has three solid phases. The delta phase is of no practical significance and need not be mentioned. For pure iron the phase change from bcc to the fcc condition occurs at 1666°F (908°C), fcc being the structure at the higher temperature. Both fcc and bcc iron can deposit carbon atoms between iron atoms in the space lattice in what is termed an interstitial solid solution. Figure 2.20 shows the interstitial positions in the iron lattice which a carbon atom may occupy. These positions are the largest gaps between positions in the lattice. In the case of bcc iron the holes have a radius of 0.36 A, while for fcc iron, the interstitial hole measures 0.52 A. The carbon atom however measures 0.70 A. Therefore, when dissolved interstitially, the carbon atom will produce some distortion in the space lattice. This distortion will be less in the case of fcc iron because the space gap is larger. As a result, fcc iron can dissolve more carbon than bcc iron.

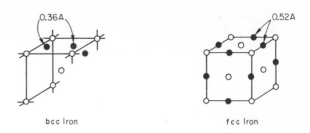

bcc Iron                    fcc Iron

Fig. 2-20   Interstitial positions in bcc and fcc iron. The largest sphere that can be inserted between iron atoms in a bcc crystal is 0.36A; in an fcc crystal is 0.52A. A larger interstitial atom will produce distortion of the space lattice.

Fcc iron can dissolve a maximum of 2 percent by weight of carbon; this amount of carbon would fill about 10 percent of the interstitial holes. The solubility of carbon in bcc iron increases from 0.008 percent at room temperature to a maximum of 0.025 percent at 1333°F (723°C).

One last type of crystal defect remains to be described, the dislocation.

## 2.20   dislocations

The theoretical tensile strength of a metal bar can be calculated as the force required to rupture the interatomic bonding forces within the crystal. If, however, the bar is tested in a tension testing machine, it will fail at a load of perhaps one-hundredth of its theoretical capacity. This is true of any metal. The explanation for the disappointing performance is the presence of *dislocations* in the lattice structure.

A dislocation is a disregistry such as the one shown in Fig. 2.21. Suppose two bricklayers work each on one end of a wall, and one mason completes the wall height with one less brick course than the other. The result will be a "dislocation" where their work joins, except that the bricklayers would call the disregistry a "hog." The presence of dislocations in large amounts results in great weaknesses in the metal crystal, and it is unfortunately not possible to cast and freeze metals in order to avoid the development of innumerable dislocations. However, dislocation-free metal whiskers have been produced in research laboratories. Since there is an extra half-sheet

Fig. 2-21   A dislocation.

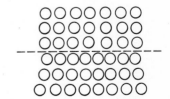

of atoms above the line of dislocation in Fig. 2.21, the upper half of the crystal is in compression, while the lower half is in tension. A shear stress exists along the line of dislocation.

## 2.21 inclusions and composites

Most metals will contain foreign particles, usually ceramic materials. If such particles are undesirable, they are called *inclusions*. If however they belong in the metal, they do not seem to be honored with a generic name. Figure 2.22 is a micrograph of cast copper showing lenticular inclusions of copper oxide. Such inclusions may enter the metal by a wide variety of methods. The oxide inclusions in the figure are of course the result of chemical combination of copper with oxygen during furnace melting. When "tapping" molten metal from a melting furnace, the metal must run out through a fire-brick-lined trough. The flow of metal will wash particles out of the firebrick; these will appear in the metal. Ceramic particles also can fall or drip from the roof of a melting furnace into the molten bath. Silicon and aluminum are added to steel to combine with dissolved oxygen, and manganese is added to take up sulfur. The end result is the appearance of $SiO_2$, $Al_2O_3$, and MnS as inclusions in the steel. Still another source of inclusions is welding slag which has been trapped in the molten metal during welding operations.

Fig. 2-22   Cast copper with $Cu_2O$ inclusions.

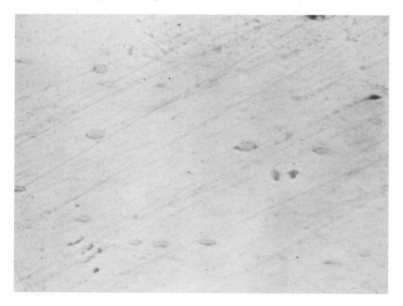

The harmful effects of inclusions are many. In the case of forming operations employed in the production of automobile bodies, the presence of inclusions may cause the sheet metal to tear while being formed between dies. The same inclusions will cause cracks to initiate in forgings or will lead to eventual failure of such machine parts as the balls in ball bearings and engine exhaust valve stems.

In other cases, inclusions may serve useful purposes and may be purposely added to a metal or other material. Carbon black is added to some plastics and rubbers for protection against ultraviolet radiation. Lead is added to metal bars to make them more machinable. High-speed steels obtain their hardness from the addition of carbides to the steel. Gray cast iron derives many of its useful characteristics from the flakes of graphite distributed throughout its structure.

The ultimate in inclusions is the blending of two or more dissimilar materials in such a way as to produce a new material with the characteristics of both materials—a blend termed a *composite*. Perhaps the most familiar composite is the group of reinforced plastics, reinforced with fiberglass, asbestos, or other filaments. Pairs of highly incongruous materials have been blended to make such useful composites as lead and polyethylene.

A highly useful type of composite is the ceramic-metal mixture, sometimes called a "cermet." One example is the carbide cutter bit used for machining metals. This is a mixture of finely ground carbides of tantalum and tungsten bonded together with cobalt metal. The structure of such a material is therefore a continuous phase of metal with a discontinuous ceramic phase. The strength and ductility characteristics of the composite material are primarily those of the continuous phase. Often, too, in a standard metal alloy there will be a distinct phase deposited at the grain boundaries of the metal, so that in effect each grain appears to be embedded in the grain boundary phase. Again, the strength characteristics of the metal will be those of the continuous phase. In general, though, such a continuous phase is undesirable in a metal, and it will be removed by a suitable heat-treating operation.

## 2.22 laminates and sandwiches

Both traditional and newer types of materials use composite construction, either fiber-reinforced, laminates, or sandwich types.

Examples of laminates include the built-up roof, which is a complex laminate of alternating layers of roofing felts and asphalts. Plywood is a laminate. Paint films are laminates. The effectiveness of such composite materials lies not in the components, but in the complete laminate—the whole is greater than the sum of the parts. Such materials are only as good as their adhesion. For example, it would be quite useless to test the effectiveness

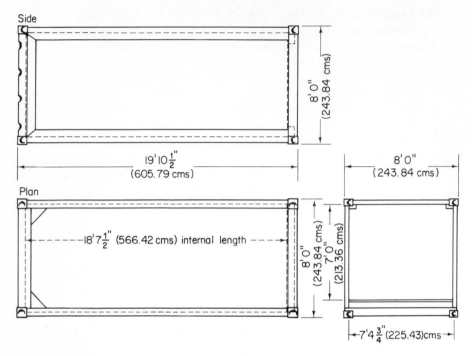

Fig. 2-23 International Standards Organization (ISO) merchandise shipping container type MLW/MLLU 24, nonrefrigerated insulated, capacity 41,260 pounds. The allowable heat leak from this container is 47 BTU per hour/degree fahrenheit.

of a detached film of paint. The paint film to be tested must be attached to its substrate—wood, metal, concrete, or other—which it is to cover and protect, and the effectiveness of the paint is the composite effectiveness of the paint and its bond to the substrate.

Figure 2.23 shows a typical container used in transportation, especially for transoceanic freight movements. The construction used is a sandwich construction: the exterior and interior faces of the walls are sheet metal separated by a core of foamed polyurethane. This foamed plastic is selected because when injected into the space between the two sheets it bonds to the sheets as it hardens. The faces of such a sandwich must provide the required load-bearing strength, stiffness, hardness, and surface characteristics. Since the core is soft and weak, it is not used to carry structural loads. Its function is to keep the sheet faces apart and to stabilize them against buckling when stressed. Frequently the core has additional functions such as heat insulation. Again, the whole is greater than the sum of the parts: neither the faces nor the core, of themselves, have any load-bearing ability, but when all are bonded into a sandwich, the strength of the sandwich is remarkably high. Other types of sandwich construction include the corrugated paper box, trailers and motor homes, prefabricated housing, and aircraft and aerospace vehicles.

1 (a) Using the American Society for Metals *Metals Handbook* or other reference, decide which of the three crystal structures, bcc, fcc, or hcp, is generally associated with high tensile strength in metals.

   (b) Which of the three crystal structures appears to be associated with high melting point?

2 What is the number of protons and of neutrons in each of the following isotopes: $_3Li^7$, $_{11}Na^{25}$, $_{16}S^{36}$, $_9F^{17}$, $_{57}La^{140}$, $_{92}U^{238}$?

3 Determine the wavelength of radiation corresponding to photon energies of 0.5 and 6000 electron volts, and state the type of radiation in each case.

4 Name devices in use for the detection of the following types of radiation: radio, infrared, visible light, ultraviolet, gamma ray.

5 Could the radiation from an X-ray tube activate the photocell on an elevator door? Why?

6 Explain why elements of Group I of the Periodic Table are suitable cathode materials for photocells.

7 The forbidden energy band between allowed energy bands in germanium is 0.72 ev wide. What photon wavelength corresponds to this bandwidth?

8 What is the general effect of a filter on the spectrum of a broad-band radiation?

9 If a metal is used as a filter for broad-band X-radiation, will such a filter suppress long-wave or short-wave radiation in general? Why?

10 Explain why in fluorescence the stimulated radiation has a longer wavelength than the exciting radiation.

11 Explain why many electrical insulating materials, such as window glass, are transparent to visible radiation.

12 What is the half-life of an element with a stable nucleus?

13 Why are radioactive sources bought by the curie instead of the pound or kilogram?

14 Suppose that radioactive Iridium-192 costs $20 a curie. If shipped by airfreight, the cost of freight is $1 per curie higher than the rail rate. Airfreight, however, saves 7.5 days' delivery time. Explain the argument for airfreight.

15 Determine the age of the earth, given the following information. Uranium-238, with a half-life of 4.5 billion years, decays to Pb-206, which is stable. U-238 always contains Pb-206 because the lead is the end element of uranium disintegration. For every 100 atoms in the lead-uranium mixture, 46 are lead and 54 are uranium.

16 A wooden ship, which turns out to be a remarkable historical discovery, is recovered from the sand dunes of Libya. The discovery party of historians must be certain of the age of the ship. Find its age, given the following information.

   A small amount of carbon in $CO_2$ is converted from C-12 to C-14 because of cosmic radiation. This carbon is taken up by plants during photosynthesis.

During the life of a plant, the ratio of C-14 to C-12 in the plant is constant, but it decreases when $CO_2$ absorption ceases at the death of the plant. Assume the half-life of C-14 to be 5600 years. The radioactivity of a living tree is 255 disintegrations per kg per sec. The radioactivity of wood from the ship is 240 disintegrations per kg per sec.

17  Why is gamma radiation more dangerous to bone than to soft tissue in the human body?

18  The half-value thickness of aluminum is to be determined for a certain X-ray tube. A sheet of aluminum 0.10 in. thick is available. When used as an X-ray barrier, this sheet reduces the radiation to 20 percent of the incident radiation. What is the half-value thickness of aluminum for this radiation?

19  If bond paper 0.002 in. thick passes 40 percent of visible light incident on it, what is the half-value thickness of this bond paper for visible light?

20  An opaque material has a reflectance of 0.20. What fraction of the incident radiation is absorbed by the material?

21  Explain the mechanism of photoelectric absorption of high-energy radiation.

22  Is the electric charge on the particle positive, negative, or zero charge in the case of (a) the neutron, (b) the electron, (c) the photon, (d) the proton? Which of these particles are found in the nucleus of atoms?

23  Explain ionization potential.

24  If the electron of hydrogen falls from the M level to the L level, what photon energy is released? What type of radiation is produced?

25  Certain quantities, such as the energy levels of an electron, are quantized or restricted to certain values. Other quantities such as lengths, distances, or weights are continuous or unrestricted. Explain quantized and continuous numbers in terms of the two types of money: cash and checks.

26  What occurs in an atom of an element for each of the following cases:
(a) an extra electron is added
(b) an extra neutron is added to the nucleus
(c) a neutron in the nucleus changes to a proton?

27  Why are helium and argon used as shielding gases for the arc when welding metals?

28  What is the meaning of fluorescence?

29  In the radiographic photography of a weld or a casting, the photographic exposure time for the film must be doubled if the number of curies of the radiographic source is reduced by half. By how much is the film exposure increased if a source of Iridium-192 ages by 1 year? The half-life is close to a fifth of a year.

30  The absorptivity of aluminum paint for visible radiation is 0.50. What fraction of light will such paint reflect?

31  What is an amorphous material?

32  What is the difference between body-centered and face-centered cubic space lattices?

33 In order to protect personnel from gamma radiation it is decided to shield them with three half-value layers of material. By how much will this protection reduce the level of radiation?

34 (a) Sugar, cream, and coffee are three distinguishable material phases. If sugar, cream, and coffee are mixed and stirred, is the mixture one phase or three phases?

   (b) Is a plastic material reinforced with fiberglass filaments one phase or two phases?

35 Explain the meaning of a eutectic temperature.

36 What is the difference between a substitutional and an interstitial solid solution?

37 Sketch a dislocation in a metal lattice structure.

38 Distinguish between a composite, a laminate, and a sandwich structure. Give an example of each.

39 What is the function of core and face in a sandwich construction?

40 If wood continues to increase in price, drafting boards may be made of synthetic materials. Is a synthetic drafting board likely to be a sandwich construction?

The technology of materials has acquired such a degree of sophistication that it is not possible to make a summary review of the significant characteristics of materials. In this book therefore material characteristics are grouped into four sections: physical, electrical, mechanical, and economic. This grouping is of course arbitrary. By "physical characteristics" we mean those more or less fundamental characteristics which may be understood or explained from the basic concepts of physics. Performance characteristics of an empirical or less fundamental nature, such as metal fatigue, coefficient of friction, or hardness, are classified here as mechanical characteristics.

Many of the physical properties of materials have already been discussed in the previous chapter, such properties including absorptivity, half-life, and others. These need not be further reviewed.

### 3.1   stress and strain

The strength of the material is the critical concern for those materials which must carry loads and forces. Ductility, not strength, may be required

# PHYSICAL PROPERTIES
# OF MATERIALS

# 3

of materials in other applications. Roofing materials must be able to "give" with movements of the building; an adhesive or a paint must accommodate thermal and other movements of the material to which it is attached.

Although the words "stress" and "strain" are often confused in ordinary conversation, each has a specific meaning. For explanation, consider the case of the 4 × 4 timber post 100 inches long of Fig. 3.1. The post supports a compressive load of 400 pounds. The stress in this post is the load divided by the cross-sectional area that supports that load:

$$\text{Stress} = \frac{\text{load}}{\text{area}} = \frac{400 \text{ lb}}{4 \times 4} = 25 \text{ psi } (172,000 \text{ N/m}^2)$$

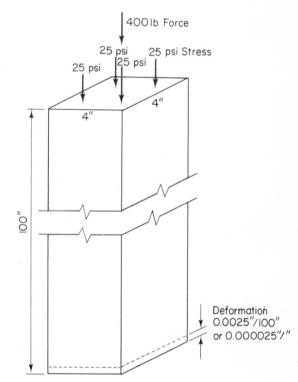

Fig. 3-1   Stress and strain in a wood post.

There are only three possible types of stresses: *tension, compression*, and *shear*. These are illustrated in Fig. 3.2. A shear force or stress tends to cut through the material, as occurs in the shearing and punching of metal plates, or it may be a twisting force or stress such as shears the head from a bolt when the torque of a wrench is excessively high. The strength of metals is lower in shear than in tension, and for most materials, especially brittle materials, the strength in tension is lower than the compressive strength.

Accompanying any stress there is always some deformation (change in dimension) of the loaded member. In our example the timber post will shorten in length under the load. Suppose that the length of the post under

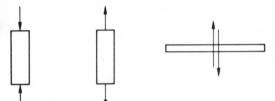

Fig. 3-2 Forces producing compression, tension, and shear stress.

load decreases by 0.0025 in. The amount of this deformation will depend on the length of the post: if the post were twice as long it would shorten twice as much under the same load or stress. It is therefore convenient to express this deformation as deformation per inch of length. Deformation per inch is the *strain* or *unit strain*, which in our example is

$$0.0025 \text{ in.}/100 \text{ in.} = 0.000025 \text{ in.}/\text{in.}$$

Materials may be strained by conditions other than stress. If the temperature of a material is increased, it will expand. This expansion is a *thermal strain*.

If the load or stress is removed from a material, two types of behavior are possible:

1. The material may remain permanently deformed by the stress. This kind of strain behavior, in which the strain is permanent, is termed *plasticity*. Wet concrete or damp soil are plastic materials. When walked on, they retain the impression of the walker's shoes.
2. The material may return to its original shape and size. This behavior, in which the strain disappears with the stress, is called *elasticity*.

Many materials exhibit elasticity under limited conditions of stress. Only a few brittle materials, such as glass, are wholly elastic when loaded to their failure stress. In general, the word "brittle" suggests a material in which strain behavior will be elastic, since such a material is capable of only a limited amount of strain. Ductile materials are those materials capable of a considerable amount of plastic deformation.

Both elastic and plastic strain are desirable characteristics. Dimensional stability is expected of manufactured industrial goods; such goods are designed so that any predictable stresses which they must carry will fall within the limited elastic range of the material. On the other hand, most forming operations during manufacture require plasticity for their execution. The raw material for an automobile fender is a flat sheet of steel. It must be deformed plastically to become a fender. But when installed on an automobile only elastic deformations are desirable (though rarely obtained in car accidents). Even hard firebrick and glass must be manufactured while they are in a plastic condition.

## 3.2 Poisson's ratio

Suppose that a bar is stretched by a tensile stress in its *x*-direction. If the volume of the bar is to remain constant, then it must experience a lateral *contraction* in the *y*- and *z*-directions. Within the elastic range, this lateral strain is proportional to the axial strain, and Poisson's ratio expresses the ratio of the two strains:

$$\text{Poisson's ratio} = \mu = \frac{\text{lateral strain}}{\text{axial strain}}$$

For metals, Poisson's ratio is in the range of 0.27 to 0.30 for elastic strains. Poisson's ratio for plastic strain is taken as 0.5.

## 3.3 stress-strain diagrams

The stress-strain diagram of a material is a most useful compilation of information about any material, since it discloses in a single diagram such characteristics as ultimate strength, limit of elastic behavior, ductility or brittleness, elasticity, plasticity, and modulus of elasticity.

A stress-strain diagram may be given for compression, tension, or shear. The most commonly used is the tensile stress-strain diagram, though compression tests may be more useful for those materials normally loaded in compression, such as concrete or plastic foam insulations. A typical sample of material used for a tension test specimen is shown in Fig. 3.3. Two center-

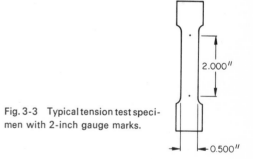

Fig. 3-3  Typical tension test specimen with 2-inch gauge marks.

2.000″

0.500″

punch marks are made in the test length of the specimen exactly 2.000 in. apart. The material between these gauge marks is the material analyzed under test, and the 2.000 in. length is termed the gauge length. A suitable strain gauge is mounted on the specimen to record the tensile strain in the gauge length under load. The specimen is then mounted in a tension testing machine such as the one shown in Fig. 3.4. This testing machine is actually a highly accurate, instrumented hydraulic press which can load materials in either tension or compression. The large dial indicates the load in pounds

Fig. 3-4  Universal testing machine. The test load is indicated on the large dial.

applied to the specimen, and the strain gauge indicates the elongation in the 2-in. gauge length. For any load the stress is found by dividing the cross-sectional area of the specimen into the load, and the unit strain is found by dividing the elongation by 2. Suppose that the results of the test are those given in the following table.

### STRESS-STRAIN READINGS FOR A STEEL TEST SAMPLE

| Load (lb) | Stress (psi) | Elongation in 2 in. | Unit Strain (in./in.) |
|---|---|---|---|
| 1000 | 1000 | 0.000066 | 0.000033 |
| 2000 | 2000 | 0.000133 | 0.000066 |
| 4000 | 4000 | 0.000266 | 0.000133 |
| 6000 | 6000 | 0.000400 | 0.000200 |
| 10000 | 10000 | 0.000666 | 0.000333 |
| 20000 | 20000 | 0.001333 | 0.000666 |
| 30000 | 30000 | 0.002000 | 0.001000 |
| 42000 | 42000 | 0.0028 | 0.0014 |
| 42000 | 42000 | 0.040 | 0.020 |
| 44000 | 44000 | 0.060 | 0.030 |
| 50000 | 50000 | 0.170 | 0.085 |
| 60000 | 60000 | 0.264 | 0.132 |
| 63000 | 63000 | 0.280 | 0.140 *Rupture.* |

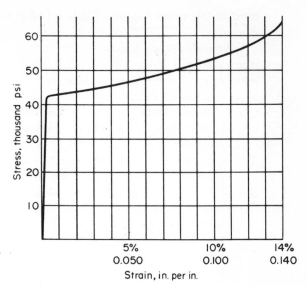

Fig. 3-5 Stress-strain curve.

Strain, in. per in.

The graph of stress (ordinate) versus strain (abscissa) is given in Fig. 3.5. Examination of either the data or the graph discloses much information. The *ultimate tensile stress* (UTS) is 63,000 psi. The maximum elongation is 14 percent in 2 in. The reduction in area (R.A.) is often measured at the break and reported with the elongation as an additional indicator of ductility, since a very ductile material will neck down considerably before fracturing. The modulus of elasticity may be calculated from the readings at 30,000 lb:

$$E = \frac{30,000}{0.001} = 30 \times 10^6 \text{ psi (200 GN/m}^2)$$

The strain is linear up to 42,000 psi. The latter value of stress is the *proportional limit*. The material may exhibit some nonlinear elasticity above the proportional limit, however; the end of the elastic range is termed the *elastic limit*. More attention however is given to the *yield stress*. This is the stress corresponding to the intersection of the stress-strain curve with a line parallel to the linear part of the curve and intersecting the strain axis at 0.2 percent strain.

The meaning of the terms "brittle" and "ductile" is disclosed by this stress-strain curve. A brittle material shows little plastic strain or none whatever, while a ductile material has an extended plastic region. The stress-strain readings for the above steel sample indicate that although not brittle, the steel is not especially ductile, since the elongation is only 14 percent. A total strain of not less than 10 percent is demanded of most metals except cast irons, but a ductile metal might give 35–50 percent elongation.

Notice that at 42,000 lb the material yields plastically without an increase in load, but at 44,000 lb this yielding ceases, and the metal takes on more stress. The metal strengthens as a result of plastic deformation. This effect is called *work-hardening* or *strain-hardening* and is found in all metals to some degree. Metals become harder and stronger when deformed plastically. If a wire must be cut, and no pliers are available, it is usual to flex the wire repeatedly in one place to cause it to break. This is strain-hardening. The repeated flexing makes the metal harder and thus more brittle.

Figure 3.6 is a tensile stress-strain diagram for a mild (i.e., soft) construction steel. As usual for steels, the specimen at first strains linearly. At about 35,000 psi the strain becomes plastic, since the material deforms at an almost constant load. At about 4 percent strain the material then takes up more load. The maximum or ultimate tensile stress is 67,000 psi, and thereafter the stress falls as the specimen necks down to a smaller diameter. The yield stress is 36,000 psi.

Fig. 3-6   Stress-strain diagram for a mild steel.

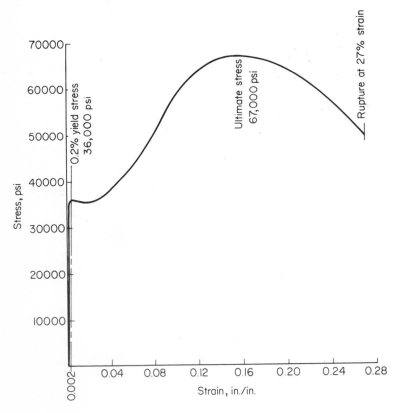

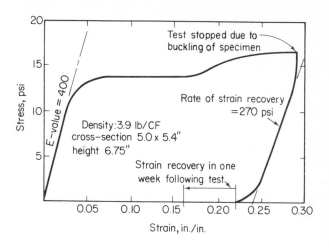

Fig. 3-7  Compression test of a block of rigid foamed insulating polyurethane 3.9 lb/cf. Almost one-half of the total strain was recovered in one week, indicating that this "plastic" is actually a rubber. Note the different slopes of the linear loading and unloading lines.

A more complex example of stress-strain behavior is given in Fig. 3.7. This diagram is for an insulating grade of rigid polyurethane foam with a density of 4 lb/cu ft. On being loaded, the material at first strains linearly as do the steel specimens. This linear strain is elastic. At about 12 psi the strain appears to become plastic, since the material deforms continually at a constant load. Actually, in this region the material is still in its elastic region, since this strain is recoverable on unloading. At about 18 percent strain the material takes up more load. The loading was stopped at about 17 psi because of incipient buckling of the block of urethane.

On being unloaded, the material recovers elastically at somewhat the same rate shown during the initial loading. This elastic recovery reduces the strain from the maximum of 29 to 22 percent. The curved portion of the strain recovery line near zero stress is the recovery during 15 minutes after the test. After one week, the permanent strain fell to 16 percent. Thus approximately half the maximum strain is elastic and half is plastic. Note that the elastic strain recovery was not instantaneous with removal of the load. Strain is not proportional to stress except in the case of a few materials, usually metals, at low stress levels and slow loading.

A material such as this polyurethane, with a maximum elongation of about 30 percent, has excellent ductility. It could extend or contract to accommodate movements of adjacent materials. On the other hand, this material is quite obviously unsuited to the support of loads: it deforms excessively at stresses which are negligible compared to those carried by the structural steel specimen just discussed. In the whole range of materials, some must be strong and stiff to support loads, and others must be flexible to accommodate movements.

## 3.4 modulus of elasticity

Most metals, some rubbers, and the elastic plastics such as bakelite, epoxy, and reinforced plastics, exhibit linear elastic behavior below the elastic limit. The *elastic limit* is the highest stress for which the strain is elastic.

If the initial portion of the stress-strain graph is linear, as in the case of the several materials just discussed, that is, strain is proportional to stress, then the material has a modulus of elasticity. The *modulus of elasticity* or Young's modulus, symbol $E$, is the ratio of stress to strain:

$$E = \frac{\text{increase in stress}}{\text{increase in strain}}$$

More informally, modulus of elasticity is referred to as $E$-value. The modulus may also be considered as the stress necessary to strain the material elastically to twice its original length, if that were possible. A very rigid material will necessarily have a very high $E$-value, since a large stress will be accompanied by only a small strain. If the stress and the strain are measured for a small rubber band, the $E$-value will be found to be about 200. The metals however have moduli of elasticity ranging from about $6.5 \times 10^6$ for magnesium to $80 \times 10^6$ for osmium. The units of $E$ are psi, or in the SI system newtons/square meter.

### MODULI OF ELASTICITY (psi)

| | | | |
|---|---|---|---|
| glass | $10 \times 10^6$ (varies) | steel | $29 \times 10^6$ |
| alumina | $40 \times 10^6$ | lead | $2 \times 10^6$ |
| graphite | $1 \times 10^6$ (varies) | magnesium | $6.5 \times 10^6$ |
| wood | $1.5 \times 10^6$ (varies) | molybdenum | $47 \times 10^6$ |
| concrete | $3 \times 10^6$ (varies) | nickel | $30 \times 10^6$ |
| polyethylene | 35,000 | titanium | $16.5 \times 10^6$ |
| aluminum | $10 \times 10^6$ | tungsten | $50 \times 10^6$ |
| beryllium | $44 \times 10^6$ | zirconium | $14 \times 10^6$ |
| copper | $16 \times 10^6$ | | |

The table of $E$-values suggests that the figure of $1 \times 10^6$ separates the high-modulus or stiff materials from the low-modulus plastics and rubbers (in the SI system $7 \times 10^9$ N/m²).

Certain very brittle materials such as brick are difficult to test in tension. For such materials the modulus of rupture, rather than the tensile strength, is determined. The modulus of rupture actually gives the tensile strength at failure in a bending test. It is unrelated to the modulus of elasticity.

In a test for modulus of rupture, the material is loaded as a beam as shown in Fig. 3.8. From the test results the modulus of rupture is calculated by the following formula:

$$\text{Modulus of rupture, psi} = R = \frac{3WL}{2bd^2}$$

where $W$ = load at failure, pounds

$L$ = loading span, actually 7 inches in the standard tests

$b$ = width of specimen, inches

$d$ = depth of specimen, inches.

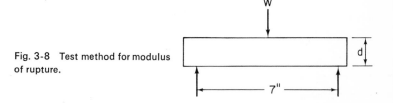

Fig. 3-8   Test method for modulus of rupture.

The tensile stress-strain curve of any such brittle material is a straight line to failure. Brittle materials are elastic materials, although their elasticity is extremely limited.

## 3.6  strain mechanisms

In elastic deformation, the interatomic spacing within the crystal lattice must increase or decrease.

Plastic deformation however proceeds by mechanisms other than change of interatomic spacing. These mechanisms are *slip* and *twinning*. Slip is shown in Fig. 3.9. Here the atoms have moved over one integral interatomic distance along the slip plane, and when the stress is removed, there will be no interatomic force to return the atoms below the slip plane to their original positions. Slip occurs along planes of greatest atomic density. Such planes are shown for fcc, bcc, and hcp metals in Fig. 3.10. There is one such plane in

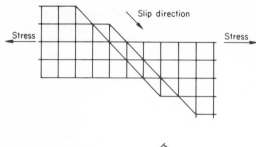

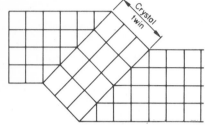

Fig. 3-9 Slip and twinning in a cubic crystal.

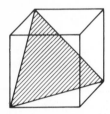

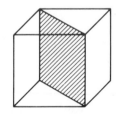

Fig. 3-10 Slip planes in fcc, bcc, and hcp metals. These planes are those of greatest atomic density. In the fcc crystal lattice there are eight such planes, including the shaded one in the diagram, in the bcc lattice six. In an hcp lattice the base plane has the greatest atomic density.

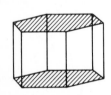

the hcp lattice. Since the face-centered cubic lattice has the greatest number of slip planes, eight in all, fcc metals are the most ductile.

Twinning, like slip, is a shear strain, in which there is a uniform tilting of the surface in the twinned region. Figures 3.11 and 3.12 show micrographs of slip and twinning.

The brittleness of the ceramic materials is explained by the atomic structure of a ceramic crystal. Figure 3.13 represents the space lattice of magnesium oxide. Slip cannot occur, for a movement of one interatomic distance will align like atoms, thus destroying the molecular structure. A sufficiently large force to execute such distortion is not physically possible.

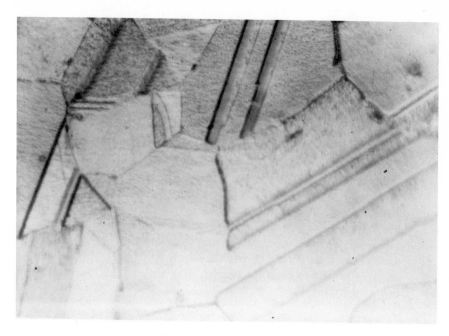

Fig. 3-11   The sets of parallel lines within the individual crystals are twins.

Fig. 3-12   A series of twins with sliplines crossing the twinning planes. This micrograph is taken at a higher magnification than Figure 3-11. Both specimens are brasses.

Fig. 3-13 Comparison of the cubic crystal structures of a metal and magnesium oxide.

If a metal crystal were a perfect crystal, without dislocations (see Chapter 2), for slip to occur it would be necessary for the bonds between all atoms across the slip plane to break simultaneously. Such a crystal would be perhaps a hundred times stronger than any metals which any experimenter has tested in a tension test. The actual strengths of metals are explained by the presence of dislocations in the crystal. Slip occurs by movement of the dislocation, one atomic bond at a time, across the slip plane as in Fig. 3.14. If the plastic deformation is continued, then dislocations are gradually consumed, and the crystal becomes more perfect and therefore stronger. This is the explanation of strain-hardening. Finally a condition is reached in which the stress required for slip to continue exceeds the stress required for rupture, and the metal finally breaks.

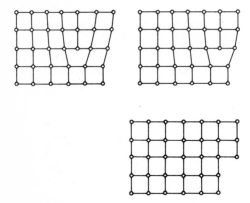

Fig. 3-14 Movement of a dislocation due to stress.

## 3.7 anisotropy

An isotropic material has identical physical properties in all three dimensions. Such a material is difficult to produce, though highly desirable. Steel sheet used in deep-drawing operations, for example, must flow uniformly in all directions while it is being shaped in the die.

Since most metals are formed by rolling, extruding, and similar processes which give a marked directionality to the metal, it would be expected that such materials would be anisotropic. The greatest strength of the material occurs in the direction of rolling, and the degree of anisotropy increases with the degree of reduction by rolling.

### 3.8 coefficient of linear
### thermal expansion

This expansion coefficient is the amount of expansion in a one-foot or one-inch length of a solid material as a result of a temperature rise of 1°. The expansion coefficient is a pure number reported as inches per inch or feet per foot per degree, and is of course a thermal strain. The convenient assumption can usually be made that this coefficient is a constant, though this assumption must be abandoned if thermal strains are to be calculated over temperature ranges of hundreds of degrees. For most materials the expansion coefficient increases slightly with temperature, and it will change with any phase change in the material.

#### COEFFICIENTS OF LINEAR EXPANSION

in./in. at 70°F.

| | |
|---|---|
| carbon steel | 0.000 006 |
| copper, brass | 0.000 009 |
| aluminum | 0.000 012 |

When laying up firebrick in industrial furnaces, an expansion allowance of $\frac{1}{4}$ in. every 4 ft is used. The silica brick used in the roofs of open-hearth furnaces for the melting of steel presents a special problem. Silica brick has a very large expansion up to about 1000°F (538°C), but almost no expansion at higher temperatures. Damage to the furnace roof can be expected if the furnace temperature falls below 1000°F. Techniques have therefore been developed to repair such roofs while hot.

Because dimensional stability is a highly desirable characteristic in finished goods, the lowest possible expansion coefficient is always sought. Since aluminum has twice the coefficient of steels, it can be expected to show twice the warpage during welding operations. In other applications the coefficient of expansion must be formulated to suit the conditions of use, as in the case of vacuum tubes for the electronics industry. If these are not to lose their vacuum, the metal and glass components of the tube must have matching expansion coefficients. The alloys of nickel and iron can be blended to give a suitably wide range of expansion coefficients.

## 3.9 melting point and boiling point

These are temperatures of phase change from solid to liquid and liquid to gas. The heat required to melt or boil unit weight of a substance is termed its latent heat.

The evaporation of water at temperatures below the nominal boiling point is a familiar phenomenon. This phenomenon occurs with all substances. The evaporated gas exerts a pressure termed the *vapor pressure*. If the pressure on the substance is reduced, then its boiling point falls to a lower temperature and its vapor pressure is increased, indicating that evaporation occurs at a faster rate. If therefore metals with low boiling points are used in high vacua or in aerospace vehicles, there may be loss of the metal through evaporation. Thus iron boils at about 5500°F (3055°C) at atmospheric pressure, but in outer space its boiling point would be reduced to the region of about 1500°F (830°C). Cadmium and zinc cannot be used in space vehicles for this reason; both metals will boil away in outer space at a temperature in the range of 100°F (38°C).

It is generally true that the oxides of metals have higher melting points than the metals themselves. There are some exceptions. The oxides of iron melt at approximately the same temperature as iron, and this is the reason why steels can be cut with the oxyacetylene torch while other metals cannot.

### MELTING POINTS OF THE METALS

| mercury | −38°F | −39°C | silicon | 2570°F | 1410°C |
|---|---|---|---|---|---|
| gallium | 85.5 | 30 | nickel | 2651 | 1455 |
| sodium | 208 | 98 | iron | 2804 | 1540 |
| lithium | 367 | 186 | titanium | 3074 | 1690 |
| tin | 449 | 232 | zirconium | 3326 | 1830 |
| bismuth | 520 | 271 | thorium | 3348 | 1840 |
| cadmium | 609 | 321 | vanadium | 3452 | 1900 |
| lead | 621 | 327 | boron | 3812 | 2100 |
| zinc | 787 | 419 | hafnium | 3866 | 2130 |
| magnesium | 1204 | 651 | columbium | 4379 | 2415 |
| aluminum | 1224 | 662 | molybdenum | 4752 | 2622 |
| copper | 1918 | 1048 | tantalum | 5425 | 2996 |
| uranium | 2071 | 1133 | tungsten | 6152 | 3400 |
| beryllium | 2343 | 1284 | carbon | 6600 | 3650 |

The material with the highest melting point so far produced is hafnium carbide.

## MELTING POINTS OF CERAMIC MATERIALS

| | | | | | |
|---|---|---|---|---|---|
| hafnium carbide | 7520°F | 4160°C | silicon carbide | 4890°F | 2700°C |
| tantalum carbide | 7025 | 3885 | zirconia | 4850 | 2677 |
| carbon | 6600 | 3650 | calcium oxide | 4660 | 2590 |
| zirconium carbide | 6370 | 3521 | beryllia | 4570 | 2520 |
| tantalum nitride | 6050 | 3343 | titania | 3870 | 2132 |
| titanium carbide | 5660 | 3127 | alumina | 3722 | 2050 |
| thoria | 5630 | 3110 | nickel oxide | 3560 | 1960 |
| zirconium boride | 5530 | 3054 | silica | 3110 | 1710 |
| magnesia | 5070 | 2800 | | | |

### 3.10 specific heat

The specific heat of any material is the amount of heat energy, measured in Btu's, that will raise the temperature of 1 lb of the material 1°F. The specific heat is therefore a measure of the amount of heat that can be stored in a material. Although the specific heat is frequently assumed to be constant, for any material it varies somewhat with temperature, usually increasing with temperature. Therefore any specific heats quoted usually apply to temperature conditions of 70°F or 20°C. These are almost the same temperature.

The light elements of the Periodic Table have the highest specific heats, and the heavy elements have low specific heats.

### TABLE OF SPECIFIC HEATS

Btu/lb/°F

| | | | |
|---|---|---|---|
| hydrogen | 3.4 | air | 0.24 |
| helium | 1.25 | steel | 0.11 |
| water | 1.00 | cast iron | 0.12 |
| lithium | 0.79 | aluminum | 0.22 |
| boron | 0.305 | copper | 0.09 |
| beryllium | 0.52 | nickel | 0.105 |
| ice | 0.5 | zirconium | 0.066 |
| steam | 0.5 (approx) | uranium | 0.028 |

The specific heat of ceramic materials is frequently taken at the rough figure of 0.2. Since organic materials usually contain considerable hydrogen, their specific heats are in the range of 0.3 to 0.5.

## 3.11  thermal conductivity

The thermal conductivity of a material is also known as the K-factor. A low K-factor, below 1.0, indicates that the material is a heat insulator. Being highly conductive, the metals have K-factors in the range 100 to 2700 Btu/sq ft-in.-°F-hr.

Figure 3.15 shows a board foot of insulating material, say fiberglass, with a K-factor of 0.3. Suppose that there is a temperature difference of 1°F between the two flat faces of the fiberglass. Then heat will flow through the fiberglass from the hot face to the cold face. The actual heat flow through the board foot of fiberglass with a 1° face difference of temperature will be 0.3 Btu/hr. The quantity of heat conducted by the material under these unit conditions is termed the thermal conductivity or K-factor, which for fiberglass will of course be 0.3 Btu per hr. Since K-factor is defined in terms of a board foot of material, the K-factor has mixed units of feet and inches.

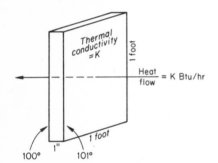

Fig. 3-15  Explanation of K-factor or thermal conductivity.

Like the specific heat and the coefficient of expansion, the thermal conductivity is assumed to be constant over short ranges of temperature. But like these other physical "constants," the thermal conductivity increases with temperature. The K-factor for many insulating concretes used for lining industrial furnaces may increase by a factor of 5 between room temperature and 1000°F (540°C).

### TABLE OF THERMAL CONDUCTIVITIES

Btu/hr-sq ft-°F-in.

| | |
|---|---|
| soils | 4–12 (increased by moisture content) |
| ice | 12 |
| snow | 5 (varies) |
| carbon steel (at 70°F) | 310 |
| stainless steel #304 | 105 |
| copper (70°F) | 2740 |
| aluminum | 1400 |

The thermal conductivity is one of the more significant properties of any material, with an influence in areas where the novice might least expect to find it. Figures 3.16 and 3.17 show micrographs of two spot welds, one made on low carbon steel with a relatively high $K$-factor and the other on stainless steel with a low $K$-factor. Because heat cannot readily flow from the area of the spot weld in stainless steel, the heat is retained in this area and the effect is an undesirable large grain growth. Such grain growth is absent in the higher conductivity carbon steel. By and large, a low thermal conductivity indicates a material which is easy to weld, since the welding heat is retained in the area of the weld. Stainless steel is easy to weld, despite unfavorable grain growth. Copper has such a high conductivity that it is virtually unweldable by many welding methods.

The following is the equation for one-dimensional heat flow:

$$Q = \frac{kA(t_1 - t_2)}{L}$$

where $Q$ = heat transferred in Btu/hr

$K$ = thermal conductivity of the material, Btu/sq ft-hr-°F-in.

$t_1$ = temperature of the hot side of the material

$t_2$ = temperature of the cold side of the material

$L$ = thickness of the material in *inches*

$A$ = cross-sectional area of heat flow path, in square *feet*.

Fig. 3-16 Spot weld joining two sheets of 24-gauge mild steel.

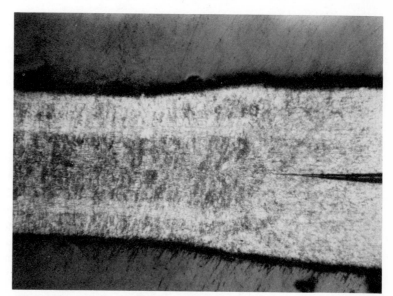

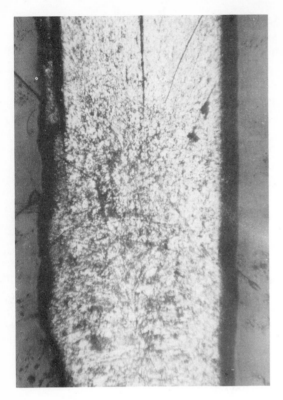

Fig. 3-17 Spot weld joining two sheets of 24-gauge type 304 stainless steel.

## 3.12 viscosity

Fluids have the characteristic that they can be pumped or otherwise made to flow. The term "fluids" therefore includes both liquids and gases. A shear stress is required to make a fluid flow. Thus if water is flowing through a pipe, the water at the axis of the pipe has the highest velocity, while a thin layer of fluid in contact with the pipe wall has zero velocity, and any layer of fluid must shear against the adjacent layer. In a solid material, shear stress is proportional to the *magnitude* of the deformation. In a fluid, the shear stress is proportional to the *time rate* of deformation. We are then led to a better definition of a fluid: a fluid is a substance that undergoes continuous deformation when subjected to a shearing stress.

One of the most important of fluid properties is *viscosity*, which may be defined as the resistance of the fluid to flow or to shear. Comparing molasses to water, molasses is a high-viscosity (viscous) fluid, while water has a low viscosity. Since fluids must be moved by pumps and fans which consume horsepower, the ideal fluid would have zero viscosity. The viscosity of most fluids does not change if the rate of deformation changes. Such fluids are termed *Newtonian fluids.*

Certain fluids however are non-Newtonian. A ductile metal will require a certain magnitude of shear stress before plastic flow occurs, and such behavior is non-Newtonian. A perfectly elastic solid must have an infinitely large viscosity. Still another and an important type of non-Newtonian fluid is a *thixotropic* fluid, of which the most familiar example is a non-sag house paint. If the paint will not sag when applied to a vertical wall, this would seem to imply an exceedingly high viscosity. But if this were the case, the paint could not be brushed out on the wall. Instead, a thixotropic fluid shows a decrease in viscosity when subject to a shear stress: the viscosity of the paint is higher when on the wall than when being brushed out. Other examples of thixotropic fluids are clay-water mixtures and firebrick mortars.

The viscosity of liquids (and solids) *decreases* with temperature; that of gases increases with temperature. The viscosity of liquids increases slightly with pressure up to about 1000 psi, but increases rapidly for higher pressures. Thus the viscosity of some hydraulic oils when used in hydraulic presses at pressure of 10,000 psi may be five times the viscosity at atmospheric pressure. To minimize leakage, higher viscosity fluids are used in high-pressure hydraulic systems, although the higher viscosity of the oil demands greater power consumption in the hydraulic pump.

It is unfortunate that the property of viscosity is measured in a bewildering variety of engineering units. Only the more commonly used units of viscosity will be mentioned here.

The fundamental equation for viscosity may be explained by reference to Fig. 3.18. Here two parallel plates are a distance $b$ feet apart, one plate being fixed and the other moving at velocity $V$ in feet per second. The fluid adheres to each plate, so that the fluid velocity is zero at the fixed plate and $V$ at the moving plate. At any point $y$ above the fixed plate the velocity $u$ is

$$u = \frac{V}{b}y$$

The shearing stress $= \lambda = \mu(V/b)$ where $\mu =$ viscosity. This equation defines viscosity. Therefore the viscosity $= \mu = \lambda(b/V)$.

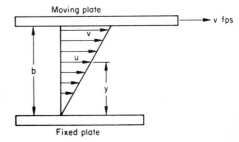

Fig. 3-18 Shear stress in a fluid between two plates.

The absolute viscosity then is a force acting on unit area of a plane surface moving at unit velocity relative to another surface at unit distance away from it. For the units of absolute viscosity, consider that

$$\mu = \lambda \frac{b}{V} = \text{stress}\left(\frac{\text{length}}{\text{velocity}}\right) = \frac{F}{L^2}\frac{L}{L/T}$$

where $F$ = force

$\quad\; L$ = length

$\quad\; T$ = time

giving units for absolute viscosity of pound-seconds per square foot, or in the metric system, dyne-seconds per square centimeter. The metric unit is called a *poise*.

$$1\frac{\text{lb-sec}}{\text{ft}^2} = 478.8 \text{ poise}$$

The disadvantage of these absolute units of viscosity is the difficulty of setting up convenient apparatus for their measurement. The most widely used instrument for determining viscosity in this country is the Saybolt Universal Viscosimeter. Using this instrument, the time in seconds for 60 cu cm of liquid to flow vertically through a small orifice is measured. This time is called the viscosity, expressed as Saybolt seconds universal (SSU). Since viscosity varies with temperature, the temperature of the test must be specified. Temperatures of 100°F and 210°F are in common use.

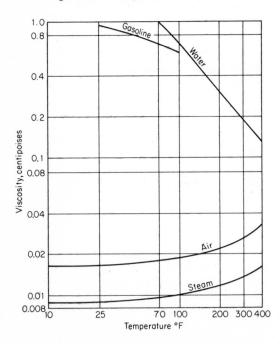

Fig. 3-19 Viscosities of common fluids.

In hydraulic systems, oils with viscosities in the range of 500–950 SSU at 100°F are used for higher pressures, 2500 psi or higher. For low-pressure systems, below 2500 psi, the viscosity is usually in the range of 150 to 300 SSU at 100°F. The latter range is approximately the viscosity of a #20 or #30 motor oil.

## PROBLEMS

1 Plot the stress-strain curve for the steel sample of Sec. 3.4 and determine the 0.2 percent yield stress. Draw a line parallel to the elastic line of the material and passing through the 0.2 percent point on the abscissa axis. The intersection of this line with the stress-strain curve is the 0.2 percent yield stress.

2 Determine the 0.2 percent yield stress and the modulus of elasticity of the materials of Figs. 11.9 and 14.22.

3 A mine timber measuring 4 in. × 4 in. carries a compression load of 8000 lb and measures 50 in. in length. Its modulus of elasticity is $1.5 \times 10^6$. By how much does it shorten under this load?

4 Determine the strain (in./in.) if a magnesium bar is stressed in tension to 10,000 psi. The modulus of elasticity is $6.5 \times 10^6$.

5 What force is required to shear a $\frac{1}{4}$-in. steel plate in punching a 2-in. diameter hole? The ultimate shear strength of the plate is 40,000 psi.

6 What is a thixotropic fluid?

7 Explain anisotropy. Is wood anisotropic? Why?

8 Suppose that a material is elastic in the solid phase under all possible conditions. What methods are available for forming it into useful products?

9 Make tensile tests of polycarbonate, polyethylene, and mylar sheets. Plot stress-strain curves and determine ultimate strength, elongation, modulus of elasticity, and yield stress.

10 Explain why hydrogen is used as a coolant in large electric motors and generators.

11 Explain why copper is a difficult metal to weld.

12 A stainless steel pipe 6.625 in. in outside diameter is butt-welded to a zirconium pipe of the same diameter. This piping system is then heated to 1100°F from 100°F. At this high temperature the butt weld cracks. What is the difference in O.D.'s between the two pipes at 1100°F? Expansion coefficient of the stainless steel is 0.000008, and of the zirconium is 0.000004.

13 A precision gauge block presumed to measure 0.75000 in. in length at 68°F is to be compared with a master block of the same length. The instructions given to the metrologist state that he must compare the two blocks while they are seated on an aluminum plate, and that they must remain on the aluminum plate for 20 min before comparison is undertaken. Explain the reason for the aluminum plate.

**14** A furnace lining of 25,000 firebrick, each weighing 8 lb, has a specific heat of 0.2 Btu/lb/°F. The furnace is heated from cold (100°F) to an operating temperature of 2000°F. At operating temperature the hot face of the brick is at 2000°F and the cold end of the brick at 1000°F. How much heat must be stored in this firebrick lining before the furnace reaches operating temperature?

**15** A 1000-gallon water tank is made of $\frac{1}{8}$-in. steel sheet and has a total surface area of 200 sq ft. The inside surface of the steel shell is at water temperature of 40°F and the outside surface of the shell at 39°F. What is the heat loss from the tank in Btuh? $K$-factor of the steel = 320 Btuh/in.-°F.

**16** The tank in the previous question is next insulated with 1 in. of foamed polyurethane insulation with a $K$-factor of 0.25. The steel shell now supplies a negligible resistance to heat flow and can be ignored. The temperature drop across the insulation is 20°F. What is the heat loss from the insulated tank?

# NATURAL
# MATERIALS

*Part* **2**

The system ice-water-steam is the most familiar and one of the most important of the industrial materials. Both water and steam are used in enormous quantities by industry, though ice is less important. A steam power plant of moderate capacity may consume 20,000,000 gallons of water per hour and produce perhaps 700,000 lb of steam per hr for the production of electric power from a steam turbine driving an electric generator.

The chemical $H_2O$ is in a number of respects a most unusual one. Its most outstanding property is its unparalleled capacity for *heat storage*. The specific heats of ice, water, and steam are all very large, and the latent heat of evaporation of water is about 1000 Btu per lb (1055 joules = 1 Btu).

The human race makes generous use of this heat storage capacity. A small amount of water or steam can remove or transport a great amount of heat, hence the use of steam and water for heating of buildings, or water cooling in various industrial processes. The heat storage of water is also sometimes a disadvantage, as expressed in the time-honored saying, "A watched pot never boils," or in its effect of depressing temperatures in the vicinity of large bodies of water such as Lake Superior.

# WATER
# AND INDUSTRIAL GASES

# 4

Fig. 4-1  Air view of a river during a snowstorm in early winter. The high heat storage of the water in the river produces a microclimate in its vicinity which is mild enough to melt the snow, while the surrounding countryside is blanketed in snow.

A second unusual characteristic of water is its *availability*. Most of the earth's surface is covered by either salt water, fresh water, or ice (permafrost). Although less than 1 percent of the earth's atmosphere is water vapor, this still represents a sizeable reservoir of water. The general availability of water gives it the economic advantage of cheapness.

A third characteristic of water is its general *corrosiveness*, although it is not one of the more aggressive chemicals. Ferrous metals can be afforded a good degree of protection from water corrosion by removing dissolved oxygen from the water in a deaeration process and increasing the alkalinity of the water. Both processes are usually necessary. High-pressure steam is corrosive to ordinary glass.

A disadvantageous characteristic of water is its expansion as it is cooled from 39°F down to ice at 32°F. This volume expansion is about 10 percent (actually 9.3 percent), a considerable amount and the cause of much maintenance expense in colder countries.

Water has a relatively *low viscosity*. This is an advantage, since motors driving water pumps do not demand high horsepower to overcome pressure drops caused by viscosity losses.

Another disadvantage of water is the high pressures required to cause water to boil at temperatures much above 212°F.

Fresh water has a specific gravity of 1.0 at 60°F and 160°F. A cubic foot of water at 60°F weighs 62.4 lb, and a gallon 8.3 lb.

## 4.2 thermal characteristics

## of water

The specific heats of hydrogen oxide are as follows:

| | |
|---|---|
| ice | 0.5 Btu/lb/°F |
| water | 1.0 at low temperatures |
| steam | 0.5 (approximately) |

The latent heat of fusion of ice is 144 Btu per lb. This figure is the basis for the capacity rating of refrigeration equipment. One ton of refrigeration is the refrigeration capacity needed to remove sufficient heat to freeze 1 ton of

Fig. 4-2   A utilidor. When building on permafrost (permanently frozen ground) in arctic areas, utilities such as sewer and water cannot be laid underground. Instead, a steam-heated and insulated trench, called a utilidor, carries sewer, water, steam heat, electric power and other utilities to all buildings of the town. The arctic is still waiting for some materials engineer to design a good utilidor, since existing installations have many defects.

water at 32°F in 24 hours. This is a heat removal capacity of $2000 \times 144$ or 288,000 Btu per 24 hrs, 12,000 Btuh, or 200 Btu/min.

Knowledge of the properties of steam and water is necessary if a person is to work in such fields as operating engineering, heating, refrigeration, air-conditioning, and internal combustion engines. The properties of steam and water, including specific heat, specific volume, latent heat, and others, are tabulated in detail in the steam tables, which are not explained here but may be found in basic texts on thermodynamics.

Sometimes we speak of steam, sometimes of water vapor. There is, however, no fundamental difference. If the steam is part of an atmospheric mixture of air and steam, it is referred to as "water vapor." If the steam is contained in a piping system, boiler, or heat exchanger, it is referred to as "steam." Again, we may use the word "boiling" or the word "evaporation." There is no difference between boiling and evaporation: both represent a phase change between the liquid and the gas state of hydrogen oxide.

## 4.3   water treatment

The solvent powers of water bring with them a host of problems that must be solved by complex and expensive water treatment processes. Water containing dissolved oxygen cannot be pumped into a boiler because the boiler tubes would be corroded by the combination of water and oxygen. Hard water cannot be used by soft drink manufacturers or breweries, or by laundries, because of damage to clothing and reasons of soap consumption. Water containing very small amounts of dissolved iron will stain utensils, plumbing fixtures, and dishes red. Most water used for industry or domestic consumption therefore requires treatment to remove or reduce dissolved gases, liquids, and solids.

There is no general-purpose water treatment method. Before any particular water treatment can be specified, an analysis of the water must be made. A standard water analysis provides data on such matters as the $p$H of the water and the dissolved and suspended solids in parts per million by weight (ppm). Ten thousand parts per million is equal to 1 percent. The $p$H value of the water expresses its alkalinity or acidity. A distilled or neutral water sample has a $p$H of 7.0. Smaller values indicate an acid water. Thus beer or soft drinks may have a $p$H of about 4. Values higher than 7.0 indicate an alkaline water. Lime or other basic compounds are commonly added to water to increase the $p$H, since acid waters are highly corrosive to piping systems.

As a comparison, two water analyses are presented below, the first one for a western city with an excellent water supply, the second an analysis of well water so poor that the well was not used.

|            | City        | Well       |
|------------|-------------|------------|
| pH         | 6.9         | 7.8        |
| calcium    | 25.2 ppm    | 1060 ppm   |
| magnesium  | 8.1         | 648        |
| sodium     | –           | 3660       |
| total cations | 33.3     | 5368       |
| bicarbonate | 107.0 ppm  | 217 ppm    |
| carbonate  | nil         | nil        |
| chloride   | 2.1         | 3817       |
| sulfate    | 9.5         | 1310       |
| total anions | 118.6     | 5344       |
| iron oxide | nil         | 0.53 ppm   |
| total solids | 125 ppm   | 6874 ppm   |

These two water analyses may be compared with the water standards of the U.S. Department of Health & Public Welfare:

|            | Maximum Permissible | Suggested Limits |
|------------|---------------------|------------------|
| lead       | 0.1 ppm             |                  |
| fluorine   | 1.5                 |                  |
| arsenic    | 0.05                |                  |
| copper     | 2                   |                  |
| zinc       |                     | 15.0 ppm         |
| iron       |                     | 0.3              |
| total solids |                   | 500–1000 ppm     |
| magnesium  |                     | 125              |

The city water analysis exceeds the standard, although its low $pH$ suggests that the water will be somewhat corrosive to piping and hot water tanks: experience confirms this. The $pH$ of a water supply should be at least 8.0 to prevent corrosion. In the case of the well water, the $pH$ is acceptable, but 6874 ppm of solids means that over $\frac{1}{2}$ of 1 percent of the supply is solid matter, or 1 gallon of solids in every 200 gallons of water. The water is also quite brackish, since the total of sodium and chloride is over 7500 ppm. Water begins to taste salty when the salt concentration exceeds 500 ppm.

The dissolved compounds that make a water "hard" are calcium bicarbonate, calcium carbonate, calcium sulfate, magnesium bicarbonate, magnesium carbonate, and magnesium sulfate. If the sum of these dissolved compounds is less than 60, the water is called soft; in concentrations exceeding

180 the water is considered hard. "Hardness" is the quality in water that prevents a soap from forming a lather, though the hardness components listed have a variety of other harmful effects. They may leave deposits on pipe walls to form a hard scale, and instances are known in which pipes were completely closed by scale. Again, if carbonates in water are heated in boilers or heat exchangers, they break down to release carbon dioxide. The combination of carbon dioxide and water forms carbonic acid, which is corrosive to iron and steel.

At 70°F and atmospheric pressure, water can dissolve 8 parts per million of oxygen. This small amount of oxygen is necessary in fresh water for the survival of fish and other animal life, but it causes corrosion in industrial equipment. This oxygen must be removed before water is fed into boilers. This is done in deaerators which heat the water to release the dissolved oxygen.

## 4.4 ideal and imperfect gases

There are three basic physical properties of gases: pressure, temperature, and volume, or specific volume. An absolute temperature scale is necessary for gas calculations, either the Kelvin or the Rankine scale; the latter is used in engineering in some English-speaking countries such as the USA and Canada. The Kelvin and Rankine scales are defined as follows:

$$°K = °C + 273$$
$$°R = °F + 460$$

The temperatures $-273$ and $-460$ are of course absolute temperatures for the Centigrade and Fahrenheit temperature scales, respectively.

Similarly, absolute pressures must be employed in gas calculations rather than gauge pressures. Gauge pressure means pressure above atmospheric pressure. Atmospheric pressure reads zero on a standard Bourdon pressure gauge, but it is actually 14.7 psi absolute at sea level. Pressure in psi absolute (psia) is equal to gauge pressure (psig) plus 14.7.

In this country a variety of other pressure scales are in use in industry. Fuel gas pressures and air pressures in ventilating systems are often measured in inches of water. Gas burners for industrial furnaces may use ounces per square inch. In instrumentation work, inches of mercury may be used for pressure readings. Finally, vacuum systems use the torr as the unit of pressure.

| | |
|---|---|
| 34 ft of water column (34 ft w.c.) | = 14.7 psi |
| 2.3 ft w.c. | = 1 psi |
| 1 ft of water | = 0.434 psi |
| 30 in. Hg | = 14.7 psi |

| 760 mm Hg | = 14.7 psi |
| specific gravity of mercury | = 13.6 |
| 1 in. Hg | = 0.491 psi |
| 1 torr | = 1 mm Hg |

Units of pressure, like most other engineering units in use in North America, are unnecessarily confusing, but practitioners seem to survive with few complaints. Consider the following example:

A centrifugal fan supplies air at a pressure of 1 in. w.c. What is the fan pressure expressed as psi?

$$34 \text{ ft w.c.} = 14.7 \text{ psi} (= 101.5 \times 10^3 \text{ N/m}^2)$$

$$1 \text{ in. w.c.} = 1/12 \text{ ft} = \frac{1}{12 \times 34} \times 14.7 = 0.036 \text{ psi} (= 248 \text{ N/m}^2)$$

Certain gases follow simple mathematical relationships between their basic properties of temperature, pressure, and volume. These relationships are called Boyle's law and Charles's law:

If the gas pressure is held constant, the volume of a gas varies directly with its temperature.

If the temperature of the gas is held constant, the volume of the gas is inversely proportional to its pressure.

Scientific "laws," like the statutory kind, are honored both in the breach and the observance. Gases whose behavior is in close accord with these two relationships are called "ideal" or "perfect" gases. Gases which do not closely obey such rules are called "imperfect" gases. In practical terms, we might say that the pressure, volume, or temperature of an ideal gas may be determined by calculation, but that this procedure is not readily possible for an imperfect gas, for which gas tables similar to the steam tables must be used. Steam is an imperfect gas; therefore the properties of steam are obtained from steam tables, not from Boyle's or Charles's law. Tables similar to those for steam are available for other imperfect gases, such as ammonia.

| Ideal Gases | Imperfect Gases |
|---|---|
| oxygen | steam |
| nitrogen | carbon dioxide |
| air | ammonia |
| argon | sulfur dioxide |
| helium | freon refrigerants |
| hydrogen | methyl chloride |
| | methane, propane, butane |

## 4.5 atmospheric gases

The gas constituents of the earth's atmosphere were tabulated in the first chapter. The most important constituent, oxygen, is of course of critical importance to both life and industry. The oxygen however is heavily diluted with nitrogen; air is mostly nitrogen. The principal uses of air are the following:

1. oxygen supply for combustion processes
2. ventilation air for environmental control
3. automatic control systems operated by air, usually at 20 psig supply pressure
4. pneumatic power systems operated at pressures up to about 80 psig, as in reciprocating air cylinders
5. compressed air for the operation of paint spray guns, pneumatic rock drills, etc.
6. pneumatic tires for highway and off-the-road vehicles

The consumption of air for these purposes runs to large figures. Over 125 lb of air are inspirated into an engine's cylinders to burn a gallon of fuel, or about 20 times as much air as fuel by weight. A small ventilating fan may handle 500 lb of air every hour. A steam power station may consume 25,000 tons of air in an hour.

1 standard cubic foot of air (scf) at 70°F and 14.7 psia = 0.075 lb

13.4 scf air = 1 lb

These two constants are of basic importance in mechanical work, and they must be memorized by practitioners who design with air. The specific heat of air at constant pressure at 70°F is 0.24, at 500°F 0.247, and at 1000°F 0.263. The viscosity of air at 70°F is 0.0000126 lb-sec/ft², increasing to 0.0000185 at 500°F.

**Nitrogen.** The properties and constants of nitrogen must necessarily be nearly those of air. Nitrogen is a rather inert gas. It boils at −320.3°F at atmospheric pressure and cannot be liquefied above −232° (critical temperature). It is available in standard gas cylinders and is chiefly used as liquid nitrogen for refrigeration of transport vehicles and as an inert or purging gas.

**Oxygen.** Oxygen is a slightly magnetic gas, weighing 11 percent more than air. Its boiling point at atmospheric pressure is −297.3°F. It is probably the most dangerous of the industrial gases. Many hydrocarbons, such as greases, will explode in the presence of oxygen.

Oxygen is available both as therapy oxygen for hospital use and industrial oxygen in 2400 psi cylinders and in larger quantities. The two types of oxygen are supplied in cylinders with a different color coding, and a cylinder of oxygen for the one purpose is never permitted to be used for the other purpose, because industrial oxygen may become contaminated with other gases through the piping system to which a cylinder of industrial oxygen is attached.

Liquid oxygen, or lox, is consumed in tonnage quantities chiefly for the following uses:

1. rocket fuel
2. steel refining. Oxygen is used to burn excess carbon out of steel and to accelerate the operation of blast furnaces and open-hearth furnaces
3. flame drilling of hard rock, using kerosene as fuel

In addition, oxygen is used as a welding gas for both gas welding and certain types of arc welding and for torch cutting of steels.

**Argon.**   Argon is a monatomic inert gas which does not dissolve in liquid metals. These characteristics make it an especially useful gas. Its boiling point at atmospheric pressure is $-302.4°F$, and its specific gravity compared to air as 1.0 is 1.38. Argon serves a wide range of uses in incandescent and fluorescent lamps, Geiger-Muller counter tubes, and inert atmosphere heating furnaces. Its largest use is in the welding and thermal cutting of metals that must be protected from oxidation, such as zirconium, aluminum, and titanium.

**Helium.**   Helium is another monatomic inert gas which does not dissolve in liquid metals. It therefore competes with argon in its industrial applications. Like argon, helium may be extracted from air, but it is also available from natural gas wells in the western states and western Canada. Its specific weight at 70°F and 14.7 psia is exactly 10 percent that of argon. Its boiling point is very close to absolute zero, $-452.1°$. The specific heat of helium is very high, 1.25 Btu/lb, second only to that of hydrogen.

Because of its small molecule, helium serves as a most sensitive leak detector. It is also used as a refrigerant in the attainment of temperatures near absolute zero.

**Neon.**   Neon is familiar from the characteristic red color of the electric discharge through such neon-filled tubes as are used in advertising signs and stroboscopes.

**Xenon.**   Xenon is also used in electric-discharge tubes. The visible spectrum from such a tube is very similar to that from the sun. Such tubes give remarkably high illumination intensity and are therefore used to activate lasers and for microscope illumination.

**Krypton.**   Krypton is another fill gas for tubes and is added to fluorescent lamps to increase the lighting intensity. Krypton-85 is a radioisotope used for leak testing.

## 4.6 carbon dioxide

Carbon dioxide gas and its solid phase, popularly termed "dry ice," is one of the most versatile chemicals of industry, with the added virtue of cheapness. It is not recovered from air but is usually produced by burning carbon compounds. Carbon dioxide gas is 53 percent heavier than air. At atmospheric pressure the phase change from dry ice to $CO_2$ gas occurs at $-109.2°$. Some of the uses of dry ice are these:

1. general purpose refrigerant
2. shrink-fitting of mating parts by cooling the inserted part
3. branding of cattle with an iron cooled with dry ice

Fig. 4-3  Cylinders of welding and medical gases at a cylinder-filling plant of Liquid Carbonic Corp.

Gaseous carbon dioxide is used in the following and many other applications in food, drug, and chemical processing:

1. fire extinguishers
2. heat transfer medium in British nuclear reactors
3. soft drinks and beer
4. removal of heat during machining and grinding operations on plastics, soft rubber, and other heat-sensitive materials
5. inert atmosphere and purging gas for removing flammable vapors
6. pressure gas for aerosol spray cans
7. greenhouse atmosphere to increase yield of tomatoes and other plants
8. shielding gas for welding

The familiar tall, small-diameter, color-coded cylinders used for welding gas supply are usually rated for a maximum internal pressure of 2400 psi, except in the case of acetylene. Such cylinders are seamless deep-drawn tubes, with a wall thickness of about $\frac{1}{4}$ in., manufactured in the same presses as drawn artillery shells. The handling of such cylinders is subject to government regulation, and every cylinder must be hydrostatically tested every five years.

The use of oxygen and acetylene for torch welding of metals is about 60 years old. However, this type of gas welding is of small importance in production welding on the factory floor, where electric arc methods dominate the scene. The most important of the production welding processes is the MIG method (pronounced "mig", an abbreviation of metal-inert gas). In this welding method a continuous spooled wire is fed into the arc at a controlled rate and is protected from the atmosphere by a flow of shielding gas from the welding gun. The most commonly used gas for MIG welding is $CO_2$. Except for spot welding, the welding done on automobiles and trucks is performed by the $MIG-CO_2$ method. However, $CO_2$ is not an inert gas,

Fig. 4-4 MIG welding of light angle iron frames using a shielding gas of 5 percent argon and 95 percent carbon dioxide. A spool of MIG wire can be seen on the right of the photograph. The extensive fine spatter of the MIG arc as seen here is characteristic of this welding method.

since it can dissociate into carbon monoxide and oxygen. Carbon dioxide therefore cannot be used for welding such readily oxidizable metals as aluminum, copper, and titanium. For such metals, either argon or helium is employed to shield the arc. Helium, being only one tenth the weight of argon, does not provide quite as good shielding of the arc in the presence of drafts or air currents.

Helium has a higher ionization voltage than argon. At a MIG welding current of 200 amps the arc potential drop will be about 25 v with argon and 30 v with helium. These voltages are not the ionization voltages of the gases but the arc voltage drop. Since the heat generated by the arc is proportional to the product of amps times volts, more heat is produced by helium. Therefore helium is preferred for welding metals of high thermal conductivity to compensate for heat conduction losses through the metal. In particular it may be impossible to weld copper with argon because of the unusually high thermal conductivity of this metal.

## 4.8   fuel gases

The most commonly used fuel gases are hydrogen ($H_2$), acetylene ($C_2H_2$), natural gas (largely methane, $CH_4$), propane ($C_3H_8$), and butane ($C_4H_{10}$). These gases are burned either with oxygen or air, the chemical reaction being called *combustion*. Combustion is an *oxidation* process occurring at a sufficiently rapid rate to produce a high temperature, usually with the appearance of a flame. If the total amount of oxidizable material oxidizes at a more or less uncontrolled rate, the process is then called *explosion*.

Gas burners, sometimes called torches, are devices for mixing gas and air in the proper proportions for combustion. The actual combustion process occurs in the visible flame after the fuel gas and the oxygen are mixed. Since the gases issue from the burner ports at a constant velocity, the flame front must continually burn back toward the burner. Therefore the size of the burner ports must produce a forward velocity of the same order as the *flame propagation velocity*, the speed at which the flame propagates through the gas mixture. Different gases have different flame propagation velocities and different air-gas ratios; this is the reason that the burner or torch must be changed if the fuel gas is changed. The flame velocity is influenced by many variables and cannot be readily stated for any fuel gas. Methane is a typical slow-burning gas: its flame velocity is only a few feet per second. Hydrogen burns at a high velocity, of the order of 20 ft per sec.

For the design of gas piping and burners the *specific gravity* of the gas must be known. When the specific gravity of a gas is reported, it is compared to air as 1.0, not water as 1.0, as is done in the case of liquids. Thus natural gas has a specific gravity of 0.65 (it varies from city to city between the limits of 0.60 and 0.70), indicating that natural gas is two-thirds as heavy as air.

The *heating value* of the fuel gas is the amount of combustion heat released per cubic foot of fuel gas measured at standard temperature and pressure and with the products of combustion cooled back to the initial temperature. The gross heating value is usually stipulated: this includes the latent heat of condensable water vapor.

The *flame temperature* of a fuel gas cannot be accurately measured. Usually a theoretical flame temperature is referred to. This is the flame temperature assuming instantaneous and perfect combustion with no heat loss to the surroundings. For a theoretical flame temperature all the combustion heat is used to heat the products of combustion, that is,

$$Q = wc(t_2 - t_1)$$

where $Q$ = Btu per pound liberated per pound of fuel

$w$ = weight of combustion products, pounds

$c$ = specific heat of combustion products

$t_2$ = flame temperature

$t_1$ = initial temperature

Although this equation represents a simple specific heat calculation, it must be noted that the specific heat of the combustion products must be the mean specific heat of the mixture of product gases over the range $(t_2 - t_1)$. This is a somewhat uncertain figure. We may attempt an example, however. Suppose we find the theoretical flame temperature for natural gas burned with air, using approximations.

We will assume natural gas to be pure methane, $t_1$ to be zero, and the specific heat of the combustion products ($CO_2$, $H_2O$, and $N_2$) to be 0.33.

The heating value of methane nearly equals 24,000 Btu per pound. There are almost 18 lb of products yielded per pound of methane.

$$24,000 = 18 \times 0.33(t_2 - 0)$$
$$t_2 = 4000°F$$

This solution is a few hundred degrees too high.

A much higher flame temperature is produced in the absence of nitrogen, which forms a substantial fraction of the combustion products when air is used. The nitrogen contributes nothing to the combustion process but acts as a heating load on the heat produced.

## PROPERTIES OF FUEL GASES

| Gas | Sp Gr | Flame Temperature | Lbs Products per lb Gas | Gross Btu/lb | Heating Value Btu/cu ft |
|-----|-------|-------------------|-------------------------|--------------|-------------------------|
| hydrogen in $O_2$ | 0.069 | 4000 | 8.94 | 61,000 | |
| acetylene in $O_2$ | 0.90 | 6200 | 4.07 | 21,570 | |
| natural gas in air | 0.65 | 3600 | 18.2 | 23,900 | 1000 |
| propane in air | 1.52 | 3660 | 16.6 | 21,800 | 2510 |
| butane in air | 1.95 | 3640 | 16.3 | 214,000 | 3100 |

*Hydrogen* has a higher specific heat than any other substance, 3.4 Btu per lb. This explains its use as a cooling gas in large electric motors and generators. Hydrogen is also used as a rocket fuel and as a reducing agent in the manufacture of parts made from metal powders, such as carbide cutting tools and self-lubricated sintered bearings. Without hydrogen, the film of oxide on the surface of the metal grains would prevent them from sintering together.

*Acetylene* is well known for its use in oxyacetylene welding and thermal cutting of metals. For welding purposes it is commonly supplied in squat black cylinders fitted with left-hand threads. Its flame temperature is the highest obtainable from the fuel gases.

Fig. 4-5 Oxyacetylene flame-cutting of steel plate with five cutting heads.

Acetylene explodes readily if compressed to pressures above 15 psig. The explosion requires no oxygen, the acetylene simply decomposing into carbon and hydrogen. Cylinders of acetylene therefore are filled by a unique method. The cylinder contains a porous plastic or other substance, and this filler is used to absorb acetone. The acetylene gas is then dissolved in the acetone. Because of this special preparation, considerable time is required to fill an acetylene cylinder to the usual pressure of 250 psig.

*Propane* and *butane* are referred to as *liquefied petroleum gases* or LPG. These are obtained from both oil and gas wells. Figure 4.6 shows storage tanks for these gases at a tank farm. The stored gas may be identified by the shape of its storage vessel. The bullet-shaped tanks store propane, and the spherical tanks contain butane. Both gases are used for heating for gas welding and cutting.

Fig. 4-6 A tank farm for liquefied petroleum gases. The horizontal tanks are for propane storage; the spherical tanks store butane.

*Natural gas* is composed of 90 percent or more methane, together with small amounts of ethane, nitrogen, and carbon dioxide. Since natural gas is odorless, a small amount of a sulfur compound is added. The smell provides warning of gas leaks.

### 4.9 vacuum

A vacuum, as used in industry, must be considered as an atmosphere or gas with special inert properties. Vacua are increasingly employed for industrial processing, so that recent years have seen the development of a separate "vacuum technology."

Industrial vacua for process uses are almost always what are called "very high vacua," that is, in the range of $10^{-3}$ to $10^{-7}$ torr ($10^{-3}$ to $10^{-7}$ mm Hg). At least two vacuum pumps in series are required to reach such low vacuum conditions. Vacua for such purposes as electron beam welding or the vacuum melting of steel are at the high pressure end of this range, but vacuum tubes, including X-ray tubes and oscilloscope tubes, must maintain a vacuum of at least $10^{-6}$ torr if they are to function properly. If the pressure is too high, there are too many gas molecules present, and the electrons of the tube current lose excessive energy in collisions with gas molecules.

A wide range of metals react rapidly with oxygen at temperatures above 1000°F and must be melted under vacuum conditions. Such metals include titanium, beryllium, zirconium, all the refractory metals, and the semiconductors such as silicon and germanium. Circumstances sometimes permit the heat-treating or welding of such metals under a protective atmosphere of argon or helium. However the welding of these metals by means of a beam of electrons under a vacuum of about $10^{-4}$ torr is perhaps more common.

The evaporation of low-melting metals under low vacua has been mentioned in an earlier chapter. Vacuum deposition of metal films has become a standard practice. The reflecting surfaces of sealed-beam headlights for automobiles are vacuum-deposited.

## PROBLEMS

1  A bale of wheat straw can absorb 50 pounds of water when saturated. Approximately how many Btu's of heat are required to evaporate this water from the bale?

2  If a crescent wrench is submerged in a few inches of water, it rusts; if submerged in a few feet of water it does not rust. Can you explain why?

3  Why is the city water whose analysis is given in Sec. 4.3 corrosive to piping systems? What kind of water treatment would make it less corrosive?

4  Foamed polystyrene weighing 1.5 pounds/cubic foot is often used as a flotation material. What is the maximum load that a foam board 4 ft × 8 ft × 2 in. will support in fresh water?

5  Determine the maximum weight that a standard 55-gallon drum when empty will support in fresh water; allow for the weight of the drum.

6  The specific gravity of polypropylene plastic is 0.90. What is its weight in pounds per cubic foot?

7  What is the absolute pressure in psia of a vacuum of 25 inches of mercury?

8  A fan delivers 25,000 cfm of air at standard temperature and pressure. What is the weight of this air in pounds per minute?

9  Why is therapy oxygen delivered in different bottles from those used to contain welding oxygen?

10  Why is helium gas such a sensitive leak detector?

11  As you drive through a strange town, you estimate the height of its overhead water tank to be 92 ft. What then is the maximum pressure available at a water tap at ground level in town?

**12** A steam heating system cuts in at 5 psig and cuts out at 8 psig. What are the corresponding absolute pressures at sea level?

**13** Define the torr.

**14** How many inches and millimeters of mercury are the equivalent of atmospheric pressure at sea level?

**15** A fan delivers air to a building at 6 in. w.c. pressure. State this pressure in psig.

**16** Why are argon and helium used as welding gases?

**17** The fuel supplied to a torch designed for natural gas is changed to hydrogen. Consult the remarks on flame propagation velocity and explain what will probably happen if this torch is lit.

**18** What is the chief constituent of natural gas?

**19** Sometimes flames with low temperatures are required for industrial operations. An excess-air burner is employed for such a flame. Why will excess air provide a cooler flame?

## 5.1 mechanical and electrical

## equipment

Mechanical engineering and electrical engineering are two broad technical areas which, though nominally distinct, overlap in such a comprehensive fashion that it is often impossible to designate a problem or project as specifically "electrical" or "mechanical." A mechanical engineer simply cannot ply his trade without comprehensive knowledge of things electrical. Similarly, in time of trouble the electrical engineer may discover that the defects in his electrical equipment are mechanical defects.

Consider first some mechanical aspects of electrical equipment. If an electric motor is down for maintenance, the reason may be either bearings or a burnout, both mechanical failures. Breakage of electric power lines caused by ice accumulation or wind vibration is mechanical failure. Fracture of the Bakelite parts of switchgear is mechanical failure. So is evaporation of the tungsten filament when an incandescent lamp fails. The dissipation of heat from electronic microcomponents is a mechanical design problem, as are the glass-to-metal seals required in all vacuum tubes Despite its name, the

# ELECTRICAL
# PROPERTIES
# OF MATERIALS

# 5

General Electric Company is famous for its mechanical engineering, both in research and in products.

The uses of electrical equipment for nonelectrical applications are almost without number. Most welding equipment is electrical. Most of the nondestructive testing methods used in manufacturing operations are electrical methods—ultrasonic, X-ray, magnetic flaw detection, eddy-current testing. An electronic induction heater used for brazing and heat-treating of metals is similar in design and components to the frequency generator of a radio transmitting station. The electrical characteristics of materials, then, are not of peripheral importance, or of interest only to electrical specialists. Ignorance of these electrical properties may prove a severe occupational handicap. The following is a case in point.

A weld shop decided to install a common ground for ten 300-amp welding machines. Without making any electrical calculations, it was decided to use a $\frac{5}{8}$-in. diameter steel bar for the ground, simply because such a bar would be "plenty big enough." It wasn't, and it had to be replaced immediately after it was installed, much to the embarrassment of the supervisor responsible.

## 5.2  electrical resistance

The one-dimensional flow of electricity through a resistance and the similar flow of heat by conduction are mathematically identical. For electric current

$$E = IR, \text{ Ohm's law}$$

$$R = \frac{\sigma L}{A}$$

where $\sigma$ = the specific resistance of the resistor material in some suitable units;

$$L = \text{length of the conductor}$$

$$A = \text{cross section of the conductor}$$

For heat flow by one-dimensional conduction, $Q = (KA\Delta t)/L$. If the formula for heat flow is rearranged, it becomes identical mathematically with Ohm's law:

$$\Delta t = Q\left(\frac{L}{KA}\right)$$

Here $\Delta t$, the temperature differential, is the potential causing heat flow and corresponding to $E$, $Q$ Btu/hr is the flow, and $L/KA$ is the heat-flow resistance $R$, where $1/K$ corresponds to the specific resistance.

For purposes of mechanical engineering, the most convenient unit for specific resistance is ohms per cubic inch, because the conductors involved in mechanical engineering may be of rectangular or irregular shape. Most resistance calculations in electrical engineering are concerned with wire and cable of circular cross section, for which the circular mil-foot is a convenient unit for specific resistance. Finally, for scientific work, the unit of ohm-centimeter is used.

If $L$ is given in inches, and $A$ in square inches, then the units of $\sigma$ are ohm-inches (ohms per cubic inch). If the circular mil-foot is used, then $L$ must be in feet, and $A$ must be in circular mils. The area of a circle with diameter $n$ mils (thousandths of an inch) is simply $n^2$ circular mils. This avoids the use of $\pi$ in calculations. A square inch of wire contains 1,273,240 circular mils. The following examples are offered for clarification.

**Example 1.** A frozen steel pipe with a 2-in. nominal diameter is to be thawed by running current from a welding transformer through it. The two welding cables are connected to the pipe at points 200 ft apart. The cross-section of the pipe wall is 1.075 sq in. If the specific resistance of steel at 32°F is 5.0 microhm-inches, what is the resistance of the pipe in ohms?

Here

$$L = 200 \times 12 \text{ in.}$$

$$A = 1.075 \text{ sq in.}$$

$$\sigma = 0.000\ 005 \text{ ohm-in.}$$

and

$$R = 0.0111 \text{ ohm}$$

**Example 2.** A nickel alloy resistance wire has a specific resistance of 800 ohms per circular mil-foot. What is the resistance of such a wire 0.1 in. in diameter and 5 ft long?

$$0.1 \text{ in.} = 100 \text{ mils}$$

$$A = 10,000 \text{ circular mils}$$

$$L = 5 \text{ ft}$$

and

$$R = 0.4 \text{ ohm}$$

Wire used as electrical conductor is usually identified either by gauge number or by circular mils. Electric wire uses the American Wire Gauge (AWG), and steel wire not for electrical conduction uses the Birmingham Wire Gauge (BWG).

## ELECTRIC WIRE TABLE

| AWG | Diameter, mils | Cross-section, circ. mils | Cross-section, sq in. |
|---|---|---|---|
| 0000 | 460 | 211,600 | 0.1662 |
| 000 | 409.6 | 167,800 | 0.1318 |
| 00 | 364.8 | 133,100 | 0.1045 |
| 0 | 324.9 | 105,500 | 0.0809 |
| 1 | 289.3 | 83,690 | 0.0657 |
| 2 | 257.6 | 66,370 | 0.0521 |
| 4 | 204.3 | 41,740 | 0.0328 |
| 6 | 162 | 26,250 | 0.0206 |
| 8 | 128.5 | 16,510 | 0.0130 |
| 10 | 101.9 | 10,380 | 0.0082 |

## SPECIFIC RESISTANCE OF MATERIALS AT 20°C

| Material | Microhm-inches | Ohms/circ. mil-foot |
|---|---|---|
| silver | 0.6 | 9.6 |
| copper | 0.7 | 10.4 |
| aluminum | 1.12 | 17.1 |
| mercury | 3.8 | – |
| nickel | 2.7 | 41 |
| iron | 4.0 | 60 |
| mild steel | 5.0 | – |
| 302 stainless steel | 29.0 | – |
| tungsten | 2.2 | 33.3 |
| carbon | 1300 | – |
| silicon | 450 | – |

In the above table of specific resistances, there are some uncertainties, since such factors as the purity of the material, the amount of cold-working, and the presence of alloying elements are not specified. Such factors may have a powerful influence on resistance.

Four methods are available for changing the properties of a material, including its electrical resistivity:

1. *Alloying.* Alloying increases both the strength of a metal and its electrical resistance, but it reduces the electrical resistance of a semiconductor.
2. *Plastic deformation.* This increases the strength and resistivity of a metal, and it decreases its ductility.
3. *Heat treatment.* Depending on the heat treatment, properties may be improved or harmed.
4. *Irradiation* by high-energy particles.

Fig. 5-1 Induction heating. The copper coil carries a high-frequency current of high amperage, inducing eddy currents in the bar inserted in it.

The effect of alloying on electrical resistance may be noted in the table of specific resistances by comparing resistance values for pure iron, mild steel, and stainless steel. Mild steel is iron alloyed with about 0.2 percent carbon and 0.3 percent of silicon and of manganese, or a total of less than 1 percent alloying additions. The effect, however, is a 20 percent increase in resistance. Type 302 stainless steel contains about 29 percent of alloying additions, chiefly chromium and nickel. As a result the resistance is increased by a factor of 7 over that of pure iron. For the same reasons, brasses have about 4 times the resistance of pure copper. About 0.1 percent of phosphorus in copper will double the electrical resistance of copper. Silicon and aluminum additions greatly increase the resistance of steel and are used when high-resistance steels are desired, as in the steel cores of transformers. The effect of alloying elements on specific resistance is so powerful that the purity of metals and semiconductors is actually evaluated by means of resistance measurements.

The resistance of most pure metals increases by a factor of 0.4 percent per degree Centigrade, or 0.23 percent per degree Fahrenheit. This statement is not true of alloys of magnesium, nickel, mercury, or of the rare earth metals, all of which may have temperature coefficients of resistance greater or less than this value.

All substances conduct electrically charged particles. According to the Ohm's law equation, $E = IR$, a nonconductor would ideally have an infinite resistance. No known material has this characteristic. However, three kinds of electrically conductive materials may be distinguished:

1. conductors
2. semiconductors
3. insulators

*Conductors* are metals and therefore have low impedance to the flow of electrons and increasing resistance to electron flow with increasing temperature. The specific resistance of conductors is in the range of $1 \times 10^{-6}$ to about $1 \times 10^{-14}$ ohm-cm. (Specific resistance in ohm-cm is about 2.4 times resistance expressed in ohm-inches.) A *semiconductor* is partially insulating and partially conducting, with a specific resistance in the range of about $10^{-2}$ to $10^6$ ohm-cm. All semiconductors have less resistance at higher temperatures and in less pure states. An *insulator* or dielectric has a specific resistance in the range of $10^6$ to $10^{20}$ ohm-cm. Of all the properties of materials, none has a wider range than specific resistance—roughly $10^{40}$.

Germanium is a typical semiconductor. At room temperature it has a specific resistance of about 60 ohm-cm. When doped with 0.0001 percent of arsenic, its resistance falls drastically to about 0.1 microhm-cm. The decrease in resistance is explained by the valence electrons of the two materials: germanium has 4 and arsenic 5 valence electrons. The extra electron of arsenic becomes free for conduction, greatly increasing the conductivity. Other elements with 5 valence electrons would have the same effect on germanium, such as phosphorus. Such elements are called "donors," and their presence in a semiconductor produces what is called an *n-type* (for negative) semiconductor. Germanium could be doped with an element such as aluminum, which has only three valence electrons. Such an element produces electron deficiencies called "holes" and is called an "acceptor." The resulting semiconductor is a *p-type* (for positive).

The *n*-type impurities contribute electrons that appear in the conduction band of the solid. Current in a semiconductor is carried by electrons in the conduction band and by holes in the valence band. The hole travels when an electron fills it, thus leaving a new hole which likewise is filled by an electron leaving another hole, and so on. Crystal imperfections such as grain boundaries promote the recombination of holes and electrons, and they must be avoided. For this reason semiconductors are produced as single crystals.

In general, a *semiconductor* must possess the following characteristics:

1. a forbidden energy gap between the valence band and the conduction band of about 1 electron volt
2. sensitivity of its resistance to minute amounts of impurities
3. decreasing resistance with increasing temperature

An *insulator* has the following characteristics:

1. a wide forbidden energy gap between the valence band and the conduction band of several electron volts
2. much higher resistance than a semiconductor
3. resistance less sensitive to impurities

Metals, which are electrical conductors, conduct as electrons move through the crystal lattice under the attraction of an applied voltage. As the electrons move along the conductor, they meet interference from ions in the space lattice: this impedance to flow is the ohmic resistance of the material. If the metal conductor is heated to a higher temperature, the ions of the space lattice vibrate over greater amplitudes, thus producing a greater collision cross-section for the moving electron. The effect is an increased resistance. In a perfect crystal without imperfections, at absolute zero, there should be no interference to electron flow and no resistance. This condition is the superconducting condition, to be discussed presently.

The resistance of a conductor is expressed as ohms or microhms (millionths of an ohm). Conductivity is the reciprocal of resistance, with the units of mhos. Mhos = 1/ohms. In certain applications, such as water treatment and water analysis, the conductivity is measured instead of the resistance.

Figure 5.2 shows the approximate specific resistance of some refractory oxides. The resistance of these insulating materials decreases markedly at higher temperatures, as in the case of semiconductors. The specific resistance of any ceramic material is sensitive to the presence of impurities such as foreign oxides.

For the transport of heavy currents in welding and the process industries, carbon and graphite are commonly used. Of the three phases of carbon,

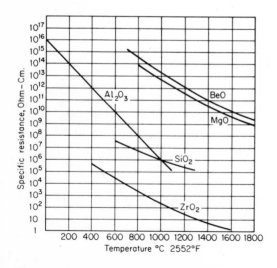

Fig. 5-2  Typical electrical resistance of refractories.

diamond is an insulator, and carbon and graphite are semiconductors. Graphite has about one-quarter the resistance of carbon, and it is the preferred material for electrodes in carbon arc welding and in arc-melting furnaces. The specific resistance of carbon ranges from about 0.0010 to 0.004 ohm-inches at 20°C, decreasing at higher temperatures.

The electrical resistance of the common plastics and rubbers ranges between $10^{15}$ and $10^{20}$ ohm-inches.

## 5.3 applications of resistance materials

*Resistance wire* is special wire for the purpose of converting electric energy into heat. Most such wire is either nickel or nickel-chromium alloy, with the properties of oxidation resistance and high electrical resistance in the range of 200–800 ohms per circular mil-foot. One such alloy with the trade name Constantan has virtually a constant resistance in the temperature range of 0–900°F. For heavy-duty heating, as in the case of electrically heated industrial furnaces, rods or tubes of silicon carbide may be used as heating elements because of their great resistance to high temperature effects. Silicon carbide has only limited resistance change with temperature.

Incandescent lamps use a tungsten filament wire which may range in diameter from 0.001 to 0.060 in. Both tungsten and molybdenum resistors are employed as heating elements in very high temperature furnaces (3000°F and higher).

For *power transmission cables* aluminum has become competitive with copper. Although the resistivity of aluminum is higher, aluminum is only one-third the specific weight of copper. For equal resistance, the weight of aluminum is 54 percent that of copper. For short cables, such as welding cables, the rule of thumb is that an aluminum cable should be one gauge heavier than the required copper cable.

*Thermistors* are semiconductor crystals with a high negative temperature coefficient of resistance of perhaps 2–3 percent per degree Fahrenheit. Thermistors are used as thermal switches. For example, a thermistor may be used within an electric motor winding to operate a relay when the motor becomes too hot.

Electrical contacts present special problems. They must be constructed of low-resistance metals, but in addition they must meet the service conditions of arcing, wear, overheating, and tendency to weld. Copper, copper alloys, carbon, and the platinum metals are the most frequently used materials in these applications. Copper oxidizes readily, and its oxide is an insulator. Frequently silver is plated over the copper contact because silver oxide is more readily reduced to metallic silver by arcing than the oxides of other metals. Platinum, palladium, and iridium are perhaps the best contactor materials for low-voltage, low-current conditions and may provide a service life of more than a billion interruptions of current.

In the field of microelectronics, the most celebrated innovation perhaps is the *printed circuit*. This consists of an insulator backing plate, the substrate, carrying a pattern of conductors and resistors etched to the required pattern from a solid sheet of conducting foil. Printed circuits are not necessarily produced by printing methods. Usually a metal foil is bonded to the insulator substrate and subsequently etched. Figure 5.3 shows printed circuitry in a control panel.

Fig. 5-3 Rear view of a printed circuit module board. The leads of the many resistors, capacitors, and solid-state components shown all pass through the substrate board and connect to the printed circuit on the other side of the substrate. This module board is one of the many elements in the Bunker-Ramo control cabinet of the milling machine of Fig. 9-10.

## 5.4 superconductivity

The resistance of a single crystal of pure metal without imperfections in its space lattice should be zero at absolute zero. At temperatures below $10°K$, the specific resistance of certain conductors falls off rapidly to zero. Such conductors are termed *superconductors*. Superconductivity is not yet well understood, but as a technology it is about to come out of the laboratory to be turned over to the engineering groups. The best electrical conductors, gold, silver, and copper, are not superconductors, that is, their resistance does not reach zero at absolute zero.

The temperature below which the electrical resistance disappears is termed the *critical temperature*. But if the magnetic field about the conductor exceeds a certain critical value, superconductivity is lost. Diamagnetism (see below—a diamagnetic material is repelled by a magnetic field) accompanies the superconducting condition, and the thermal conductivity is very low, though not zero.

Technetium has the highest critical temperature, $11.2°K$, followed by niobium (columbium) at $9.2°K$ and lead at $7.2°K$. The critical temperature of the other superconducting metals is generally in the range of $0.35-5.0°K$. In general, a superconducting metal will have 2 to 5 valence electrons outside the closed energy levels, will be a poor conductor at room temperature, and will not be ferromagnetic. Certain chemical compounds are also superconducting, with transition temperatures as high as $18°K$. Superconductivity at considerably higher temperatures may be a development of the near future.

Applications of superconductivity have been considered in the following areas:

1. lossless power distribution
2. space satellites
3. computers
4. magnetohydrodynamic generation of electric power

The benefits of power transmission without $I^2R$ loss are obvious, but they present equally obvious difficulties in refrigerating a long transmission line. The development of magnetohydrodynamic (MHD) power generation is already well under way. The explanation of this process is not a proper topic for this book: suffice it to say that MHD is a brilliantly imaginative application of Fleming's rule for direction of magnetic and electric fields as discussed in any high school physics text. The use of superconductivity for switching circuits is found in *cryotrons* or cold switches. If a superconductor has a coil wound about it, a sufficiently high current passed through the coil will create a magnetic field which will destroy superconductivity in the conductor. This is an example of a simple cryotron switch.

## 5.5 magnetism

The theory of magnetism is a rather complex subject which is still being unravelled by physicists, and only its simpler aspects can be discussed here. Three kinds of magnetism must be noted.

**1. Ferromagnetism.** This is the type of magnetism shown by materials which are strongly attracted to a magnet. Ferromagnetism is a property of iron, cobalt, nickel, $Fe_3O_4$, gadolinium, dysprosium, terbium, and certain metal alloys and ceramic alloys. Ferromagnetism, like superconductivity, is found only below a certain critical temperature, which in the case of ferromagnetism is called the *Curie temperature*. The Curie temperature of gadolinium and dysprosium is below room temperature; that of iron is 1435°F. Above the Curie temperature ferromagnetic materials become paramagnetic. Unlike paramagnetism and diamagnetism, ferromagnetism is a property only of solids.

**2. Paramagnetism.**  Paramagnetic materials, which may be solid, liquid, or gas (oxygen exhibits strong paramagnetism, for example), are weakly magnetic and orient themselves parallel to a strong magnetic field. Unlike ferromagnetism, paramagnetism is not limited by temperature.

**3. Diamagnetism.**  This is a weakly negative type of magnetism, that is, diamagnetic materials are repelled from the strongest part of a magnetic field to the weakest. The rare gases and the metals bismuth, beryllium, copper, and zinc are examples of diamagnetic materials.

An electron may be considered as a spinning negative electric charge. Since any such spinning charge is magnetic, an electron is in effect a minute magnet. There are two possible rotations of spin, clockwise and counterclockwise. In any completely filled energy level of the atom, each electron pairs with another electron of opposite spin. The electron pair is nonmagnetic, since the magnetism of one spin cancels that of the opposite spin. When all the electrons in an atom are thus paired, there can be no paramagnetism, and diamagnetism results. An unpaired electron will produce a paramagnetic condition. Thus any atom such as a silver atom (one atom in the outermost populated energy level) must be paramagnetic. There are other electron configurations which can produce paramagnetism, but these will not be discussed.

The three types of magnetism may also be defined in terms of *permeability*. Permeability measures the relative ease with which magnetic flux or magnetism may be developed in a material, or it is the relative conductivity of a material for magnetic lines of force. A vacuum is assigned a permeability of 1. Diamagnetic materials have a permeability of less than 1; paramagnetic materials have a permeability greater than 1. The maximum permeability of the common ferromagnetic steels ranges up to about 20,000, though special alloys may have permeabilities exceeding 1,000,000.

Suppose that a material is enclosed in a wire coil and that the coil carries an electric current. This will produce a magnetizing effect on the material within the magnetic field of the coil. The magnetizing effect will be proportional to the ampere-turns of the coil. The flux density $B$ (lines of magnetic force per unit cross-sectional area) in the magnetized material will be some

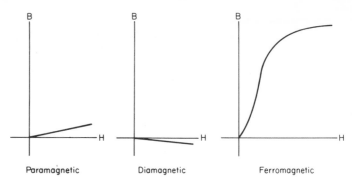

Fig. 5-4 Flux density and magnetizing force.

function of the amp-turns per unit length $H$:

$$B = f(H)$$

For both diamagnetic and paramagnetic materials, $B = \mu H$, where $\mu$ is the permeability. See Fig. 5.4.

As Fig. 5.4 shows, the relationship between $B$ and $H$ for a ferromagnetic material is nonlinear, and the permeability, $B/H$, is not a constant. There is a limit to the flux density of any ferromagnetic material; when this point is reached the material is said to be saturated. Magnetic amplifiers and saturable reactors use this phenomenon of saturation for their operation.

Soft steels are used in such electromagnetic equipment as motors, transformers, and relays. The steel is magnetized when current is switched to the magnetizing coil. Such steels instantly lose their magnetism when the current is switched off. Hard steels and nickel-steel alloys are used for permanent magnet applications.

The steel core of a transformer presents special problems which are solved by the use of laminations of so-called "electric" steels. The purpose of a transformer is the induction of a voltage or current in the secondary winding. Unfortunately the steel core is also a secondary, and a current is likewise induced in it. This current is termed an *eddy current*, and it represents an $I^2R$ loss of power, usually referred to as an iron loss, just as the $I^2R$ losses in the transformer windings are called copper losses. Soft steels have a relatively low specific resistance and would produce large eddy currents and powerful heating effects in a transformer core. To reduce such eddy currents, the core is laminated to increase the resistance by reducing the cross-section carrying current, and the steel used is a silicon steel. Up to about 5 percent silicon is added. The silicon atoms replace iron atoms in the bcc space lattice. The effect is to produce distortions in the space lattice and thus to increase the specific resistance of the steel (by a factor of about 5 times for 3 percent silicon). Electric steels, however, have a lower saturation induction. Carbon and nitrogen reduce permeability in such steels and are therefore kept to as low values as possible; the permeability is improved if the steel is coarse-grained.

The following steels are not ferromagnetic:

1. type 300 stainless steels
2. steels containing about 12 percent manganese (Hadfield manganese steels)
3. any steel above its Curie temperature

The paramagnetic manganese and stainless steels are termed *austenitic* and have pronounced work-hardening characteristics.

Certain nickel-iron alloys containing about 50 percent nickel, such as Perminvar, have a constant permeability. The most popular permanent magnet materials are the *Alnicos*. These are iron-nickel-aluminum alloys containing about one-quarter nickel. Like all permanent magnet materials, they are very hard.

Since World War II the *ferrites* have come into extensive use for electronic circuitry. These are complex ceramic oxides exhibiting strong paramagnetism. Typical ferrites are $NiO.Fe_2O_3$ or $MnO.Fe_2O_3$. Since such materials are insulators, no eddy current losses are involved in their use.

## 5.6 magnetostriction

The low-pitched hum of a 60-cycle transformer is a familiar sound. When the magnetic field of the steel transformer core builds up to the maximum in its positive or negative half-cycle, the core strains slightly under the influence of the magnetic field. This cyclical strain is the cause of the acoustic wave from the transformer. Since the steel core increases in length on both half-waves of alternating current, the sound wave is 120-cycle instead of 60-cycle. This strain that occurs when a ferromagnetic material is magnetized is called *magnetostriction*. The magnetostrictive effect is reversible: if a stress (and therefore a strain) is imposed on the material, the magnetization of the material will be altered.

Magnetostriction produces a positive strain (elongation) in iron and a negative strain (shortening) in cobalt and nickel. The effect is greatest in nickel, of the order of 30 millionths of an inch per inch maximum.

The famous antisubmarine Sonar was an early application of magnetostriction. Similar devices are now applied to ultrasonic inspection of welds and castings, ultrasonic detection and counting in automatic conveying,

Fig. 5-5 Ultrasonic transducers and amplifier-relay for such operations as counting packages on a conveyor belt, detecting pile-ups on a belt, etc. One transducer serves as ultrasonic transmitter and the other as receiver. (SONAC equipment of Delavon Manufacturing Company)

ultrasonic cleaning, ultrasonic welding, depth soundings in oceanography, phonograph pickups, and ultrasonic homogenization of milk, peanut butter, and other food products. Figure 5.5 shows an ultrasonic instrument for the detection of pile-ups on conveyor belts.

## 5.7 piezoelectricity

Piezoelectricity is analogous to magnetostriction. Certain ceramic materials such as quartz and titanates produce the piezoelectric effect. When a crystal of such material is compressed, a small voltage is generated across the opposite faces of the crystal, and conversely, a potential difference across the opposite faces will cause the crystal to expand or contract. Piezoelectricity is the preferred method of instrumentation for recording pressures in engine cylinders and gun barrels. For this purpose a small mounted crystal is screwed into the wall of the cylinder as a spark plug is, and the pressure is read out as a voltage on an amplifier. Although piezoelectricity as a method of instrumentation is competitive to magnetostriction in general, the piezoelectric effect is not capable of the high power that can be produced by the magnetostrictive effect. Ultrasonic cleaning, for example, requires considerable power in the hundreds of watts, and the transducers to supply such power must be magnetostrictive.

Figure 5.6 shows a piezoelectric transducer searching a material for flaws, which show as "pips" on an oscilloscope, also shown in the figure.

*electrical properties of materials | 115*

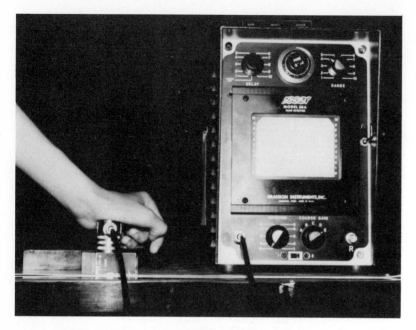

Fig. 5-6   Piezoelectric flaw detector. The detecting crystal under
the operator's hand sends an ultrasonic wave into the workpiece
underneath. Defects are disclosed by means of their ultrasonic
echoes on an oscilloscope.

## 5.8   electron emission

Probably most people, particularly students of electrical, electronic, and
mechanical engineering, are somewhat overwhelmed by the myriad types of
electrical equipment in use in industry and science. This jungle of equipment
can be reduced to intelligible order by reducing it to three basic types:

1. resistors—including wiring
2. inductors—including motors, generators, transformers, reactors, and solenoids
3. diodes

A *diode* is any device composed of a negative electrode, called a *cathode*,
which is a source of electrons, and a positive electrode, called an *anode*,
which collects the supply of electrons released by the cathode. The commonly
encountered diodes include vacuum tubes, gas-filled rectifiers such as the
mercury-arc rectifier, X-ray tubes, photocells, and welding arcs.

Release of electrons from the material of the cathode is possible by the following methods:

1. heating, called thermionic emission
2. photon capture, called photoelectric emission
3. electron bombardment, called secondary emission
4. a high-voltage electric field, called field emission

The minimum energy required for an electron to escape from the metal cathode is called the *work function*. The work function is measured in electron volts and is analogous to an ionization potential, though the two are not the same thing. This work function may be supplied by any of the four methods enumerated above.

The work function of metals lies in the range of 1 to about 6 electron volts. For electron emission efficiency the work function must of course be as low as possible. Metals such as cesium or rubidium, with a single outer electron, have work functions of about 2 ev. This low work function can be satisfied by photon energy in the visible light range, so that such one-electron metals are suitable cathodes for photocells. Platinum has a work function of about $5\frac{1}{2}$ electron volts, so that this metal might be used as an anode, but it is unsuitable as a cathode material.

When heated to the usual operating temperature of 2500°K, tungsten is a copious source of electrons, with a work function of 4.5 ev. The work function of thorium is 3.4 ev. By adding 1–2 percent of thoria (the oxide of the metal) to the tungsten cathode, the work function is reduced to 2.7 ev. Thoriated tungsten electrodes are used both in vacuum tubes and in welding arcs for the tungsten inert gas (TIG) welding method used for much aerospace welding. Other oxides such as strontium oxide and barium oxide are also used to reduce work functions of cathode materials such as tungsten, tantalum, and nickel.

## 5.9 photoconductivity

The photoconductive effect occurs when a semiconductor is irradiated with a sufficiently high photon energy to cause a pronounced reduction in its electrical resistance. A semiconductor is distinguished by a relatively small energy gap between the valence band and the conduction band. The absorbed photon energy lifts an electron across this gap to the conduction condition. There are many photoconductive materials, but the more commonly used ones include selenium, silicon, cadmium sulfide, lead telluride, lead sulfide, and thallium sulfide. The photoconductive material is usually a deposited

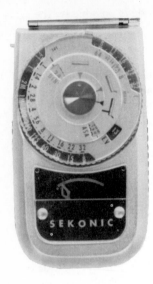

Fig. 5-7  A light meter for photography employing a silicon photoconductive cell.

thin film on a substrate material. Such materials may be used as detectors or switches to record the presence of visible light, ultraviolet, or X-radiation. Cadmium sulfide is preferred for X-ray detection.

## 5.10  *thermoelectricity*

If two dissimilar metals are joined and this junction then heated, a small voltage in the millivolt range is produced, known as the *thermoelectric effect*. This is a method of producing direct-current electric power directly from heat. Currently the thermoelectric effect is being developed for a variety of uses.

There are actually three thermoelectric effects, but only the following two have industrial applications:

**1. The Seebeck Effect.**   If the two junctions of two dissimilar metals are held at different temperatures, a voltage is produced that is closely proportional to the temperature difference. This Seebeck arrangement of wiring is known as a thermocouple and has been in use for the last hundred years as a temperature-measuring device.

**2. The Peltier Effect.**   If current flows through two junctions of two dissimilar metals, one junction will be heated and the other cooled. In terms of practical use, the Peltier effect indicates that one junction is a tiny furnace and the other a tiny refrigerator.

Since Seebeck's discovery there have been repeated attempts to produce electricity directly from heat by his method. So long as metal conductors were employed, these attempts failed because of the tiny currents produced and the abysmal efficiencies—less than 1 percent. But more recently, advances in solid-state physics have led to the use of semiconductor thermoelectric devices with efficiencies closer to 10 percent. Bismuth telluride and lead telluride are the favored semiconductors for this purpose. Thermoelectric power generators with up to 5 kw capacity are now coming into use, and the thermoelectric refrigerator is nearing the end of its research and development stage. Figure 5.8 is a diagram of a semiconductor thermoelectric device, which operates in the following manner.

Fig. 5-8   Semiconductor thermo-electric device.

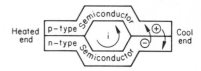

The heat energy applied at one end raises electrons from the valence band to the conduction band. For each electron thus transferred, a hole is left behind. The electrons travel to the cold junction through the $n$-type material, while the holes travel to the cold junction through the $p$-type material. The two unite at the cold junction, the electrons dropping into the holes and releasing energy to the cold junction. Since the charges carried by the holes and electrons are opposed to each other, there is a flow of electric current as indicated in the diagram. At the hot junction both holes and electrons are moving away from the junction. To make up for the loss of these charges, electrons are raised from the valence band to the conduction band to create new pairs of electrons and holes. Since energy is required for this operation, heat is absorbed at the hot junction.

For temperature-indicating thermocouples, only four combinations are in common use, although other combinations must be used for very high temperature work:

1. copper-constantan,— 300–700°F, 2% accuracy
2. iron-constantan, 0–1200°F, 4% accuracy
3. chromel-alumel, 1000–2200°F, 4% accuracy
4. platinum-platinum plus 4% rhodium, 2000–3000°F, 5% accuracy

The first metal in the pair is always the positive metal. The iron-constantan pair provides the highest electromotive force per degree. Chromel-alumel is the choice for oxidizing conditions, and platinum-platinum rhodium for reducing conditions. Chromel, alumel, and constantan are all very high nickel alloys.

For greater outputs, a number of thermocouples may be connected in series. Such a series is termed a *thermopile*. Many domestic gas furnaces contain a separate electric power system using a thermopile direct-current generator heated by the pilot gas flame of the furnace. If the pilot flame should go out, no thermopower is generated by the thermopile. The load on the little generator is a solenoid gas valve, and when the thermopile shuts down, the solenoid will cease to hold open the gas valve, which will then be shut by a spring. The house is thus protected against the possibility of an explosion caused by raw gas flooding the furnace without being ignited by the pilot flame.

## 5.11 corrosion

Corrosion is any undesired chemical action that attacks industrial materials in such a way as to degrade their properties or reduce their economic value. The combustion of steel during oxyacetylene cutting is not corrosion, but the same chemical action during rusting of steel is corrosion. Although corrosion is largely confined to metals, corrosion is an equally serious problem with firebrick linings of metal-melting furnaces. The chemical attack of a molten slag upon a firebrick may destroy the brick at a rate of an inch a day in an extreme case, necessitating replacement of the firebrick lining every week at great expense. Corrosion of concrete too is a common problem. Concrete may be corroded by sulfates in soils, by coal ash, or the concrete floor of a dairy may be corroded by the lactic acid in milk.

Corrosion then is one of the facts of life for industrial materials. Unfortunately, however, not all corrosion problems have textbook solutions; while extensive knowledge of materials and processes is certainly necessary, diagnostic skill and even intuition may often be required. Panacea solutions to corrosion problems are a snare for the unwary. People at large, engineers included, have great faith in the healing powers of the stainless steels for corrosion; unfortunately these expensive steels will solve only a restricted range of corrosion problems. Corrosion problems must be understood before they can be solved, although understanding may be hard to gain in some cases. It will not be our intention here to deal comprehensively with the whole field of corrosion, but merely to suggest some general approaches to the commoner kinds of corrosion in the commoner kinds of equipment and materials.

The beginning of wisdom in corrosion is to be convinced that no metallic, ceramic, or organic material is completely corrosion resistant. So-called corrosion-resistant metals such as aluminum, stainless steels, and titanium obtain their corrosion resisting characteristics by the establishment of a thin adherent oxide surface film developed by an actual corrosion process with oxygen. This oxide film passivates the surface and prevents further corrosion—sometimes. Thus the surface of any sheet of aluminum is actually aluminum oxide. The oxides are perhaps the most stable of all chemical compounds, and of course chemical stability simply means corrosion resistance. In general it must be observed that metals tend to revert to their ores by means of corrosion processes, or, in other words, an ore is a corroded metal.

Aluminum or stainless steel, being protected by surface oxides, will not normally be attacked by oxidizing chemicals. But a reducing chemical will attack them, since such a chemical can break down the protective surface oxide layer. Such metals therefore are readily attacked by chlorides, ferric chloride for example. Indeed, few chemicals are more corrosive than chlorides. The thought of using salts on icy roads, standard practice on this continent, is in principle a rather staggering metallurgical notion. Not even a stainless steel automobile would be proof against road salt.

One of the most frequently occurring types of corrosion is *galvanic* corrosion, the cause of which is electrical. In any wet battery or storage battery there is an anode and a cathode connected by a liquid called an electrolyte. The electrolyte is partially ionized and thus can carry an electric current. The metal of the anode is corroded, and the process of corrosion in these circumstances is termed galvanic corrosion. In an electric storage battery or dry cell, galvanic corrosion is employed as a means of generating electric power.

Now suppose that a small rectangular water tank is welded together by using steel angle iron for the frame and sheet aluminum for the sides. Here we have the conditions for galvanic corrosion. The water in the tank will act as an electrolyte, one of the metals as an anode, and the other as a cathode. In this case, the aluminum will act as the anode and will corrode, probably in areas close to the angle iron. The corrosion proceeds by means of small electric currents circulating through the metal from the aluminum to the steel and back to the aluminum by way of the water. As in the case of a storage battery, there will be a difference in voltage between the anode and the cathode in the presence of the electrolyte. These electrode potentials are measured in the following fashion.

If a metal is placed in a solution containing ions of the same metal, metal ions will transfer either from the solution to the metal or vice versa. In the latter case, the metal becomes negative with respect to the electrolyte, and in the former case, positive. This follows from the fact that the metal ions are positive and unlike charges attract. After a certain period of time an equilibrium condition will be reached, with the potential difference between metal and electrolyte equalizing the rate of transfer of ions back and forth, so that there will be no further transfer. This potential difference is known as the electrode potential and depends on the nature of the metal, the temperature, and the concentration of the solution.

For a reference standard a so-called "hydrogen electrode" is used. This is actually a platinum electrode immersed in an acid solution through which is bubbled sufficient hydrogen gas to produce saturation. The potential between the electrode and electrolyte is defined to be zero when a normal acid solution is used. A normal solution is one which contains 1 gram per liter of hydrogen ions at 25°C.

The standard electrode potential of a metal is taken to be the potential difference between the hydrogen electrode and an electrode of the metal immersed in a normal solution of one of the metal's salts. For example, an aluminum electrode under these conditions will lose positive ions until the electrode reaches a steady negative potential of $-1.70$ v relative to the solution. Similarly, a copper electrode will receive ions until it reaches a positive potential of $+0.34$ v relative to the solution.

These electrode potentials can be arranged in a series, known as a galvanic series, thus

| | |
|----|--------|
| Mg | $-2.375$ |
| Al | $-1.70$ |
| Zn | $-0.762$ |
| Cr | $-0.71$ |
| Fe | $-0.44$ |
| Ni | $-0.23$ |
| Pb | $-0.126$ |
| H  | $0.00$ |
| Cu | $+0.34$ |

This series gives the relative corrodability of metals under conditions of galvanic corrosion. If two metals are coupled in a galvanic pair, the most anodic will be corroded. This will be the one with the highest negative electrode potential. Thus if zinc is plated on iron as galvanized iron, in the presence of rainwater as an electrolyte, the zinc will corrode. Corrosion is caused by loss of ions to the electrolyte.

In any strict sense a galvanic series must be set up for different electrolytes, but the differences are rather minor. The following galvanic series for seawater is usually considered universal in application:

*Anodic, or Corroded End*

| | |
|---|---|
| magnesium | high copper brass |
| zinc | copper |
| aluminum | aluminum bronze |
| cadmium | nickel alloys |
| aluminum alloys | nickel |
| carbon and low alloy steels | stainless steel (passive) |
| cast iron | monel |
| stainless steel (active) | titanium |
| high zinc brass | *Cathodic, or Protected End* |

Note that the position of the stainless steels is equivocal.

The usual methods of combating corrosion are these:

1. Use of high-purity metals. An alloy is always much more vulnerable to corrosion than a pure metal.
2. Proper design and fabrication. Examples are either the avoidance of crevices into which an electrolyte can penetrate or the use of aluminum rivets of the same alloy as the aluminum sheet joined by the rivets.
3. Cathodic protection, such as the insertion of a magnesium anode in a domestic hot water tank or the burying of a magnesium anode beside an underground gas pipeline.
4. Use of inhibitors, such as phosphate coatings on steel, chromate treatment of aluminum.
5. Surface coatings such as paints.

## PROBLEMS

1 In many metal-working processes, the workpiece is heated by inducing eddy currents to flow in the workpiece itself. Pieces of tubing stock having a 3-in. length, 2-in. outside diameter, and a wall thickness of 0.06 in. are being induction heated. The eddy currents circulate in the plane of the cross-section of the work. The specific resistance of the workpiece material is 6.0 microhm-inches, and the induced current 1000 amps. Calculate the resistance of the workpiece and the induced power in watts. The workpiece can be considered a conductor with a cross-section $3.0 \times 0.06$ in.

2 300 amps of welding current is being carried through a #00 copper welding cable 50 ft long. For proper operation of the welding arc not more than a 2-v drop is allowed in this cable. Is the #00 cable size sufficiently large?

3 Corrosion occurs in an aluminum sheet around the rivets fastening the sheet because of the condensation of water vapor. Can you diagnose the problem and suggest a cure?

4 Low-temperature tempering furnaces operate at temperatures up to 1200°F; hardening furnaces operate up to 1800°F or higher. Specify thermocouple combinations for these two temperatures.

5 Differentiate between magnetostriction and piezoelectricity.

6 What is a superconductor?

7 What is resistance wire?

8 Differentiate between paramagnetism and ferromagnetism.

## 6.1  points of view

By and large, the story of materials begins with the digging of a ceramic material, usually a mineral, out of the earth for processing in an industrial furnace. In the steel industry, for example, iron ore, a ceramic material, is extracted from open-pit mines and reduced to metal in a blast furnace. Cement, perhaps the most important manufactured ceramic product, also is manufactured in a type of industrial furnace called a rotary kiln from the two ceramic raw materials clay and limestone.

These two instances illustrate the basic operations of the materials industries. The primary product is usually a ceramic material; the primary process requires an industrial furnace. Virtually all other industry is dependent on the massive and dirty primary processes of refining the basic ceramic materials.

Ceramic materials are unglamorous, and therefore their importance is little understood. But as finished products, they are indispensable throughout industry. Perhaps the most critically important ceramic products are the refractories that confine the great process heats of the primary industries.

# CERAMIC
# MATERIALS

But a full roster of everyday ceramic products includes the grinding wheel that puts the fine finish on ground steel shafting, the carbide bits used to machine metals, the piezoelectric crystal that reads the pressure in an engine cylinder, the mica spacer in a vacuum radio tube, the glass envelope of a photocell, the ruby crystal in a laser, the titanium oxide in paints, the porcelain enamel on domestic appliances, the ferrites in the memory of a large digital computer, the clay and sodium sulfate in paper, and even the talc in baby powder and the calcium carbonate in toothpaste.

Ceramic materials did not phase out of technology when the Stone Ages of man (ceramic ages) ended. New uses are continually being found for these materials, and in recent years ceramic materials have even been used as lubricants (molybdenum disulfide) and fuels (uranium oxide nuclear reactors and boron hydride for solid propellant rockets). Most of the ceramic materials are silicates, aluminates, oxides, carbides, borides, nitrides, or hydrides. A general classification of products is difficult because of the great versatility of these materials, but the following list includes the major groups:

1. abrasives
2. whiteware—toilet tanks, sinks, etc.
3. chemical stoneware
4. electrical porcelain
5. porcelain enamel
6. glass
7. brick and tile (such as clay sewer pipe)
8. cements and concretes
9. mineral ores
10. aggregates and grogs
11. refractories
12. slags and fluxes
13. insulations

The field of ceramics has a terminology of its own. Here are some of the more important definitions.

A *glass* is a ceramic or ceramic component which is amorphous. Thus most firebrick contains glass, but the term here obviously is not restricted to window glass.

A *slag* is a ceramic formulation used in the refining of molten metals. Slags usually serve two purposes:

1. They absorb undesired impurities from the metal, such as sulfur from steel.
2. They protect the metal from atmospheric gases.

A *flux* serves the same purpose in a ceramic mixture that an alloy serves in metals, that is, it lowers the melting or softening point of the mixture. Fluxes are usually necessary in the welding and brazing of metals in order to remove the oxide surface of the metal, because oxides usually have high melting points. Particularly in welding it is often difficult to differentiate a flux from a slag, but this distinction is probably unimportant.

A *refractory* is a high-temperature ceramic used as a lining for a heating furnace.

A *frit* is a low-melting ceramic mixture, usually the raw material for porcelain enamel.

Refractories are subject to chemical action with metal-melting slags. Slags are either acid or base (alkaline), and refractories must be of the same type as the slag. Thus an acid refractory will be rapidly attacked by a basic slag made up of limestone. Silica is acid, magnesia is basic, and graphite and chromium oxide are neutral refractories.

## 6.2 pyrometric cone equivalent

Ceramic materials, like metals, are rarely used in pure form and therefore do not have a definite melting point, particularly since most ceramic mixtures contain some amorphous glassy components.

A true melting point is the temperature at which solid and liquid phases of the same composition exist in equilibrium. Most of the commercially used ceramic materials soften progressively over a range of temperatures, and in this range both crystalline solid materials and glassy liquid coexist. To evaluate the high-temperature behavior of such materials, the usual method is to determine the "Pyrometric Cone Equivalent" or PCE. In this procedure, a ground sample of the material to be tested is molded into a test cone of slim pyramid shape, mounted with a series of standard pyrometric cones which have known softening values. The group is heated at a controlled rate until the test cones soften and bend. The number of the standard cone that softens at the same time as the test cone is reported as the PCE value of the material.

### SELECTED PYROMETRIC CONE EQUIVALENTS

| PCE | °F | PCE | °F | PCE | °F |
|---|---|---|---|---|---|
| 6 | 2246 | 11 | 2417 | 20 | 2845 |
| 7 | 2282 | 12 | 2435 | 26 | 2950 |
| 8 | 2300 | 13 | 2462 | 28 | 2939 |
| 9 | 2345 | 14 | 2552 | 30 | 3002 |
| 10 | 2381 | 15 | 2605 | 32 | 3092 |
| | | | | 40 | 3425 |

The term *spalling* is usually defined as the thermal cracking of ceramic materials. Most of these materials will crack if rapidly heated or cooled. Familiar examples of spalling are the cracking of concrete in barbecues and fireplaces. Spalling is an especially important problem in the case of firebrick.

Thermal expansion is a strain, and this strain is proportional to the thermal expansion coefficient. If the expansion is unimpeded, no stress accompanies this strain, according to Newton's third law. But frequently materials are not free to expand. Thus the hot face of a firebrick may be restrained from expansion by the cooler portion of the brick behind the hot face. Such restraint produces stress. The magnitude of these thermal stresses will be influenced by the thermal conductivity, the modulus of elasticity, and the expansion coefficient of the material. The stress will be proportional to the expansion coefficient, for a material with zero expansion will have zero thermal strain. Likewise the stress will be proportional to the modulus of elasticity, for a material with zero modulus of elasticity will not be stressed when it is strained. Now consider a material with an infinite thermal conductivity. There must be uniform temperature throughout the volume of such a material, and the bulk of the material, being at the same temperature as its heated surface, will not restrain the expansion of the heated surface.

Spalling of course will not occur in materials with considerable ductility. Nor, from the above argument, will it occur in materials with low expansion coefficients, low moduli of elasticity, and high thermal conductivity. This is borne out by experience. Metals rarely spall, because of their usual ductility and their high thermal conductivity. (Thermal stresses can be very high in stainless steels however because of the low thermal conductivity of these materials.) Graphite brick and crucibles cannot be spalled because of high thermal conductivity. Glass and silica brick will spall readily because of a very high expansion coefficient between room temperature and 600°F, though above this temperature range the thermal expansion is virtually zero and there is no risk of spalling. Pyrex glass, well known for its resistance to thermal cracking, contains boric oxide to provide a low expansion coefficient.

It is often possible to determine by inspection whether a ceramic mate-

Fig. 6-1 Spalling of refractory brick due to cooling (tension) and heating (shear).

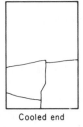

Cooled end

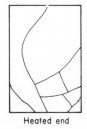

Heated end

rial has spalled because of sudden heating or sudden cooling. Sudden cooling produces failure by tensile stresses, shown by fracture lines at right angles to the edges of the material. See Fig. 6.1. Sudden heating results in shear failure, indicated by cracking at 45° to the edges of the material.

## 6.4 general characteristics
## of the ceramic group

Before discussing individual members of the broad group of ceramic materials, we shall first summarize those characteristics generally shared by the whole group.

**1. Chemical Composition.** The group consists chiefly of oxides, carbides, and silicates, though some borides, hydrides, and nitrides are in use in industry. In their chemical reactions with each other at high temperatures, such as reactions between slags and refractories, it has been noted that they may be acid, basic, or neutral. They are however chemically stable and generally uncorrodable at room temperature. In general these materials are somewhat impure, the commonest impurity being iron oxide. Most of the ceramic materials occur naturally and are extracted by open-pit mining, quarrying, or excavating methods. But beginning in this century, an increasing number of these materials have been synthetically formulated. The synthetic ceramics include the carbides and most of the abrasives.

**2. Stress and Strain.** The ceramic materials are brittle and therefore unworkable. When mixed with water in the unburned state, raw materials such as clays are plastic. All the ceramic materials become increasingly plastic or less viscous at high temperatures. The only applicable machining method for these materials is grinding.

**3. Joining Methods.** The only joining method generally workable for ceramic materials is that of cementing with either inorganic mortars or organic cements. A few special solders are used with glass, and electron-beam welding will join ceramics to ceramics or ceramics to metals, although this method is little used.

**4. Cost.** The naturally occurring ceramics are inexpensive, with costs generally in the range of $\frac{1}{4}$ of a cent to 5 cents per pound, but synthetic ceramics are considerably more expensive than this, as anyone who has bought a carborundum sharpening stone knows. Because these materials are heavy and of low value, they are not usually shipped over great distances.

**5. Physical Characteristics.** Except at high temperatures, these materials are electrical insulators, though carbides will conduct at room temperature. Thermal conductivities are in the range of 5 to 100 Btu per in., making ceramics less thermally conductive than the metals, which have conductivities above 100 Btu per in. thickness. Most ceramics are good heat insulators when highly porous and are used as insulating materials. Specific heat is usually about 0.2 Btu per lb. Almost all are hard and abrasive, with very high melting or softening points.

## 6.5 the common oxides

Six common oxides are the basic materials for a wide range of industrial products:

| | |
|---|---|
| silica | $SiO_2$ |
| alumina | $Al_2O_3$ |
| lime | $CaO$ |
| magnesia | $MgO$ |
| iron oxide | $Fe_2O_3$ |
| rutile | $TiO_2$ |

*Iron* has two oxides. The red oxide of iron rust is $Fe_2O_3$; the black oxide found on hot-rolled steel sheet is $Fe_3O_4$. Except as iron ore, iron oxide has few industrial uses, its chief one being as a flux in the manufacture of cement. It is an ever-present fluxing impurity in clays, minerals, and other deposits. White clays such as kaolin are white because of the absence of iron oxide, and this characteristic makes such clays valuable. The iron oxides are unusual in that they melt at approximately the same temperature as iron, 2800°F.

*Lime* and *magnesia* are basic (alkaline) in their chemical reactions. Lime is produced from limestone, $CaCO_3$, and magnesia from magnesite, $MgCO_3$. In addition the mineral dolomite is a solid solution of about half limestone and half magnesite. Magnesia is chiefly employed as a furnace refractory in metal-melting and cement-burning furnaces.

The chief uses of lime are these:

1. major ingredient in portland cement manufacture
2. component of window glass
3. metal-melting slag
4. chemical used in the manufacture of pulp and paper
5. chemical used in various water treatment processes and to raise the $p$H of water
6. manufacture of sand-lime brick
7. component of sand-lime-cement mortar
8. ingredient in masonry paints

This list by no means exhausts the manifold uses of lime.

Three simple chemical reactions are of great importance in the various uses of lime. Quicklime or hot lime is made by dissociation of calcium carbonate at 1830°F (1000°C):

$$CaCO_3 \rightarrow CaO + CO_2$$

This process is called lime burning. Hydrated lime or slaked lime is produced by reacting hot lime with water:

$$CaO + H_2O \rightarrow Ca(OH)_2$$

Slaking produces considerable quantities of heat. Slaked lime hardens by a slow reaction with carbon dioxide in the air:

$$Ca(OH)_2 + CO_2 \rightarrow CaCO_3 + H_2O$$

Lime (CaO) has a cubic space lattice with alternating Ca and O atoms in the lattice.

*Silica* is found naturally in its quartz phase, most often encountered as quartz rock, quartz sand, or sandstone. At temperatures above 1600°F the phase changes to tridymite, and at 2678°F to cristobalite. The melting point of pure silica is 3130°F (1722°C). Silica may be used at temperatures very close to its melting point, for like carbon, it has the unusual property of maintaining a very high strength almost to its melting point. Silica brick is used in the arched roofs of open-hearth steel-melting furnaces operated at temperatures of 3000°F. The high expansion and spalling characteristics of silica at low temperatures have already received mention.

When silica is melted and resolidified, a silica glass or quartz glass is formed (not the type of glass used as window glass) with two industrially significant properties:

1. transmission of ultraviolet radiation
2. low coefficient of expansion at low temperatures and remarkable spall resistance. Quartz glass is used in ultraviolet and infrared lamps.

Of these common oxides, *alumina* is the all-round best in its properties and has a wide range of uses in ceramic technology such as nickel does in metals technology. The melting point of pure alumina is 3720°F (2050°C). This material has good hot strength, reasonable spall resistance, and good corrosion resistance against molten slags. Its many uses include spark-plug bodies, ceramic cutting tools for machining metals, grinding wheels, and firebrick. The equilibrium diagram for $Al_2O_3$-$SiO_2$ is given in Fig. 2.19 on page 51. A great many minerals and clays, many types of firebrick, and portland cement are compounds of alumina and silica.

Rutile is one of the more common constituents of the earth's crust. It is a white oxide used to provide opacity and whiteness to paints and porcelain enamels. Rutile and lime are the most useful flux coatings (so called; they are really slag coatings) on manual arc-welding rods.

## 6.6 ceramic insulating materials

Ceramic insulating materials give excellent service up to hot-face temperatures as high as 3000°F. The $K$-factor of these materials may be as low as 0.3 Btu per in. at room temperature, though at 1000°F it will be in the range of 1 Btu per in. thickness. Most of the foamed plastics, such as polyurethane and polystyrene, have lower $K$-factors but generally cannot be used at temperatures above 150°F.

Insulating materials are available in several forms: rigid board or block insulation, blankets or batts, formed insulation to fit piping, and granular loose fill in bags or bulk. Loose fill materials may also be used with cements to make insulating concretes or trowelled cements.

*Asbestos* is a fibrous mineral mined in Canada. Though its chief use is in porous insulating materials, it is also molded into hard asbestos millboard. Asbestos "shorts" are used for loose fill. Asbestos does not have unusually high temperature resistance and tends to crumble at temperatures much above 1000°F.

*Fiberglass* is made of spun fibers of glass with a melamine or other plastic binder and is used as pipe insulation as well as in insulation blankets. Its temperature limitations are even lower than those of asbestos. The binder burns out at about 600°F, indicated by a change of color from yellow to white.

*Rockwool* is another gray fibrous insulation made by blowing high-pressure steam at a stream of molten rock composition. Rockwool is usually supplied in blankets or rolls.

85-*percent magnesia* (with 15 percent asbestos) is a white powdery insulation usually formed into pipe insulation in three- or four-foot lengths for low-temperature applications.

*Calcium silicate* is white and powdery also, but it has the higher temperature rating of about 1200°F.

*Diatomaceous earth* is the geological deposit of the skeletal remains of the marine alga known as diatoms. This material is almost pure silica and is used as loose fill. Its fusion point is about 2450°F.

*Vermiculite*, often known by its trade name Zonolite, is a foliated aluminum-iron-magnesium silicate somewhat similar in appearance to fine mica. When heated at 2000°F it expands into accordion-like granules with excellent properties as an insulating material or a light-weight aggregate for concretes. The expansion may be as much as 30 times the original size. Vermiculite is supplied in bags of 4 cu ft, a bag weighing about 25 lb. It is possible to make a vermiculite concrete light enough to float on water.

*Perlite* is a white silica mineral which also can be expanded at a temperature of about 1600°F to provide a bulk material which is formed into wallboard and tile for sound insulation. Light-weight concretes made of perlite are not so tough or crack resistant as vermiculite concretes.

*Haydite.* Certain types of clays when heated release gases which bloat the clay. This expanded clay is called haydite. Haydite is used as an aggregate in light-weight concrete blocks and in low-heat duty refractory concretes. However, a haydite concrete is heavier than concretes using perlite or vermiculite.

## 6.7 glass

In ceramic science, the word "glass" signifies any amorphous component of a ceramic mixture. More generally, glass is a transparent silica product which may be amorphous or crystalline, depending on heat treatment.

Glass has a remarkable range of uses. The architect Le Corbusier, a pioneer in the use of glass as an architectural material, once suggested that the story of architecture is simply the search for the window. Today the glass curtain wall on high-rise buildings is a part of the skyline of any large city. To the architect or the civil engineer, glass is a structural material. To the electrical engineer or the instrument engineer, glass is an electrical insulator or a tube envelope with transmission characteristics required for radiation. To the mechanical engineer, glass may be many things; he uses glass pipe, fiberglass insulation, and even glass springs. To the packaging engineer, glass is a container for such liquids as milk and beer.

Glass is a solid insofar as it has the rigidity and hardness customarily associated with solid bodies. However, it lacks a continuous space lattice. Any crystalline ordering is on a submicroscopic scale in glass, and in this respect, therefore, glass is a hard liquid. When molten glass is cooled, it tends to crystallize, but its high viscosity is a strong deterrent to crystal formation. As a result, the liquid state becomes "frozen in." Glass properties are unfavorably affected if crystals develop in the glass.

A convenient rule of thumb says that the "melting temperature" of glass (it does not possess such a characteristic) is that temperature at which glass has a viscosity of $10^2$ poises. Although the many different grades of glass have considerable variation in their viscosity, Fig. 6.2 is a typical viscosity curve for glass.

A brittle material, glass at room temperature is completely elastic up to its ultimate tensile strength. It always fails in tension. See Fig. 6.3. The strength of glass depends to a very great extent on the residual stresses set up in the glass as a result of cooling or other thermal treatments, the size and shape of the part, the surface condition, and the surrounding atmosphere. A network of microscopic cracks covers the surface of a glass, and these are a

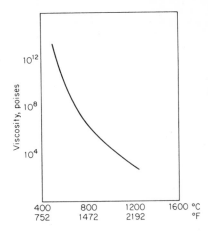

Fig. 6-2 Typical viscosity-tem-
perature curve for a glass.

Fig. 6-3 A baseball may be used to prove that ceramics are
brittle and have low tensile strength. The circle of fracture is a
line of constant stress, the ultimate tensile strength of the glass.

major factor in reducing the strength. Water and water vapor are mildly
corrosive to glass when adsorbed within such cracks and assist in reducing
the apparent strength. Maximum strength is found when a virgin surface
exists on the test specimen. Freshly drawn glass fibers with extremely small
diameters have given tensile strengths as high as $10^7$ psi; larger specimens
break under stresses in the range of 10,000 to 20,000 psi.

The effect of water vapor on the strength of glass may be noted in the
method that an experienced glass worker uses to break a glass rod. He
scratches the glass with a cutter, then briefly exhales on it or wets it.

The modulus of elasticity of most glasses is in the range of 10,000,000 psi, about the same rigidity as aluminum.

Neglecting glasses tailored for special transmission characteristics, glass transmits visible light well, up to 90 percent, but it is opaque to ultraviolet and gives limited transmission in the infrared region.

## 6.8  types of glass

Four general classes of glass may be differentiated:

1. *Soda-lime glass.* This is the cheapest glass, used for windows and doors.
2. *Borosilicate glass* (Pyrex glass). This glass has a low expansion coefficient (about $20 \times 10^{-7}$ in./in./°F), and therefore has good spall resistance.
3. *Fused silica.* Outstanding thermal shock resistance.
4. *Pyroceram.* A crystalline glass invented by Corning Glass Works.

Window glass is a soda-lime-silicate, typically analyzing

| | |
|---|---|
| silica | 74% |
| alumina | 1 |
| soda ($Na_2O$) | 15 |
| lime | 10 |

These proportions may be varied slightly. Fiberglass fibers for example may have 10% or more alumina with a reduction in silica.

Silica sand is the basis of glass. This is a remarkably pure white sand, virtually pure silica, usually consisting of fine grains of 50 to 100 mesh cemented together with kaolin. For glass manufacture a maximum iron oxide content of about 0.25 percent is allowed, since the presence of excess iron oxide will give the glass a brown color similar to a beer bottle. Most of the silica sand mined for glass manufacture in the United States and Canada comes from the St. Peter formation, a sand deposit of vast extent in the central states. The formation extends from northern Michigan to Minnesota and south to the Ozark area of Arkansas, and it is mined chiefly in Illinois, Minnesota, Missouri, and Arkansas. This silica sand was formed from Pre-Cambrian granites in Canada and displaced to the United States by movements of the Ice Ages.

Silica and soda combine chemically to produce a glassy substance called sodium silicate (water glass), which is soluble in water and used in large quantities as an adhesive in paper bag manufacture. If lime also is added to this chemical reaction, the solubility of the product in water is reduced. When sufficient lime is added, a relatively insoluble glass product results which however may still be attacked by water under certain circumstances.

Fused silica glass (quartz glass) has excellent spall resistance, good transmission of ultraviolet radiation, and the best resistance to chemical attack of any type of glass.

Quartz glass has a high melting point and is difficult to work. To reduce the melting point and the viscosity, in most glass compositions the silica is fluxed with soda ($Na_2O$ or $K_2O$), lime, borax, phosphorous oxide ($P_2O_5$), and occasionally other oxides. Phosphate glasses will transmit in the ultraviolet range, while lead oxide provides shielding against X-radiation. Borosilicate glass contains about 80 percent silica, 10 percent borax, and 10 percent or less of alumina, soda, and other oxides. In general the corrosion resistance of glass may be considered to be proportional to the silica content.

Figure 6.4 shows typical electrical resistivities of glass. The electrical resistance is reduced by the addition of soda, potash, and alumina, particularly soda; borax and lime increase resistance. Sodium ions in glass are carriers of electric charge. However lime-soda glass has sufficiently high resistance to be suitable for the envelopes of electron tubes, except for photocells, which may require transmission of ultraviolet radiation.

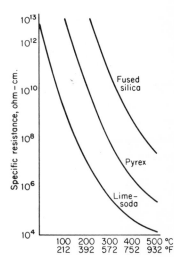

Fig. 6-4 Electrical resistance of glass.

The pyrocerams are polycrystalline glasses with softening temperatures above 2000°F, low specific heat of about 0.1, thermal conductivity of about 30 Btu/°F/in., and a modulus of elasticity of $17 - 19 \times 10^6$. The thermal expansion can be formulated from slightly negative to 0.000003.

## 6.9 rock and stone

The weight of all structures must be carried either on soils or on rock. Both soils and rocks are also used as raw materials for manufactured products.

Concrete is a manufactured rock, made from limestone and clay. Rock and concrete are similar in some of their physical properties. The *E*-values of rocks however are greater than those determined for concrete. Tensile strength of both concrete and rock specimens are about 10 percent of the compressive strength. However it is not possible to produce concretes that are as strong in tension or compression as the stronger rocks. Strength tests of rock specimens are not necessarily representative of the properties of the rock mass, since the extracted rock specimen is relieved of the pressures and the continuity of the rock deposit in which it lay.

Rock is classified by its geological origin into three types:

igneous
sedimentary
metamorphic

*Igneous* rock was formed by the solidication of molten material, usually at some depth in the earth's crust or at the earth's surface by volcanic or other action. Granites are examples of igneous rock and are chiefly composed of quartz, feldspar, orthoclase, and smaller amounts of mica. The high quartz content of granites makes these rocks very hard.

*Sedimentary* rocks are formed by the deposition, usually under water, of mineral matter chiefly produced by the destruction of pre-existing igneous rocks. Sandstone for example is a sedimentary rock of cemented quartz particles, the cementing material usually being calcite. Silica sand is sedimentary, being bonded usually with kaolin, a clay. Shale is deposited from clay. Limestone is a calcium carbonate, often containing fossils. Dolomite is a mixture of calcium carbonate and magnesium carbonate. All these sedimentary rock types tend to be weaker than igneous rock and are usually jointed and stratified.

*Metamorphic* rock is either igneous or sedimentary rock which has been changed from its original structure by extreme heat or pressure at some period in its geological history. Marble, quartzite, and slate are examples. Slate is formed from clays and shales; marble is a recrystallized limestone or dolomite.

Rocks are basically aggregates of mineral particles. Hardness is an indication of strength in rock as in other materials, though the strength of a rock is in part dependent on particle size, smaller grain size resulting in a stronger rock. The strength of a rock is also influenced by structural imperfections such as porosity, cracks, or weak particles. In order to drill rock, the drill points must be harder than the rock. Rock drills use tungsten carbide inserts, which are harder than all common rocks.

For commercially used stone, the hardness and the strength are proportional to the silica content. The cost of stone is also roughly proportional to the silica content, because the difficulty of drilling, cutting, polishing, and working of the stone increases with silica content. Thus limestone, which contains little silica, is easy to drill or process; it is called soft rock in the mining industry. Quartz and quartzite are largely silica and thus very hard and abrasive, or hard rocks. Granites also are siliceous and thus more expensive.

In these times, when seemingly anything can be made from plastics, mention should be made of synthetic stone, now a common material. It is less expensive than natural stone. Synthetic stone is used as thin veneer on the face of buildings, for counter tops, bathroom sinks, and many other purposes. It is cast from ground mineral and a polyester plastic binder. As an example, a black marble formulation would consist of the following:

| | |
|---|---|
| polyester resin | 20 lb |
| black pigment | 3 |
| fine silica sand | 110 |
| | 133 lb |

Note that synthetic marble contains no marble. Synthetic granite however contains about 40 percent ground granite.

## 6.10 clays

Clays and soils have their origin in the mechanical and chemical disintegration of rock. Clays are complex aluminum silicates containing attached water molecules. If only alumina, silica, and water are present in the clay, the clay is white after firing. Most clays contain iron and manganese oxides. These fluxing ingredients give color to the clay and reduce the refractoriness or PCE of the clay.

A wide range of products is produced from clays. These include pottery, chemical stoneware, electrical porcelain, brick, tile, haydite, drilling mud for oil wells, filters, portland cement, and catalysts for the "cracking" of petroleum. Clay through the ages has appealed to the artistic impulses of men. Every civilization has produced beautiful pottery, perhaps the finest being the ceramic work of the Sung dynasty of China about a thousand years ago.

The civil engineer distinguishes three broad groups of soils:

1. granular or cohesionless soils, including sands and gravels
2. fine-grained soils, including silts and clays
3. organic soils, including peat, muskeg, and muck

The gravel component of a soil is that fraction with constituents larger than 2 mm ($\#10$ U.S. standard sieve). The sand component is that fraction with constituents between 2 mm and 0.1 mm ($\#140$ U.S. standard sieve). The individual particles of a fine-grained soil cannot be distinguished by the naked eye. Clays are plastic when blended with suitable water content, but silts cannot be plasticized. Silt particles are in the range of 0.05 to 0.005 mm, and clay particles are still smaller.

A large number of clay minerals have been distinguished, classified according to their silicate structure. The following clays are used as industrial raw materials:

**1. Kaolin.**   A white clay, nearly pure kaolinite, $Al_2O_3.2SiO_2.2H_2O$. Kaolins are used for china, electrical porcelain bodies, sanitary ware, and in paper, rubber, and firebrick.

**2. Montmorillonites.**   This is a large group of common clays. The group includes talc, which is used in the production of the electrical dielectric known as steatite, valued for high-frequency applications.

**3. Aluminous Clays.**   These are not widely distributed over the earth's surface; there are for example deposits in the United States but none in Canada. Bauxite, $Al_2O_3.3H_2O$, is found in the West Indies and the southern states and is the mineral from which aluminum is most cheaply extracted. Diaspore, $Al_2O_3.H_2O$, is a beautiful and highly refractory clay which is the basis of the remarkable fireclay industry of eastern Missouri.

**4. Vermiculite.**   Found in Montana and elsewhere. This material was mentioned in Sec. 6.6.

**5. Flint Clays.**   These are hard clays almost devoid of plasticity, which are added to more plastic clays. The flint clay may be likened to the skeleton of the clay structure, with the plastic clay acting as cementing binder.

**6. Feldspar.**   Feldspar is an alkali-aluminum silicate rock, not a clay. It is an ingredient of many industrial clay mixtures, however, serving as a flux.

**7. Fireclays.**   These are plastic clays of high refractoriness, cone 27 (2920°) or higher. Deposits of fireclay are distributed over the United States and Canada.

**8. Ball Clays.**   These are fine-grained, highly plastic clays chiefly used in whiteware and electrical porcelain.

## 6.11 manufacture of
## clay products

Few clay products are made of a single type of clay. Electrical porcelain for example is a blend of feldspar, flint, ball, and china clay, all prepared to required sieve sizing. A vitreous sanitary ware, such as is used in sinks and bathtubs, may have the following composition:

| | |
|---|---|
| flint clay | 30% |
| feldspar | 26 |
| ball clay | 18 |
| kaolin | 26 |

In order to process such clay mixtures into finished products, the mixture must be tempered with water. The several clay manufacturing methods may be classified in order of decreasing water content as follows:

1. slip casting      12–50% water (percent of weight of clay)
2. soft mud process      20–30% water
3. stiff mud process      12–15% water
4. dry press process      4–12% water

In *slip casting*, the water-clay mixture, called a slip, is poured into a mold of plaster of Paris, which takes up water and causes a solid shell to form. The slip in the center is then poured out and the mold removed after further drying.

The *soft-mud process* is chiefly used in the manufacture of building brick, and the *stiff-mud process* chiefly for firebrick. These processes extrude the clay products through an extrusion die in a continuous length, the long extrusion being cut into brick by means of wires pulled through the extrusion.

The *dry press process* forms the clay product in a steel mold. The stiff-mud extrusion process, because of the smearing action of the extrusion die on the product, produces a brick of lower porosity than the dry press method. Brick made by either method can be identified by surface appearance, the extruded brick showing longitudinal extrusion marks. The lower porosity of the stiff-mud process is often desirable for firebrick, which must resist the penetration and corrosive attack of metal-melting slags.

Building brick is either sand brick or clay brick, in the grades of *common brick* and *face brick*. Face brick is darker, denser, stronger, and held to closer dimensional tolerances than common brick. The greater density of face brick implies lower porosity with better resistance to frost damage. Sand-lime brick contains about 10 percent hydrated lime in sand, the brick being cured in steam autoclaves to a monocalcium silicate.

Burning temperature for building brick may range from 1800–2200°F. Firebrick are burned at temperatures above 2400°F and as a result are considerably denser than building brick. A common brick weighs about 4 lb, as compared to 6 to 10 lb for firebrick. Standard size for a common brick is usually $8 \times 4 \times 2$ in.; a standard firebrick, called a *nine-inch straight*, is $9 \times 4\frac{1}{2} \times 2\frac{1}{2}$. The reason for the difference in size is that often the two kinds of brick must be built into a wall, with the firebrick on the hot face and common brick on the outside of the wall. Firebrick does not use mortar joints in

Fig. 6-5  A brick mural.

the usual meaning of the term. But building brick, because of their dimensional irregularities, require a thick bed of mortar. If building brick are laid with half-inch-thick mortar joints, and firebrick with joints of zero thickness, courses of common brick and firebrick can then be bonded together and the courses will match.

Certain clay products, such as chemical stoneware, must have zero porosity. This condition is obtained by burning the clay product in the *vitrification* range of temperature. This is the temperature range at which the most fusible or glassy ingredients of the product become suddenly less viscous, that is, they "melt." The liquid thus formed begins to dissolve the unfused material and flows into the pores to reduce porosity and increase density.

*Radial chimney brick* are made by the same methods as building brick. Such brick are segments of circles, used only in the construction of large radial brick chimneys.

*Acid-proof brick and tile* are used for building storage tanks which must contain acids (except hydrofluoric acid) and other corrosive chemicals used by such industries as pulp mills and water treatment plants. To resist chemical attack, such tile is very hard and has a smooth vitreous working face. Brick sizes correspond to those used in firebrick construction, and the mortar is a special acid-proof cement.

The ceramic products which line the nation's industrial furnaces are called refractories. These materials must be numbered among the most critically important of the materials that support an industrial civilization, for primary industries depend heavily on high-temperature processes. Such processes include the manufacture of abrasives, extraction of metals from their ores, heat-treating and forging of metals, generation of steam power, and the production of carbon products, glass, cement, brick and other materials.

Harbison-Walker Refractories Co., in the book *Modern Refractory Practice*, defines the place of refractories in the following quotation.

> Refractories are defined as nonmetallic materials suitable for the construction or lining of furnaces operated at high temperatures. Stability at high temperatures—both physical and chemical—is the primary requirement for refractory materials. They may be called upon, while hot, to withstand pressures from the weight of furnace parts or contents, thermal shock resulting from rapid heating or cooling, other stresses induced by temperature change, mechanical wear resulting from movement of furnace contents, and chemical attack by heated solids, liquids, gases, or fumes.

Because of the severe operating conditions to which they are exposed, the life of refractory linings may be brief. In extreme cases a refractory lining may have to be replaced weekly.

The range of refractories employed in industry is literally enormous. Therefore only the more commonly used items can receive mention here. To reduce this complex range of products to simple order, we may first divide all refractories into three types:

**1. Firebrick.**   These are modular burned refractory shapes, laid in an overlapping half bond with each other, and usually, but not always, bonding with thin refractory mortar.

**2. Refractory Castables.**   These are refractory concretes, poured in place.

**3. Plastic Refractories.**   These are unburned blocks of refractory in a plastic state. The soft slabs of plastic are piled up and pounded with a hammer into a monolithic wall, being burned in place when the furnace is lit. Certain types are referred to as ramming mixes, and these are installed with an air rammer.

Refractory castables are usually supplied in 100-lb sacks. Construction concretes are made by blending sacks of cement with sand, gravel, and water. A sack of refractory castables, however, includes all the required ingredients except the water. The castable is a mix of refractory cement, usually the high-

Fig. 6-6 Refractory installation in a downdraft kiln. The burners are in the round turrets, and the exhaust flues are the holes in the floor brick being installed. A portion of the 40-foot diameter roof dome may be seen, dark-colored brick at the bottom of the dome and white brick at the top. The triangular patches on the white brick result from dipping the brick in the mortar.

alumina cement called lumnite cement (lumnite cement is discussed in the next chapter) and fireclay aggregate, called "grog." The grog may be crushed and burned firebrick. The use of sand and gravel, or of ordinary portland cement, is not possible in refractory concrete except for very low temperature applications.

In addition to these differences in raw material mix, there are other differences between construction concrete and refractory castables:

1. Refractory castables are more expensive. A 100-lb sack of some special refractory castables may cost as much as a cubic yard of concrete, though this is an extreme case. Castables are not poured in the large quantities that concrete is.

2. Refractory castables set very quickly and are frequently mixed only one sack at a time for this reason. Often they are simply mixed with water in a wheelbarrow.
3. Hardened castables weigh about 125 lb per cu ft. Concrete weighs 150 lb per cu ft.
4. Refractory castables are rarely trowelled. They are too rough to trowel well, and trowelling may promote spalling of the trowelled face at high temperatures.
5. Castables do not have the tensile or compressive strength of concrete. Castable strength is moderate at room temperature, decreases to low values at about 1000°F, then increases to a maximum above 2000°F because of the formation of a ceramic bond. See Fig. 6.7.

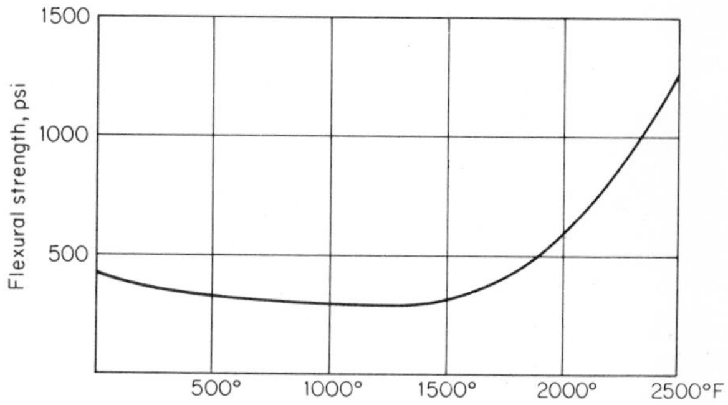

Fig. 6-7 Strength curve of a typical refractory castable. Strength is a minimum in the range of 1000–1500°F for all castables.

For light-duty applications, castables are sometimes made of ordinary portland cement, using for aggregate such materials as vermiculite, haydite, or perlite, but not sand, gravel, or crushed stone.

## 6.13 firebrick

Because of the varying demands of different types of industrial furnaces, firebrick shapes are available in many types of materials.

**1. Fireclay brick.** Except when special conditions dictate the use of special materials, fireclay brick are employed. There are however many applications that call for special materials. Fireclay brick are not generally suitable for metal melting, for slag and fume attack, or for conditions of severe abrasion. Burned fireclay shapes are available in several grades of refractoriness: superduty, high heat duty, medium duty, and low duty. Such fireclay brick contains about 40 percent alumina and 60 percent silica. Both the stiff-mud and the dry press process are used in their production.

**2. High alumina brick.** For more severe conditions such as in the manufacture of cement, high alumina brick is used, in grades of 50 percent, 60 percent, 70 percent, up to 99 percent alumina, or mullite brick, 72 percent alumina. Alumina is resistant to the attack of many slags and fumes and has higher strength than fireclay.

**3. Silica.** It has been noted that silica is an unusual material in that it maintains its strength at high temperatures. Silica brick therefore is used in applications where great strength is necessary, such as in the arched roofs of metal-melting furnaces. It is an acid brick and therefore must be used in contact with acid slags in metal meltings. Silica brick are rarely cooled below 1200°F. Below this temperature spalling is almost certain because of a high coefficient of expansion. Silica brick therefore cannot be used in intermittently operated furnaces.

**4. Graphite.** Because graphite oxidizes readily in the presence of air, graphite brick is used below the hearth line of metal-melting furnaces such as blast furnaces. Graphite is an excellent refractory since it cannot be spalled (because of high thermal conductivity) and its strength increases with temperature. If protected from oxidation it may be used at 6000°F. Few refractories are usable above 3000°F (1700°C).

**5. Magnesite and chrome ore.** Magnesite brick is magnesium oxide, the brick being a dark brown color. Chrome refractories, which are a rich green color and surprisingly heavy, are largely chromium oxide with liberal amounts of alumina and iron oxide. Large tonnages of magnesite refractories are made from the magnesium chloride in sea water. The principal products of the sea are fish and firebrick.

These are basic refractories and must be used to resist the attack of basic slags in metal melting. Actually most slags are basic, especially for steel, copper, and nickel; therefore the linings of melting furnaces are largely magnesite brick (this is not true of aluminum melting). Chrome is usually employed as a plastic refractory, being then referred to as "chrome ore." Both materials have poor hot strength and poor resistance to spalling.

The mortar for magnesite brick is a strange one indeed. For mortar a thin steel plate is used, and, surprisingly, it makes the strongest of all mortars. The magnesite brick are laid with thin steel plates between them, then the furnace is heated, and the steel oxidizes to ferric oxide which combines chemically with the magnesia. After this chemical reaction it is hardly possible to break the brick joint with anything less than a pneumatic hammer or explosives.

**6. Insulating Firebrick.** These are light-weight porous refractories. While an ordinary firebrick may weigh from 6 to 10 pounds, an insulating firebrick may weigh $2\frac{1}{2}$ to 3 lb. Most insulating brick is made of fireclay. To obtain decreased density, fine coke or sawdust is added to the raw mix. The combustible additive burns out, producing a large amount of gases, which develops the required porosity in the firebrick.

Insulating firebrick offers little resistance to abrasion or slag attack. However, they are commonly employed in what are termed "reheat furnaces," which include forge furnaces, heat-treating and brazing furnaces, etc. This brick is used simply to save on fuel costs. A large amount of heat is required to bring an industrial furnace up to temperature, if the furnace is lined with hard firebrick. Insulating brick has the same specific heat, but it weighs only a third as much, so that only one-third the heat is required for warm-up.

The ingenious applications to which materials may be put are best illustrated by refractories. The use of steel plates with basic firebrick is one such example. Another is the use of ladle brick to line the holding ladles for molten steel. The molten steel may have a temperature of 2900°F or higher, and it would be expected that a highly refractory brick would be used for the ladle lining. On the contrary, the softening point of the ladle brick may be *lower* than the temperature of the steel because it is manifestly impossible to make a liquid-tight brick lining. Even if this were possible, the brick would probably crack. The molten steel will run into any small openings in the brick lining and may then melt through the steel ladle. The worst possible accident in a steel mill is a metal breakout in an area where men are working. The possibility of such a breakout is circumvented by using a low heat duty ladle brick. The skin of such a brick softens and runs, thus forming a closed glassy surface over the ladle to prevent a breakout. As an extra safeguard, ladle brick are formulated for a high coefficient of expansion.

Building brick is laid in thick joints of sand-cement-lime mortar. If such a mortar were to be used with firebrick, the mortar would spall out under the attack of heat, and the brick wall would collapse. This is occasionally proved by the rare masonry contractor who attempts to lay firebrick with the wrong mortar. Except for basic brick and the linings of rotary kilns, firebrick are bonded in the following fashion: The mortar used is a fireclay-base mortar. It is watered down to a soupy consistency. The brick is simply dipped into the mortar and laid on the wall. Note from this description that firebrick work is not a trowel trade.

Firebrick walls are given an expansion allowance of 1 in. every 16 ft, horizontally. Firebrick masons generally use pieces of corrugated cardboard for the expansion joint. The cardboard burns out when the furnace is fired.

A wide variety of shapes is required in firebrick construction. All these shapes are produced to a dimensional tolerance of $\pm 2$ percent, since there is no thick mortar joint to compensate for variations in size from brick to brick. The standard $9 \times 4\frac{1}{2} \times 2\frac{1}{2}$ firebrick is referred to as a "9-inch straight." The arch, wedge, and key shapes are used for building arches and for turning circles. These brick are illustrated in Fig. 6.8.

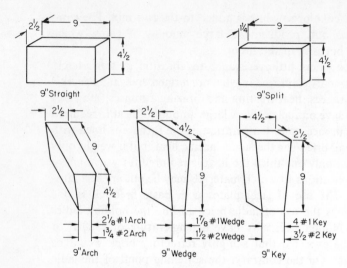

Fig. 6-8 Standard 9-inch firebrick shapes.

Both building brick and firebrick are laid in several styles of courses. These are illustrated in Fig. 6.9. Header and stretcher courses are most commonly used. With both types of brick, a half-bond overlap is used with alternate courses. Sometimes in building construction using either brick or concrete block, no half bond is employed, and all courses are registered on the bottom one. This arrangement is known as *stack bond*. It is not used with firebrick.

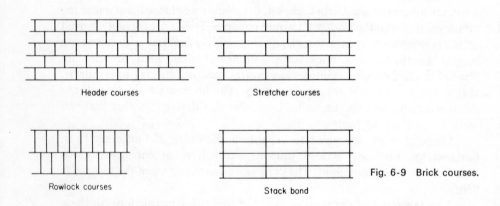

Header courses

Stretcher courses

Rowlock courses

Stack bond

Fig. 6-9  Brick courses.

## 6.14 *ashes, slags, and fluxes*

Ashes, slags, and cement are chemically similar. Most contain silica, alumina, lime, magnesia, and iron oxide, though in varying amounts. Portland cement used in the making of concrete consists chiefly of silica and alumina and lime, with smaller amounts of the other oxides. Coal ash and furnace slags, which have nearly the same composition as cement, are often added to reduce the cost of the cement. An *ash* is the inorganic residue remaining after the combustion of some material such as coal, oil, or wood.

*Slags* are inorganic oxides necessary in the melting and extracting of metals. In steel melting, for example, phosphorus and sulfur, which are always present in steels, are undesirable and must be reduced to the lowest possible levels. Both substances oxidize during the melting of the steel. To extract them from the molten steel, a basic slag is made by charging lime into the melting furnace. The lime slag floats on the surface of the metal. In a manner of speaking, we may say that the oxidized impurities prefer the company of other oxides rather than the molten steel and are attracted into the slag, reacting chemically with it. The molten metal will oxidize too, but the slag is so formulated and controlled that metal losses to the slag are minimized. This, however, may be difficult to arrange in the case of metals that oxidize more readily. Such a metal is chromium. Stainless steels contain approximately 20 percent chromium, and in melting these steels there is some loss of chromium into the slag. Similar chromium loss occurs when welding stainless steel with flux-coated welding rods. In the case of metals that oxidize extensively at high temperatures, such as titanium or zirconium, air-melting under a slag blanket is quite out of the question, for such metals would be completely lost to the slag as oxides.

*Fluxes* are actually inorganic solvents used to dissolve and lower the fusion temperature of other oxides. A familiar example is a soldering flux. This is a substance which can dissolve and remove the surface oxide on the metal to be soldered. Soldering (brazing and welding also) is not possible until surface oxides have been removed. Chlorides and fluorides are the most powerful fluxes, but they are corrosive to metals and must be removed after application.

## 6.15 abrasives

With the exception of diamond, all the abrasives used in large volume in industry are synthetic. Natural abrasives still have minor uses: garnet and flint are used for sandpaper, emery paper is used for polishing metals, rottenstone and pumice for wood finishing, and walnut shells for cleaning aircraft engine parts.

An *abrasive* is a hard material used to wear away a softer material. The hardest materials are the ceramics; hence their selection for abrasive purposes.

*Diamond*, the only significant natural abrasive, is the hardest material known. It outlasts other abrasives by factors of 10 to 100 or more. Diamond therefore is used for those applications in which the life of other abrasives is poor. Typical applications are in brick and concrete saws, wiredrawing dies, diamond drills for drilling hard rock (brass cores set with diamond grit), and cutters for machining hard metals. Like any abrasive, diamond is quickly ruined by impact.

Next to diamond, the hardest abrasive in common use is silicon carbide, SiC, perhaps better known by one of its trade names, carborundum. This highly versatile material, already mentioned for its uses as a refractory, also serves as a heating element in electrically heated industrial furnaces, as pipe, and is even formed into pumps for pumping sand. It is blue-black in color. Both beryllium carbide and boron carbide are harder than silicon carbide, but they have not yet found extended use as abrasives.

*Silicon carbide* was discovered about 80 years ago as a result of unsuccessful attempts to make artificial diamond. The raw materials for silicon carbide are cheap, but the manufacturing process is not. Sand, coke, and sawdust are mixed, and a high-temperature (4500°F) electric arc is passed through the mixture for a considerable period of time. The sawdust has the function of burning out to provide porosity for the escape of gases from the mass. At the conclusion of the process the center of the mixture is converted to silicon carbide, which is then crushed.

*Aluminum oxide* is known by various trade names, such as Aloxite. This material is easily distinguished from silicon carbide by its lighter color, which may be various shades of white or brown. It is made from bauxite, the ore of aluminum, by fusion of the bauxite and crushing.

Aluminum oxide is not quite so hard as silicon carbide, but it is tougher and thus more resistant to impact. It is preferable to silicon carbide for such applications as floor sanding machines—wooden floors contain nails, which are damaging to the abrasive grit. Aluminum oxide is the preferred material for grinding the harder metals, since it wears away faster than silicon carbide, thus exposing new edges for cutting. Silicon carbide is selected for cast iron and the softer types of nonferrous metals.

In general, only these two abrasives are used in grinding wheels. The grit in the wheel is bonded either with a ceramic material such as sodium silicate or with an organic material such as a rubber or plastic cement.

## 6.16  carbides

The properties of silicon carbide are typical of other carbides: great hardness, high melting point, high compressive strength, and brittleness. The highest melting point so far discovered is that of hafnium carbide, 7520°F. Unfortunately these remarkably high temperatures cannot usually be exploited, because the carbides will oxidize above 2500°F in air, producing carbon dioxide and monoxide.

Aside from very special applications, only a few carbides are as yet in general use. Silicon carbide has abrasive and other applications. Tungsten

Fig. 6-10 Small melting crucibles for induction melting of metals. Graphite and clay crucible on the left, silicon carbide crucible on the right, sillimanite (alumina-silica) crucibles in the rear.

carbide and tantalum carbide are used in cutting tool tips such as masonry drills and machining tips. The nuclear age has given birth to other carbides such as uranium carbide and boron carbide. Boron carbide, $B_4C$, 75-80 percent boron, is used as a shielding material against neutrons, deriving its effectiveness from the high neutron capture cross-section of boron. Though harder than silicon carbide, it is not used as an abrasive.

A most interesting employment of carbides, which is discussed in Chapter 9, is their use as a hard constituent in steel alloys such as tool steels for cutting other metals. Such carbides appear as microscopic constituents in steels as a result of suitable heat treatment, and they are the ingredients that give tool steels their hardness, wear resistance, and cutting capacity.

The cemented carbide cutting tool materials consist of hard carbides, usually tungsten carbide, titanium carbide, and tantalum carbide embedded in a binder metal, usually cobalt. They are often given the name "carboloy" in conversation, though this is the trade name of only one of many manufacturers. Such carbide bits are supplied in two types: tungsten carbide and complex carbide (WC + TaC + TiC). Though the tungsten carbide type is stronger for a given hardness, the machining chip tends to weld to this type if steels are cut. The complex carbides are therefore used for machining steels, and the tungsten carbide type is used when machining nonferrous metals, cast irons, and nonmetallic materials.

The carbide cutter bits contain about 90 percent carbides. A high-speed tool steel will contain a maximum of 20 percent of carbide constituents. The carbide bits therefore considerably outwear high speed steels and other cutting steels and retain their hardness much better at the high temperatures of metal cutting, which may exceed 1500°F.

## 6.17 carbon and graphite

The phases of carbon are included in this chapter because most of the uses of carbon are refractory uses, deriving from its high melting temperature. Like silicon carbide, the forms of carbon have a wide range of uses: graphite is used as a lubricant because of its softness, and diamond is used as a cutting tool because of its hardness.

Carbon and graphite offer the properties of chemical inertness (except against oxygen above 1000°F), absolute spall resistance, good electrical and thermal conductivity, low thermal expansion, and most important of all, *increasing* strength with increasing temperature. The carbon phases can be employed at temperatures as high as the sublimation point of the material, 6600°F. At atmospheric pressure, carbon does not melt but changes from solid to gas, that is, it sublimes, at 6600°F. But in metal-cleaning operations, as every housekeeper knows, carbon has the dubious honor of being the most difficult type of soil to remove from metal surfaces, since no standard metal cleaner will attack it chemically.

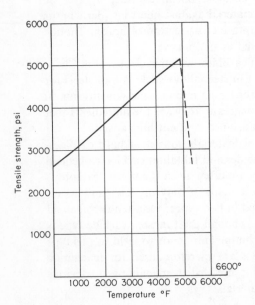

Fig. 6-11 Typical strength curve of graphite.

The tensile strength of carbon materials varies with the type, grade, and method of manufacture but may range from 1000 to 3600 psi at room temperature. The tensile strength approximately doubles when the temperature is raised above 4000°F. The spall resistance of carbon materials is so outstanding that it is difficult to devise an experimental method of cracking them by thermal stresses only.

The various forms and grades of carbon embrace a wide range of products. *Carbon black* or amorphous carbon is a finely divided pigment produced by the thermal decomposition of oil or natural gas. Soot is a noncommercial example of carbon black. Such blacks are most familiar for their use in automobile tires and polyethylene water pipe. When carbon black is heated for- a prolonged period at a high temperature, *graphite* is formed, a crystalline phase of carbon with properties markedly different from those of carbon black. Graphite is easily machined and has a relatively high thermal and electrical conductivity compared to other forms of carbon. Graphite is produced by heating carbon black to 5000°F approximately, but if instead the processing temperature is only 2750°F, *industrial carbon* is produced. Industrial carbon is extremely hard and has lower thermal and electrical conductivity than graphite. Both industrial carbon and graphite weigh approximately 100 lb per cu ft.

Graphite has a lower specific resistance (about 0.0004 ohms per in. cu), than other phases of carbon: it is therefore preferred as an electrical conductor, despite its higher cost. By far the greatest tonnage of carbon and graphite is used in the metallurgical and electrochemical industries as anodes in arc-melting furnaces, particularly in the steel and aluminum industries. Such anodes carry many thousands of amps and range in diameter up to a maximum of about five feet. The arc-melting of steel, described in Chapter 9, requires electrodes with smaller diameters of 1 to 2 ft. The raw material for such products is usually *petroleum coke*, produced from crude oil at oil refineries. Other electrical uses for graphite and carbon include brushes for electric motors and anodes for large electronic tubes such as rectifiers and ignitrons. These materials are not used as cathodes because they are poor emitters of thermal electrons.

### 6.18   porcelain enamels

These enamels are actually ceramic surface coatings applied to steel, iron, and aluminum for protection against corrosion and for decorative appearance. The enamel frit is a low-melting mixture of oxides resembling small particles of glass. Quartz and feldspar are blended with fluxes such as borax or soda ash (sodium oxide) and with materials such as titanium oxide or inorganic colors for color or opacity.

Fig. 6-12 Porcelain enamelled stoves.

Enamel is applied after first cleaning the metal surface with an alkaline cleaner. The glassy frit is ground fine, mixed with water, and applied by dipping or spraying. The enamel coat is allowed to dry and is then "burned" or fired at temperatures in the range of 1000–1500°F.

Porcelain enamel is applied to a wide range of household appliances and equipment. It is also used for architectural purposes such as curtain wall sheets on high-rise buildings and automotive service stations, often with highly interesting effects.

A variant of the enameling process is the application of *ceramic coatings* to parts requiring abrasion resistance or protection against high temperature or oxidation. Materials such as silica or tungsten carbide may be applied. Parts of aerospace vehicles, rocket motors, and gas turbines may require ceramic coatings.

## 6.19 ceramic lubricants

Ceramic materials are used as lubricants under conditions either of extreme high pressure or high temperature. Graphite is perhaps the best known of these lubricants. Lime is used as a lubricant in wire-drawing, and molten glass in the extrusion of steel shapes. Mica, soapstone, and talc are occasionally used as lubricants.

Molybdenum disulfide, $MoS_2$, is a more recent addition to the list of ceramic lubricants. This material is gray in color, with a high specific gravity of 4.9 and a melting point of 2200°F. Its coefficient of friction is very low, but it is capable of sustaining remarkably high bearing pressures.

The silicone oils are lubricating oils with a silicon base instead of a carbon base. They are serviceable to 500–600°F, well beyond the useful range of petroleum oils, and at low temperatures do not congeal as do the petroleum lubricants.

## PROBLEMS

1 Explain the influence of thermal conductivity on the spall resistance of ceramic materials.

2 Why is pure silica not used for window glass?

3 What is the difference between a silt and a clay?

4 Why does standard firebrick measure $9 \times 4\frac{1}{2} \times 2\frac{1}{2}$ in.?

5 Explain the meaning of vitrification. What clay products do you know that are vitrified?

6 What is a "glass"?

7 What is meant by the terms grog and castable?

8 Explain the purpose of a slag.

9 Why would crushed limestone not be used as aggregate in refractory castable?

10 How many 9-in. straight firebrick are required per square foot of furnace wall when laid in header courses?

11 Explain the reason for the following selections:
(a) silica brick for an arch roof on a high-temperature furnace
(b) graphite brick below the molten bath level in a metal-melting furnace
(c) magnesite brick at the slag line in a furnace using a lime slag.

12 What is refractory? A refractory castable?

13 What is the meaning of spalling?

14 Name one use for each of the oxides silica, alumina, iron oxide, lime, magnesia, and titanium oxide.

15 What is the difference between quicklime and hydrated lime?

16 Why is glass weak in tension?

17 What advantage distinguishes borosilicate glass?

18 Explain the difference between igneous, sedimentary, and metamorphic rock. State an example of each type.

19 What mineral constituent determines the relative difficulty of drilling rock?

20 What is synthetic stone?

21 What is the origin of clays?

22 What is the basic difference in the manufacturing method between the stiff-mud and the dry press process of making brick?

23 Consult Sec. 6.5 and state what would happen to a refractory concrete using limestone aggregate when heated to 2000°F.

24 Why are insulating firebrick frequently used in furnaces?

25 What is the difference between a header and a stretcher course of brick?

26 Which abrasive is selected for grinding: (a) hardened steel, (b) brass, (c) gray cast iron?

27 Why is tungsten carbide not used as a cutting tool for machining steel?

28 What type of carbon is employed in automobile tires?

29 Name the softest and hardest forms of carbon.

30 What is petroleum coke?

31 What is a porcelain enamel?

The words "cement" and "concrete" are often confused in conversation, as in referring to a concrete sidewalk as a "cement sidewalk." Cements, whether organic such as rubber cements or inorganic such as the portland cement used in road and airport paving, are adhesive materials. In a concrete, the cement must coat the surface of all the particles of aggregate (bulk or filler material) in order to bind the whole into a monolithic mass.

Cements then are adhesives, as are brick mortars. Whether there is any difference in meaning between the words "cement" and "adhesive" is uncertain, though the word cement seems to imply a thicker adhesive joint than is used with an adhesive. An adhesive used to bond ceramic materials is almost always called a cement, but if the same adhesive material is used to bond metals or organic materials it may well be termed an adhesive.

The cement used in civil engineering construction is termed portland cement. This is a hydraulic silicate cement and is produced in several types. Refractory castables and corrosion-resistant concretes commonly use an alumina-base cement called lumnite cement. Lumnite cement is more expen-

**CEMENTS**
**AND CONCRETES**

**7**

sive than portland cement and is used only where its superior properties justify the additional cost.

Inorganic cements and mortars harden by taking up carbon dioxide or water in chemical reactions. The ancient Romans used a brick mortar resembling our own sand-lime-cement mortar by using quicklime, CaO, and burned clay in the form of crushed brick. Such mortars, ancient or modern, harden very slowly as the lime combines with carbon dioxide in the air to form rock-hard $CaCO_3$ or limestone. Portland cement is a more complex mixture of chemicals and has a more complex chemistry of hardening. But compared to lime, portland cement hardens much more quickly and attains considerably higher strengths mainly because of the presence of tricalcium silicate, $3CaO.SiO_2$, which is not found in lime mixtures.

Fig. 7-1  An all-fuel prefabricated chimney designed by the author, insulated with Perlite Lightweight Concrete.

## 7.2  the raw materials for

## portland cement

The method of manufacturing portland cement from its raw materials is largely determined by the fact that tricalcium silicate can only be produced within a restricted temperature range and only in a solid-state chemical reaction. If tricalcium silicate is melted, it decomposes to lime and dicalcium silicate. The melting point is 4064°F; the temperature of manufacture must be in the range of 2300–3450°F.

The raw materials for portland cement are clay and limestone. The chemical composition of ordinary clay is such that its silica content is about twice its aluminum content, giving it the approximate chemical composition $Al_2O_3.2SiO_2.2H_2O$. The two basic ingredients, clay and limestone, must be blended in suitable proportions and "burned" together under conditions that produce an appreciable amount of tricalcium silicate. This solid-state reaction is promoted by including in the raw material mix a substance which serves chiefly as a flux: ferric oxide. Without the fluxing action of ferric oxide a much higher temperature of reaction between the clay and the limestone would be required.

Pure limestone, $CaCO_3$, without some dolomite, $MgCO_3$, is virtually unknown. Dolomitic limestones, which are about half limestone and half dolomite, are unsuitable for cement manufacture. Magnesia does not combine with the acid silica and remains as free magnesia in the finished product. If free magnesia, or even free lime, were present in the cement, it would "slake" or combine with water to produce $Mg(OH)_2$. This reaction produces an increase in volume, and the effect would be to crack and disintegrate the concrete. Expansion caused by magnesia is more dangerous than expansion caused by lime, because its rate of development is quite slow, and its first effects might not appear until the passage of a few years. For this reason most standard specifications for portland cement limit magnesia to a maximum of 5 percent. There are always limited amounts of magnesia in any cement, and it is this material which gives cement its gray-green color.

## 7.3 the manufacture of cement

A typical raw feed by weight for a portland cement might be

| | |
|---|---|
| $SiO_2$ | 15.5% |
| $Al_2O_3$ | 2.5 |
| $Fe_2O_3$ | 2.0 |
| $CaO$ | 42.0 |
| $MgO$ | 2.5 |
| $CO_2$ | 35.5 |
| | 100.0% |

with $CO_2$ combined as $MgCO_3$ and $CaCO_3$. The raw materials are ground to 200 mesh in a ball mill and burned in the cement kiln either dry or mixed with water as a slurry.

The rotary kiln in which the raw material is reacted is one of the big industrial furnaces which, like the blast furnace, plays an indispensable role in our industrial civilization. Rotary kilns are used to process a wide range of products, from alfalfa and alumina to brucite, chrome ore, dolomite, ferrochromium, garbage, haydite, iron ore, lime, and manganese ore, to pozzolans, pyrites, perlite, and plaster of Paris, to vermiculite and zinc oxide. The kiln is a cylinder made of steel. A small laboratory or pilot-plant kiln may have an inside diameter of 3 ft. A large cement kiln may be 16 ft or more in diameter, though the usual cement kiln is 11 or 12 ft in diameter. The length of the kiln shell may be from 400 to 600 ft. The steel plates of the kiln are not less than $\frac{7}{8}$ in. thick for a cement kiln. The kiln is inclined at an angle of a few degrees, so that raw material fed into the high end can travel slowly down toward the low end as the kiln rotates. The kiln is supported on large rollers and rotated at about 0.6 rpm by a large electric motor of 125 hp or more. A standby diesel engine is available in case of electric power failure, for if the kiln does not rotate it will warp because of its great weight and length. The kiln is driven through a geared bull-ring that encircles the kiln at the middle of its length. Expansion of the kiln is in both directions from this fixed point. Figure 7.2 shows an interior view of a cement kiln.

The discharge end of the kiln is enclosed in a housing called a firing hood. Figure 7.3 is a view of the inside of a firing hood with its refractory lining of firebrick and plastic refractory. So that repair crews may have

Fig. 7-2 A cement kiln under repair. The large globules of clinker adhering to the refractory lining are characteristic of kiln operation. The steel shell of this kiln is 11 ft in diameter, 7/8 in. thick, and 400 ft long.

Fig. 7-3  The firing hood at the end of the kiln shown in Fig. 7-2. The large gas burner may be seen in the middle of the hood.

access to the interior of the kiln, the firing hood may be retracted on a set of railroad rails. A large gas burner about 12 in. in diameter supplies heat for the process through the end of the firing hood. The operating crew controls kiln operation from the firing end, on what is called the "burning floor"; here the panel of automatic control instruments is mounted.

The raw kiln feed is metered into the kiln at the high or feed end and is slowly tumbled down to the discharge end in a journey that takes a few hours to complete. The tumbling action produces a product in pellet form. As the raw materials are heated, the following chemical actions occur in sequence along the kiln.

1. Water is driven out of the feed.
2. At 630°F, $MgCO_3$ decomposes to $MgO$ and $CO_2$.
3. At 1632°F, $CaCO_3$ decomposes to $CaO$ and $CO_2$.

One barrel of cement (or 4 sacks) weighs 376 lb. Because of loss of carbon dioxide, about 583 lb of solid feed are required to produce 376 lb of kiln product. The heat requirement is nearly 1,000,000 Btu per barrel of cement produced.

4. Finally in the "burning zone" of the kiln, which is about the last 70

ft, the components undergo a solid-state reaction at a temperature of about 2700°F to produce calcium silicates. This is an exothermic reaction, releasing heat. The reaction that is most difficult to complete is the combination of the last trace of lime with previously formed dicalcium silicate to form tricalcium silicate. The amount of lime remaining in the kiln product after burning is usually less than 1 percent.

5. The resulting product, a black slag in small globules, is dropped through the floor of the firing hood into a cooler below the kiln. This product is called "clinker."

Following the burning operation, the clinker is ground to about 325 mesh (44 microns), blended with about 3 percent gypsum, and bagged or stored in large cement silos.

An interior view of an 11-ft cement kiln shut down for repair is shown in Fig. 7.2. The kiln is lined with 9 in. of firebrick, and the clinker adhering to the firebrick can be seen in the figure. At the time the photograph was taken, the kiln had cooled for three days, yet the clinker, though relatively cool at its surface, was still red hot underneath. The two repair men in the kiln are drilling holes into the refractory lining to determine the extent to which the lining is worn away. Ordinarily the lining of the burning zone will last 8–9 months.

Lining a cement kiln with firebrick is an interesting process that few persons have the fortune to witness. The lining is formed from segments of a circle, called kiln liners, with 35 brick liners to a circle for an 11-ft kiln. Each brick is 4 in. thick (in the direction of the axis of the kiln), which gives three courses per foot of kiln length. Since each kiln liner weighs 16 lb, the refractory lining weighs 1700 lb per ft of kiln, or almost a ton. About 8 ft of kiln are lined at a time, and the bricklaying is carried on continuously day and night until completed—this equipment is too expensive to be tied up with the casual effort of dayshifts only.

The firebrick are laid without mortar and stay in place simply by fitting tight together. The last brick in each circular course must be carefully cut to make a tight fit in closing the circle.

The brick for about 8 lineal ft of kiln are laid in the bottom half of the kiln. Figure 7.4 shows the bricking sequence. A timber is run along the top of the courses on each side of the kiln, and trench jacks installed every 2 ft, jacking against these timbers with maximum pressure to prevent any movement of the brick. The kiln is then rotated a quarter turn. Another quadrant of the kiln wall is bricked, two more timbers set up, and another set of jacks installed, again at 2-ft intervals. There is now a jack every foot, making for crowded working conditions. The kiln is rotated another quarter turn, and the circular courses completed. The jacks are then carefully removed, the crew watching the bricks at the top of the kiln to see if they slip (since a kiln liner weighs 16 lb, the crew members wear hard hats). Occasionally the

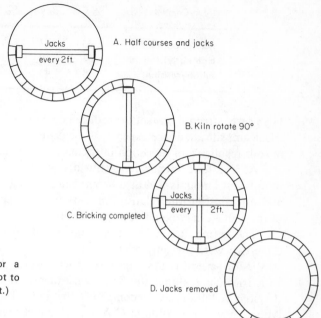

A. Half courses and jacks

Jacks every 2 ft.

B. Kiln rotate 90°

C. Bricking completed

Jacks every 2 ft.

D. Jacks removed

Fig. 7-4 Brick sequence for a rotary kiln. (The drawing is not to scale nor to correct brick count.)

bricking is unsuccessful; if the bricks slip slightly, then the courses must be rebricked.

## 7.4 portland cement compounds

Portland cements are composed of four principal chemical compounds:

1. tricalcium silicate, $3CaO.SiO_2$, abbreviated $C_3S$
2. dicalcium silicate, $2CaO.SiO_2$, abbreviated $C_2S$
3. tricalcium aluminate, $3CaO.Al_2O_3$, abbreviated $C_3A$
4. tetracalcium aluminoferrite, $4CaO.Al_2O_3.Fe_2O_3$, or $C_4AF$

The strength of these cements is controlled by the content of the two calcium silicates, which together constitute at least 70 percent of the cement constituents.

Three types of portland cement are in common use, though certain industries and operations require special types such as oil-well cement. The three most used types are normal portland cement (Type I), high early strength cement (Type III), and sulfate-resistant cement (Type V). These cements have the following typical compositions:

| Cement | % C$_3$S | % C$_2$S | % C$_3$A | % C$_4$AF |
|---|---|---|---|---|
| normal portland | 45 | 27 | 11 | 8 |
| high early strength | 53 | 19 | 10 | 10 |
| sulfate-resistant | 38 | 43 | 4 | 8 |

High early strength cement is produced by employing a raw mix giving more tricalcium silicate at the expense of dicalcium silicate. Resistance to sulfates in soils results from reduction in tricalcium silicate.

All the four chemical constituents in portland cement take up water in different amounts and at different rates during the setting of the cement, and by adjusting the relative amounts of these constituents the properties of the cement may be altered appreciably.

The setting rate of both tricalcium and dicalcium silicate is slow, and if these two were the only constituents of cement, the cement would remain plastic for several hours after mixing with water. But the opposite happens with tricalcium aluminate. This constituent, when contacted with water, develops a "flash set" accompanied by the release of considerable heat. Although the proportion of C$_3$A in portland cement is not high, its presence would lead to flash setting of the whole cement mass. This explains the use of a small amount of gypsum, CaSO$_4$, in cement. Gypsum prevents such flash setting and regulates the setting time of the cement. It does this by quickly combining with C$_3$A to form a needle-like compound, calcium sulphoaluminate, which contains a considerable amount of chemically bound water or "water of crystallization."

## 7.5 hydration of portland cement

Portland cement, like any hydraulic cement, sets and hardens by taking up water in complex chemical reactions, a process called *hydration*. Provided that a source of water is constantly present (water vapor in the air is a suitable supply), portland cement may continue to harden over a period of months. The reaction completed first is the combination of gypsum with C$_3$A and water to form calcium sulfoaluminate. This chemical action is completed within 24 hours. The other compounds in cement will hydrate by taking up about 25 percent of their weight in water. The amount of water added to cement is not gauged from this relationship, however.

C$_3$S hydrates in the following reaction

$$2(3CaO.SiO_2) + 6H_2O \rightarrow 3CaO.2SiO_2.3H_2O + 3Ca(OH)_2$$

while $C_2S$ hydrates thus

$$2(2CaO.SiO_2) + 4H_2O \rightarrow 3CaO.2SiO_2.3H_2O + Ca(OH)_2$$

Thus the final product of the hydration of both calcium silicates is $3CaO.2SiO_2.3H_2O$, a compound which is given the name *tobermorite*. But $C_3S$ in its hydration releases three times as much calcium hydroxide to the concrete mix: this has a significance for the corrosion resistance of portland cements, which will be dealt with presently. $C_3A$ hydrates without taking up or releasing any lime. Some of the lime set free by the hydration of the calcium silicates is taken up in the hydration of the $C_4AF$.

The workability of cement paste does not change for some time after water is mixed with it. In this apparently dormant period, however, there are rapid and complex chemical reactions, including the $CaSO_4$ and $C_3S$ reactions just discussed. Setting of the cement is completed in a matter of hours, while hardening is not completed for months, although there is no fundamental difference between these two processes. Rapid setting, however, is not synonymous with rapid hardening. Gypsum retards setting; the use of hot water accelerates setting. The finer the cement powder is ground, the greater the surface exposed to water and the faster the set. Rapid setting and high early strength are provided by an increase in $C_3S$, as shown in the analysis of high early strength cement given earlier. Fast-setting cements, however, release considerable quantities of heat in their short setting times.

The ultimate compressive strength developed in a portland cement results from the hydration of the two calcium silicates. $C_3A$ and $C_4AF$ contribute little to strength of the concrete, although they reach their maximum strength levels in a comparatively short time.

The complex chemistry of cement setting just discussed occurs under atmospheric conditions of temperature and pressure. A variant of this method is used in the manufacture of sand-lime brick, and to a degree, in the manufacture of concrete blocks. These products are cured with steam under heat and pressure. Sand-lime brick uses a mixture of fine quartz sand and 5–7 percent hydrated lime. This is compacted, then cured for several hours at a steam pressure of 100 psi and a temperature of about 350°F. At this temperature the silica reacts readily with lime to form calcium silicate hydrates of high strength.

## 7.6 heat of hydration

When hot lime (quicklime) is slaked with water, the reaction with the water is exothermic, that is, heat is released. The generation of heat may be so pronounced that the water may boil. The lime in portland cement is eventually converted to calcium hydroxide also, through the formation of calcium silicate hydrates. The evolution of heat is therefore to be expected when cement sets. Most cements set at a sufficiently slow rate that this evolution of heat is not a source of trouble. However the heat of hydration in massive concrete structures such as dams could cause cracking because of expansion and subsequent contraction. Such pours may have to be refrigerated. The cement manufactured for large pours is formulated for low heat of hydration. $C_3S$ releases twice as much heat of hydration as $C_2S$, and $C_3A$ three times as much.

## 7.7 corrosion of concrete

Water attacks concrete to a slight degree. Part of the hardened cement paste is free calcium hydroxide, a chemical that is fairly soluble in water and therefore can be leached out in the presence of water. This leaching action may produce the efflorescence occasionally observed on concrete and brick structures. Efflorescence is water-soluble matter such as salts or calcium carbonate which is deposited on the surface of the concrete through capillary action. Such efflorescence may be removed by wire brushing or treatment with dilute hydrochloric acid.

Sulfates are often found in soils and water. Sulfates attack cement by reacting with calcium hydroxide in the cement to produce first gypsum and then water-rich calcium sulfoaluminate. This chemical action is harmful because of the volume expansion that accompanies it. Sulfate-resistant cement therefore is low in tricalcium aluminate. It may be noted here that sea water contains significant amounts of magnesium sulfate.

Occasionally damaging chemical reactions occur between the cement and the aggregate materials of the concrete. This is termed an *alkali-aggregate reaction*. The actual chemistry is complex and imperfectly understood, but in brief a high-alkali cement attacks susceptible aggregates containing silica (which is acid) and develops disruptive pressures in the concrete. Pozzolans are used to control this reaction.

## 7.8 pozzolans

Pozzolans are siliceous materials, not cements in themselves, which combine with lime in the presence of water to form compounds with cementing properties. Pozzolans are named from the fact that the ancient Romans used volcanic siliceous materials as pozzolans obtained from deposits at Pozzuoli near Naples. A wide range of materials offer possibilities as pozzolans: calcined clays and shales, pumicites, blast furnace slags, brick, and flyash from boilers. Pozzolans, like cement, must be finely ground.

As much as 40–50 percent of pozzolans may be added to cement, and they generally produce the following effects on the concrete:

1. improved workability
2. lower heat of hydration
3. improved watertightness
4. improved resistance to sulfates
5. reduced alkali-aggregate reaction
6. better resistance to leaching
7. reduced cost

Weathering resistance however is lessened somewhat, and the modulus of elasticity of the concrete may be slightly reduced. Because of the lower heat of hydration, pozzolan additions are made to concrete used in large pours such as dams.

One of the most interesting of the pozzolanic materials is flyash. When pulverized coal is burned in a large steam generator, most of the ash is carried up the boiler toward the stack and collected in hoppers at the rear top of the boiler. This flyash is even finer than portland cement and consists of small glassy spheres of rather complex chemistry. Most flyashes show about 30–50 percent silica, up to 10 percent carbon, 10–20 percent iron oxide, about 20 percent alumina, and lesser amounts of calcia, magnesia, and alkalis.

## 7.9  lumnite cement

In the previous chapter we referred to the use of high-alumina cement in the blending of refractory concretes. Such aluminous cements are usually termed *lumnite cement*. Lumnite cement finds most of its applications in the field of mechanical engineering, just as most of the applications of portland siliceous cement are in civil engineering. Lumnite cement has additional uses in refractory castables, however. It was first produced in France in 1913, and its French name, Ciment Fondu, is sometimes used in this country. Compared to portland cement, it is more expensive, highly resistant to chemical attack (a characteristic of alumina), and fast-hardening, developing great strength in 24 hours.

Unlike portland cements, the chemical composition of a lumnite cement is somewhat variable, but approximates the following analysis:

$SiO_2$       3–11%
$Al_2O_3$     33–44%
$CaO$        35–44%
$Fe_2O_3$    4–12%
$FeO$        0–10%

The aluminates in this cement are chiefly monocalcium aluminate, CA, but include also $CA_2$ and $C_{12}A_7$. The silica may be combined as $C_2S$ or in more complex compounds.

The limestone-clay raw materials used in the manufacture of portland cement are too low in alumina and too high in silica to be suitable for lumnite cement. Pure limestone and bauxite are employed. Bauxite, the usual ore of aluminum, is rather expensive because it is relatively scarce. Lumnite cement cannot be produced by the clinkering of the raw mixture in a rotary kiln, as is the practice for portland cement. Instead, the mixture is melted, often in electric arc furnaces; hence the French name Ciment Fondu—molten cement.

The hydration of lumnite cement is virtually complete in 24 hours of curing. More water is used than in portland cement, and a higher-strength cement results. Hydration releases free alumina instead of free calcium hydroxide, and this in part explains the excellent corrosion resistance of this cement. Temperatures above 70°F in the curing process cause a significant drop in the strength of the concrete. Unfortunately the rapid hydration means a much more rapid release of heat than occurs in the case of portland cement. Concrete pours even as thin as two inches become noticeably hot to the touch. Sometimes water-cooled metal formwork is employed on lumnite concrete, and the forms are removed as soon as possible so that the concrete can be hosed with cold water.

Having no free calcium hydroxide, lumnite cement is resistant to sulfate attack.

Quick-setting cements for repair of water leaks are sometimes made by mixing lumnite cement and lime or lumnite cement and portland cement. Such mixtures do not produce a strong cement, but they do serve the purpose intended, that is, flash-setting.

With the use of suitable aggregates such as burned fireclay materials, lumnite cement is used in refractory castables for use at temperatures as high as 2700°F. It is commonly employed in chimney linings, castable shapes, blast furnace foundations, furnace doors, rocket-launching pads, and taxi areas for jet aircraft. Finer aggregates are used in refractory castables than are customary in portland cement concretes. For high-temperature work, reinforcing rods must not be used in refractory concretes. Expansion of the rods can crack the concrete, and in any case there is no bond between the reinforcing material and the concrete as there is in constructional reinforced concrete. As with portland cement, the lower the ratio of water to cement, the higher the strength of the concrete. Besides refractory uses, lumnite cement is used for floors which are subject to the attack of aggressive chemicals, such as acids in fertilizer plants, lactic acid in dairies, and organic acids in meat-packing plants.

Ordinary concrete should not be subjected to temperatures much higher than 500°F. Above this temperature, some type of spall and heat-resistant concrete should be used.

In a standard concrete about 75 percent of the volume of the mix is occupied by aggregates—crushed rock, gravel, and sand, with heavier pours using larger pieces of aggregate. Any cubic yard of concrete must be filled either with cement paste or with aggregate. Since the cement is the most expensive ingredient, the amount of cement must be minimized, consistent with the strength and other quality requirements of the concrete. The minimum cement requirement is met by suitably grading the aggregate particles so that small particles can fill the voids between the larger particles. Each aggregate particle, small or large, is completely embedded in cement paste, and all spaces not filled by aggregate must be filled by cement paste, except for entrained air in the concrete.

The hardened concrete will weigh about 150 lb per cu ft. Most concretes are batched to provide an ultimate compressive strength of 2500 to 4000 psi after 28 days, the standard curing interval before the strength of the concrete is tested. Figure 7.5 shows a typical strength curve of concrete with the passage of time. Higher strengths than 4000 psi are possible but are not commonly required. The tensile strength of concrete is not high but is proportional to the compressive strength. The modulus of elasticity of concrete is about 1000 times the ultimate compressive strength, though strain is not proportional to stress in concretes.

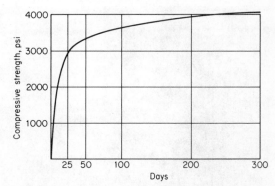

Fig. 7-5   Strength curve of a 3000 psi portland cement concrete cured in air. Strength after 28 days is 3050 psi.

The strength of concrete depends chiefly on the water-cement ratio, with a low ratio producing a strong concrete. While only a small amount of water is required to complete the chemical reactions of setting concrete, more than this is used to make the concrete more workable.

A great many special admixtures have been developed for concrete. The use of accelerators such as calcium chloride has already been mentioned. Air-entraining mixtures are used to improve the resistance of concrete to frost action and salt corrosion. About 1 percent of the volume of air-entrained concrete will be minute voids from 10 to 1000 microns in diameter (1 micron = 1/25,000 in.). Hardeners are added to industrial concrete floors to improve the resistance of the floor to abrasion and dusting. Cast iron and emery particles are two of the many materials used as hardeners.

Light-weight concretes are made by using perlite, haydite, vermiculite, or other expanded minerals as aggregates. In addition to the advantage of less weight, light-weight sections offer heat insulation and a degree of acoustic absorption. Light-weight concretes, however, have greater shrinkage on curing and need more care in proportioning the required water. They require more water than standard aggregates because the expanded aggregates themselves absorb water.

Concrete is used as a biological shield for protection of personnel from radiation. Since all the elements in concrete, H, O, Al, Si, Ca, etc., are light elements, none has high absorption for gamma rays. Such absorption is provided by adding various materials to the concrete, including magnetite (iron ore), barite (containing barium, a heavy metal), or steel punchings.

*Grout* is a special concrete mix for seating machine bases, levelling base plates, and for closing up fissures in underground rock formations or dams. Since grouts must be thin, they are made up of nonshrinking mixes of cement and sand and occasionally other additives.

A terrazzo floor is shown in Fig. 7.6. Some remarkably beautiful floors are made in terrazzo, which is produced by laying mixtures of concrete

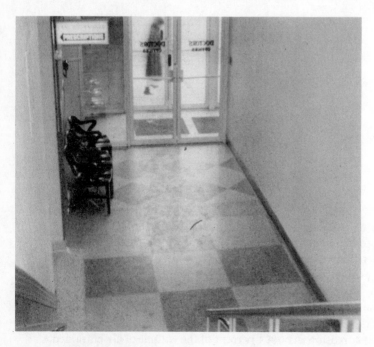

Fig. 7-6   Terrazzo floor in a clinic building.

containing colored marble chips, usually in black, white, red, yellow, or green. Terrazzo floors require special crews for their installation.

## 7.11 asphalt concrete

Asphalts, which are discussed more fully in Chapter 13, are residues from the distillation of crude oil. Although asphalts are also used for roofing and waterproofing, about three-quarters of all asphalt production is used for highway and airport paving. Asphalt refined to specifications for paving purposes is called asphalt cement.

Asphalt cements are produced in several grades. The relative hardness or "consistency" of asphalt cement is measured by the penetration of a weighted needle (100 grams) into the asphalt. The depth of penetration is measured in units of 0.1 mm at 77°F after 5 seconds. The test is similar to the types of hardness tests performed on rubber, plastics, and metals discussed in the following chapter. The deeper the penetration, the softer the material.

Asphalt makes a flexible pavement. Such pavements are mixtures of asphalt cement and aggregates. Although the pavement must be flexible, that is, elastic, it must be resistant to permanent displacement under traffic loads. The surface of the pavement must be sufficiently rough to provide traction under adverse weather conditions; otherwise the familiar "slippery when wet" signs must be posted on the highway. With these requirements in mind, the function of the aggregate is to carry the traffic loads imposed on the pavement and to provide rough texture, while the asphalt cement binds the aggregate and provides waterproofing.

## PROBLEMS

1   What purpose does iron oxide serve in the manufacture of portland cement?

2   What is the chief chemical compound in normal portland cement?

3   Why can magnesium oxide not be substituted for lime in portland cement?

4   What is the difference between rotary kiln clinker and finished portland cement?

5   Why is gypsum added to portland cement?

6   Why is mortar not used to bond the firebrick lining in a large rotary kiln?

7   Which of the several chemical compounds in portland cement is fast-setting?

8   What is a pozzolan?

9   Why are pozzolans added to the cements used in the heavy concrete pours of large dams?

10  Apart from the matter of cost, why would lumnite cement have few applications in the field of civil engineering?

11  What is meant by hydration of cement?

12  What is a hydraulic cement?

13  What reaction is the first to occur when cement sets?

14  Consult suitable references and describe the procedure and purpose of the slump test for concretes.

# METALS

*Part* **3**

The many mechanical characteristics of materials cover a wide range, chiefly empirical in nature, and few of them are clearly and fundamentally understood. We shall deal first with particle size. Next we shall consider a group of characteristics which may be termed *serviceability* factors, including purity, brittle fracture, and others. Finally, we shall conclude the chapter with a discussion of *processability* factors such as machinability, formability, and weldability. In general it could be said rather ironically that the length of the word describing the characteristic is directly proportional to the depth of our ignorance surrounding it. Although we can put numerical values on the characteristic known as machinability, for example, we cannot really measure this factor, and the numerical values we have are the subject of some dispute and dissatisfaction.

## 8.1 particle size

Small particle size is measured either in mesh or in microns. Mesh size refers to sieve sizes. Thus a 200-mesh sieve has 200 openings per lineal inch, or

## MECHANICAL
## CHARACTERISTICS
## OF MATERIALS

## 8

40,000 per sq in. A micron is 1/1000 mm or 1/25,000 in. Portland cement is ground to an average of about 325 mesh, which is the same as 44 microns; 200 mesh corresponds to 50 microns and 100 mesh to 149 microns.

Dusts consist of particles in size from 1 to 50 microns. Smoke is carbon particles smaller than 0.1 micron. Diameters of gas molecules range from 0.055 micron down.

It is not ordinarily desirable for particle size to be uniform in any aggregate of particles to be processed into finished materials. Just as in the blending of concrete aggregates, small particles are needed to fill the interstitial spaces between the larger particles. Particles for foundry sand or for sintered carbide cutting tools therefore must be graded over a range of sizes, with a proper proportion in each grade. Figure 8.1 shows a typical fine aggregate grading for a concrete. Notice that 100 percent of these fine particles will pass a #4 sieve (4 openings per inch) but virtually none will pass a #100 sieve (100 openings per inch).

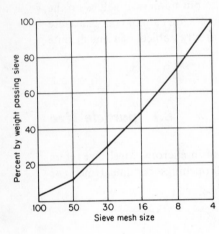

Fig. 8-1 Fine aggregate grading for a concrete.

Usually several steps are required to reduce bulk material to a smaller size. The first reduction will be done by a primary crusher, such as a jaw crusher, gyratory crusher, or hammer mill, which may reduce the material to lump of about 4-in. size maximum. The next step may be a rod mill. This is a rotating heavy steel barrel which is loaded with the bulk material and long 2-in. diameter steel rods. As the material is tumbled the rods reduce the material perhaps to $\frac{1}{4}$-in. size. The rod mill makes a more uniform reduction of particles than other crushing or grinding methods because the grinding rods are supported by material all along their length. The final grinding stage will be a ball mill, using steel balls about 4 in. in diameter. The ball mill can grind to nonuniform powder sizes. The crushed and ground material will finally be sorted by screening.

Finer sizes require more power for their reduction. Horsepower requirements for reduction are approximately proportional to the surface area of the particles produced. This means that if 10 hp is required to reduce to 1-in. size, then 20 hp is required to reduce to $\frac{1}{2}$ in.

At the same time as the number of alloys and composite materials increase, industry and science make ever more exacting demands for purity of materials. Standards of purity that were impossible twenty years ago are now routine in some types of manufacturing.

Transistor semiconductor materials such as germanium and silicon must be prepared to standards of almost absolute purity—a part or two per billion by weight. After purification by a remarkable vacuum melting method known as zone refining, the doping materials required to produce an *n*-type or a *p*-type transistor are added in amounts of a few parts per billion. Unless such purity standards are met, the characteristics of the transistor cannot be controlled within proper limits.

Steel technology can now control very closely the elements that dissolve interstitially in the iron space lattice, such as oxygen and nitrogen. These elements contribute to brittleness of various types in steel and are held to maximums of a few thousandths of a percent by weight in airmelted steel, or less in vacuum-melted steel.

Small amounts of alloying impurities have a pronounced effect in increasing the electrical resistance of copper. About 0.02 percent arsenic or aluminum will reduce the conductivity of copper by 6 percent, though other alloying ingredients have less powerful effects on conductivity.

Nuclear fuels are frequently enclosed in zirconium tubes. Zirconium is selected for this application because of low neutron absorption, and therefore it must be relatively free of impurities of high neutron absorption cross-section. Specifications for nuclear-grade zirconium therefore limit boron and cadmium to maximums of half a part per million by weight, because of their high absorption. Ores of zirconium contain about $2\frac{1}{2}$ percent hafnium, another neutron-absorbing element which must be limited to fewer than 50 parts per million in the finished zirconium. Even 50 ppm is only 0.005 percent.

Not even water escapes these stringent demands for purity, especially if it is used in the generation of high-pressure steam or in nuclear power generation. Since dissolved solids in water can absorb neutrons, cooling water for nuclear reactors may contain no more than about one part per million of dissolved solids.

Argon gas used for welding purposes is about 99.995 percent pure. Zinc for die-casting purposes must be at least 99.99 percent pure (four nines zinc) and sometimes is 99.999 percent pure (five nines). Without this rigorous purity, die castings develop various troubles, including dimensional instability and corrosion.

## 8.3 brittle fracture

The tension test described in Sec. 3.3 is a simple and reliable indication of the strength and ductility to be expected of a metal under many service conditions. However, this is true only when a metal exhibits ductile behavior in service. There are circumstances under which certain ductile metals behave in brittle fashion. Two important cases of brittle failure of ductile metals in service must be examined. These are called brittle fracture and fatigue. Fatigue will be examined in the next section.

Under the conditions of a standard tensile test the metal is subjected to stress in only one dimension. An inherently ductile metal, such as a structural steel, will elongate by slip along the planes in the crystal inhabited by the greatest number of atoms. The metal may also twin. These two mechanisms of ductile behavior were described in Sec. 3.6.

If now a saw cut is made in the specimen at right angles to the direction of stress, the metal will maintain its ductility during the tension test. It will not be sensitive to the presence of this notch. However, at the root of the saw cut stress conditions are complex and are no longer one-dimensional, but two-dimensional.

Under the impact conditions of the Charpy test, now to be discussed, it is possible for the ductile metal to lose its ductility and become remarkably brittle. The Charpy test uses a notch in the specimen to produce a complex stress concentration at the root of the notch, and therefore it is a test of notch sensitivity.

The Charpy test is one of the most revealing tests for disclosing low-temperature brittleness in otherwise ductile metals. While face-centered cubic metals such as copper, aluminum, and austenitic stainless steels remain ductile under all circumstances, the ductility of body-centered cubic and hexagonal metals is sensitive to the temperature of deformation, and the purpose of the Charpy test is to disclose such lack of ductility in the presence of a notch. In such brittle fracture the metal breaks in tension along the faces of the space lattice, but it does not twin or slip in shear.

For a Charpy test of a metal, samples of the shape and size shown in Fig. 8.2 are prepared. The figure shows a vee-notch milled across the specimen, but other types of notches may be employed, such as a keyhole type. The specimen is supported horizontally at its ends in the impact testing machine and struck on the side opposite to the notch by a swinging hammer that begins its swing with 264 ft-lb of potential energy. When the hammer strikes the specimen at the bottom of its swing, it has 264 ft-lb of kinetic

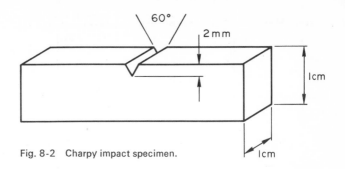

Fig. 8-2   Charpy impact specimen.

energy. This energy is partly dissipated by fracturing the specimen, the remaining energy carrying the hammer up on the backswing. The Charpy tester records on a scale the amount of energy in foot-pounds required for fracture—actually it measures the amount of backswing potential energy and subtracts this from the original 264 ft-lb to obtain the fracture energy. Figure 8.3 shows an impact tester with the pendulum hammer ready for release from the 264 ft-lb position.

Suppose that a number of Charpy specimens of a particular steel are tested over a range of temperatures from +200 to −100°F, and the results are plotted on a graph of foot-pounds against temperature. A typical S-

Fig. 8-3   Charpy impact testing machine with pendulum hammer raised to the 264 ft-lb position.

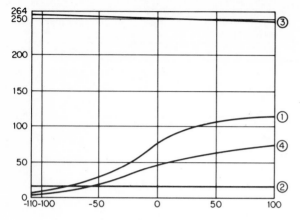

Fig. 8-4 Charpy impact test results: (1) a notch-tough rimmed steel, (2) 2024 aluminum alloy in hard temper, (3) pure nickel, (4) a 0.35-percent carbon steel plate $2\frac{1}{2}$ in. thick.

curve results, similar to the curve 1 in Fig. 8.4, which is a plot of the results for a notch-tough rimmed steel. At the high end of the temperature range the steel offers great resistance to fracture, giving a torn appearance at the break in what is termed a *ductile failure.* The torn appearance of the break is caused by shear failure along the slip planes within the grains of the metal. See Fig. 8.5. But as the temperature is dropped, a temperature range is reached in which the notch-toughness of the metal falls drastically, the fracture appearance changes to a bright crystalline break, and the steel has become brittle. Brittle fracture indicates a tension failure, as indicated in Fig. 8.5.

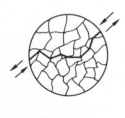
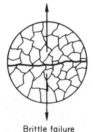

Ductile failure                    Brittle failure

Fig. 8-5  Mechanics of Charpy impact failure. Above the transition temperature failure occurs by shear along the slip planes within the individual crystals. Below the transition temperature failure occurs by tension forces separating the interatomic planes.

The minimum impact strength allowed for many applications is 15 ft-lb per specimen. The temperature corresponding to this impact value is called, rather arbitrarily, the transition temperature between ductile and brittle failure. Other temperature values are sometimes called the transition temperature, and indeed any of the following temperatures might be termed the fracture transition temperature:

1. the temperature corresponding to 15 ft-lb
2. the average temperature between ductile and brittle failure
3. the inflection point on the $S$-curve (this would be almost the same as No. 2.)
4. the lowest temperature at which the specimens show 100% ductile failure
5. the temperature at which 50% of the area of the break is ductile and 50% is brittle
6. the highest temperature for which none of the area of the break is ductile

Catastrophic failures of structures caused by brittle fracture have been fairly common until recent years. Changes in steel specifications and more knowledgeable methods of fabrication have now greatly reduced the incidence of brittle fracture. But minor failures still occur and are a source of mystery to many fabricators unacquainted with the brittle fracture phenomenon. The author has had reported to him a considerable number of brittle fractures of small steel parts used in cold storage rooms held at temperatures as high as $+30°F$. Brittle fracture has been especially evident in welded structures. The thermal strains of fusion welding build a great deal of residual stress into structures. "Big inch" gas pipelines have cracked over great lengths under very little stress, and even ships have split completely in two.

Such brittle failures occur only below the transition temperature of the steel and are prevented by the simple expedient of using steel with a lower transition temperature than any service temperature likely to be met. Some degree of prevention can also be exercised by good fabrication practice. Undercuts, notches, and arc burns caused by welding operations, fillet welds, hammer marks, and other such defects can serve as fracture initiation points for a fulminating collapse, for these defects have the same influence as the notch in the Charpy specimen.

Several methods are available for lowering the transition temperature of a steel. Perhaps the cheapest method is that of killing the steel with aluminum or silicon. This killing operation is discussed in the following chapter. In brief, it removes oxygen from the steel, with the effect of lowering the transition temperature. An even more powerful method is to reduce the carbon content of the steel. Brittle fracture considerations are a part of the explanation why manual arc welding rods for mild steels do not contain more than 0.1 percent carbon. Both carbon and oxygen dissolve interstitially in the space lattice of iron: brittle fracture characteristics must therefore be affected by interstitial elements.

Fine-grained steels, other things being equal, will have lower transition temperatures than coarse-grained steels. Further, the thicker the plate, the higher the transition temperature, and this consideration often dictates adjustment of the chemistry of thicker plate.

Brittle fracture does not occur in face-centered cubic metals, such as nickel. Another method of lowering transition temperature in steels is to alloy the steel with nickel. A 3.5 percent nickel low-carbon steel may be safely used at $-150°F$; for cryogenic vessels to store oxygen or nitrogen at $-320°F$, 9 percent nickel is required. Manganese is also beneficial, but it does not have the powerful effect of nickel. The austenitic stainless steels, copper, and aluminum, all being fcc, are safely used in refrigeration and cryogenic equipment.

The nuclear age has disclosed new aspects of brittle fracture. The transition temperature of a bcc metal is raised if the metal is heavily irradiated.

## 8.4 fatigue of metals

Fatigue failure is another type of brittle failure in otherwise ductile metals, but unlike brittle fracture, is not influenced by crystal structure, that is, it cannot be prevented by selecting a face-centered cubic metal. Fatigue failures are produced by sufficient stress levels repeated thousands or millions of times and therefore are found in shafts, gear teeth, hold-down bolts, vehicle frames, and other machine components. Fatigue failure is undoubtedly the most common type of failure in metals.

Fatigue failures in any metal occur at stress levels well below the yield stress of the metal. The failure begins as a slowly deepening crack at some region of stress concentration, such as the root of a gear tooth, a shoulder or keyway in a shaft, an impression stamped into a part (such as an identifying part number), a machining mark, or a change of section such as a fillet weld in a truck body or piece of farm machinery. With constantly repeated stress cycles the crack deepens, leaving progressively less metal to carry the load. Finally, without warning, fracture occurs.

The maximum repeated stress which can be applied indefinitely under fatigue conditions is called the *endurance limit* of the material. Obviously, for one cycle only the fatigue strength of a material will be its elastic limit (since plastic deformation is usually not acceptable). As the number of cycles is increased, this allowable maximum fatigue stress decreases. Finally the endurance limit is reached, the stress at which no fatigue failure occurs at an indefinite number of stress cycles. Figure 8.6 shows a fatigue curve for a steel. For steels, the endurance limit will be about 40 percent of the ultimate tensile stress and will be reached at about 10,000,000 cycles of stress. These figures are based on the assumption of complete reversal of stress in each cycle. Note from Fig. 8.6 that a fatigue curve is plotted as linear stress against logarithmic cycles. Such a curve is really a plot of "fatigue elastic limit" against cycles.

Nonferrous metals in general do not exhibit an endurance limit (Fig. 11.5). The maximum allowable fatigue stress for such metals falls constantly, even to $100 \times 10^6$ cycles. For nonferrous metals it is usual to determine the safe stress for a finite life of $N$ cycles, which may be $10^8$ cycles. Since fatigue stresses are a vital feature of aircraft operation and since aircraft are built

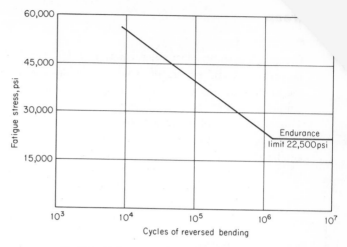

Fig. 8-6 Fatigue curve for a machinery steel with a sharp stress concentration machined into it. The reversed bending fatigue test was performed on the testing machine of Fig. 8-7.

chiefly of nonferrous metals, great care is taken in the determination of allowable fatigue stresses in aircraft.

Figure 8.7 is an illustration of a fatigue testing machine. A small specimen of metal is flexed by the machine as it is rotated at 10,000 rpm. The machine has a revolution counter to indicate the number of cycles of stress reversal. When the specimen fails, the machine automatically stops.

Fig. 8-7 Fatigue testing machine. The fatigue sample is gripped by a collet at each end (a wrench is one collet) and flexed as it is rotated at 10,000 rpm.

g-term plastic deformation. It is a familiar effect in lead, wood, ?, and paper, all of which will sag under relatively light loads if period is sufficiently long. Only a few metals will creep at room temperature, but creep is a cause of metal failures at elevated temperatures. Thus if the time of operation is sufficiently prolonged or if the combination of temperature and stress is sufficiently severe, the blades of a gas turbine engine may fail by striking the engine casing as they elongate from creep effects. Creep of course is accelerated by higher temperatures.

Figure 8.8 shows the typical progress of creep in a metal. In the primary

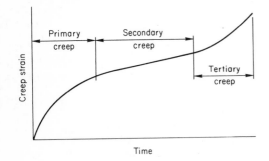

Fig. 8-8    Elongation due to creep.

stage of creep, the metal elongates at a diminishing rate. This rate reaches a minimum which persists for a considerable period of time. A linear rate of creep thus is characteristic of the secondary stage of elongation. In the final or tertiary stage, the creep rate accelerates toward failure.

Large steam turbines are expected to last for 20 years while driven by steam at temperatures above 1000°F. For such machines the stress levels must confine creep to 1 percent elongation in 100,000 hours of operation.

## 8.6   high-temperature

## resistance

Throughout the twentieth century, and particularly since World War II, developments in the technology of materials have followed each other in rapid succession, accelerated by the relatively new science of solid-state physics. This remarkable progress is nevertheless marked by a few failures. Competent researchers have been unable to produce a ductile ceramic or a cutting tool bit which will meet the demands of machine tool operators. Perhaps the most notable failure is our inability to develop materials with an extended life under service conditions exceeding 2000°F.

There is no serious lack of materials with high melting points. Refractory metals such as tungsten and molybdenum, natural oxides such as beryllia and thoria, and the element carbon all have melting points in the range of 5000–6000°F or slightly higher. For still higher melting points we have developed such synthetic carbides as hafnium carbide. Unfortunately, all these refractory materials fail in some important property at temperatures far below their melting points. Carbon begins to oxidize rapidly above 1500°F. Moduli of elasticity of all the metals fall to lower values at several hundred degrees Fahrenheit (see Fig. 8.9), and so do

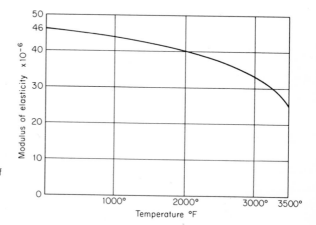

Fig. 8-9  Modulus of elasticity of molybdenum.

yield strength, hardness, and ultimate strength. The highly refractory metals such as tungsten oxidize readily above 1000°F. A number of oxides such as beryllia are serviceable at temperatures of 3000°F, but such materials are too brittle to be used as structural components of equipment and are confined in their applications to furnace linings and melting crucibles.

Since ductility as well as high-temperature resistance is a mandatory requirement for most materials in high-temperature applications, the only usable materials for high-temperature work are generally metal alloys. The most useful of the high-temperature metals is also the best low-temperature metal—nickel. This metal offers the best all-round combination of tensile strength, ductility, creep strength, and oxidation resistance, and many of the high-temperature alloys (the superalloys) are nickel-base. Some of these superalloys may be used at temperatures as high as 2200°F, though the useful life is rather short at such a temperature. Even the stainless steels with nickel contents from 8 to 35 percent have reasonably good high-temperature properties.

High-temperature resistance is of course not a property but a complex of highly demanding properties: strength, stiffness, hardness, ductility, dimensional stability, and oxidation resistance chiefly. An especially severe demand on materials at these temperatures is chemical inertness. Above 2000°F virtually all materials become chemically active, and it must be expected that the material will react with its environment.

One solution to the high-temperature problem is an obvious one and is frequently used with a degree of success. That is to use a ceramic coating on a metal substrate. This solution may raise problems of its own, however. The coefficients of expansion of the two materials must be compatible, and at high temperatures the metal and the ceramic must not engage in chemical action or mutual diffusion into each other.

Aerospace vehicles require materials for short time exposure to temperatures as high as 15,000°F, in heat shields and rocket nozzles. Such a temperature is twice as high as the highest melting point known. Ironically, the low-melting plastic materials offer perhaps the best solution to this service problem because they possess certain other characteristics. There are actually three possible methods of resisting such temperatures.

**1. Heat Sink Method.** A material of high thermal conductivity is used to distribute the heat developed throughout the entire mass of material. The material will of course reach a high temperature; therefore a metal must usually be employed. A metal such as beryllium has suitable properties: high specific heat, sufficiently high melting point, and low specific weight. Carbon also has these properties.

**2. Refractory Metal Method.** The refractory metals such as tungsten, melting at 6152°F, have the properties of high melting point, rapid oxidation at higher temperatures, volatile oxides of low melting point, and high thermal conductivity. In use as a heat shield at ultrahigh temperature, the metal will oxidize. The oxide will melt or vaporize, and the latent heat can be used to cool the material.

**3. Ablation Method.** This has been the most successful solution to the problem. The method relies on degradation and decomposition of a reinforced plastic material. Cooling action is produced by the following effects:

1. high specific heat of the material
2. high latent heat of the degradation products
3. heat required for degradation
4. radiation of heat
5. loss of the decomposed and volatile materials, which flow into the adjacent air layer

and keep the heat shield cool. The material ultimately chars to a porous carbon, and this acts as an insulating layer which also protects the heat shield

Both phenolic and epoxy reinforced plastics make excellent heat shields.

## 8.7 low-temperature resistance

Low-temperature resistance is as vague a term as high-temperature resistance. Here we shall interpret the term to mean suitability for use below $-250°F$, which is the cryogenic temperature range. The problems of creep, oxidation, low strength, and low $E$-value do not exist at low temperatures. Here the problem is largely that of notch brittleness.

Both strength and modulus of elasticity of metals increase somewhat at low temperatures, with sometimes some loss of ductility. In general, face-centered cubic metals must be used for their notch toughness, such as the austenitic stainless steels or aluminum. Figure 8.10 shows the change in properties of type 316 stainless steel at low temperatures.

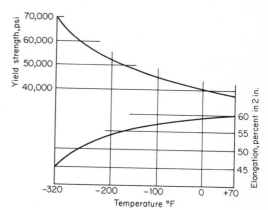

Fig. 8-10 Properties of stainless steel type 316 at low temperature.

## 8.8 corrosion resistance

The subject of corrosion was briefly dealt with in Sec. 6.11. The subject of corrosion resistance of materials is too detailed to be dealt with in a single section of this book. Instead, specific aspects of corrosion will be noted where they arise. Outstanding resistance to corrosion is given by such materials as tantalum, the plastic Teflon, and the ceramic alumina. However,

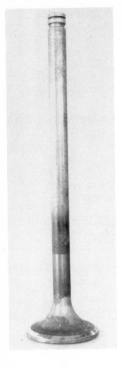

Fig. 8-11 An exhaust valve from a large stationary diesel engine. Service conditions subject the valve to heat, impact, and fatigue. The head of the valve is made of stellite.

all of these materials may be corroded by hydrofluoric acid. There is no panacean material which is resistant to all corrodents.

## 8.9 abrasion resistance
## and impact resistance

*Abrasion* is a wear or loss of material caused by movement of an abrading material, such as sand, parallel to the surface of a machine component and causing scouring of the component. *Impact* is the wear that results from hammering effects perpendicular to the component. Wear by abrasion removes material by scratching; impact removes material by peening, mushrooming, cracking, chipping, or spalling.

These two types of wear call for different types of metal to resist them. Resistance to impact requires toughness, hardness, and strength, and often it is best provided by some metal with pronounced work-hardening characteristics. Resistance to abrasion is provided by a very hard surface, such as is provided by carbides. An alternative method of providing abrasion resistance is the use of a suitable soft and elastic but tough material that will deform rather than score; rubber and nylon are two such materials which in many applications provide better wear resistance than harder materials. Pumps for pumping sand may be lined with silicon carbide; others may be lined with rubber.

The constant loss of metal caused by both abrasion and wear are corrected in the process industries by the use of hard-surfacing rods deposited by arc-welding methods. The selection of the best type of hard-surfacing material is a vexing problem that can sometimes be solved only by trial and error. Tungsten carbide and chromium carbide provide best resistance to abrasion; austenitic stainless steels and austenitic manganese steel with 12 percent manganese offer best resistance to impact. Switch frogs on railroad lines must take the battering of train wheels and are cast from 12 percent manganese steel.

Fig. 8-12 A cattle stunner. This component is fired by compressed air into the medulla oblongata of the cow to stun it before killing. Service conditions subject this product to fatigue, impact, and corrosion from the saline body fluids of the animal. The bar of the stunner is a hardenable stainless steel.

## 8.10 hardness

There are a great many hardness tests currently in use. Among the reasons for this multiplicity of methods is that a single hardness method cannot span the great range of hardnesses from the soft plastics to the hard ceramics. Almost all hardness tests measure indentation hardness after the principle of asphalt testing outlined in Sec. 7.11. Some type of indentor is pressed into the specimen; a small indentation indicates high hardness and a large indentation indicates low hardness. However, in an alternative method, the scleroscope test, a steel ball is bounced off the specimen to indicate the hardness by the height of the bounce.

Everyone early in life acquires the intuitive notion that hardness is related to strength. This of course is a generally correct idea. Metals are generally hard and strong, plastics usually soft and weak. Harder metals are stronger metals. Alloys are harder than their base metals. Hardness and strength may be increased by suitable heat treatments in the case of some metals and plastics, and in the case of plastics, by increasing molecular weight or length of polymer chain. Plastic straining (strain-hardening or work-hardening) increases the hardness of all metals, some to a greater degree than others. Figure 8.13 shows the effect of cold reduction of mild steel on its hardness.

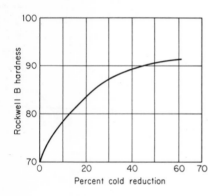

Fig. 8-13  Strain-hardening curve for a mild steel.

Since a standard tension test occupies considerable time in the operations of machining the specimen and then testing it, Brinell, a Swedish engineer, devised a hardness test which indicates ultimate tensile strength as well as hardness. This test is commonly used for metals and provides a Brinell Hardness Number, or BHN.

The Brinell hardness tester is shown in Fig. 8.14. A hardened steel ball 10 mm in diameter is forced into the metal specimen under a 3000 kg load (500 kg for nonferrous metals). The average pressure on the spherical surface of the indentation is the BHN, or by formula

$$BHN = \frac{P}{\frac{\pi D}{2}[D - \sqrt{D^2 - d^2}]}$$

where $P$ = load in kilograms

$D$ = diameter of indentor ball

$d$ = diameter of impression

Fig. 8-14 Brinell hardness tester.

The diameter of the impression is determined by a low-power measuring microscope.

The ultimate tensile strength of a carbon steel is given closely by the simple formula

$$UTS = 500 \times BHN$$

This formula applies only to steels.

The Brinell test is not as quick or convenient a hardness test as the Rockwell test. Further, it leaves a rather large impression, which may be damaging to the part tested, and it cannot test material less than about $\frac{5}{16}$ in. thick or small objects. However, the large impression gives a good average reading over a considerable area and is relatively immune to the effect of scratches and machining marks.

The disadvantages of the Brinell hardness test led to the development of other indentation hardness methods. The Vickers Pyramid Hardness tester uses a very small four-sided pyramid indentor. The most popular indentation tester, however, is the Rockwell tester of Fig. 8.15. There are several Rockwell scales using different indentors and loads. These are summarized in the following table. Only the B and C tests are in widespread use.

Fig. 8-15    Rockwell hardness tester.

## ROCKWELL SCALES

| Scale | Used for | Load, kg | Indentor | Hardness Read on |
|-------|----------|----------|----------|------------------|
| A | steel sheet | 60 | brale | black numbers on dial |
| B | standard test | 100 | 1/16" ball | red numbers on dial |
| C | standard test | 150 | brale | black numbers on dial |
| D |  | 100 | brale | black numbers on dial |
| E | soft metals | 100 | 1/8" ball | red numbers on dial |
| F | brass | 60 | 1/16" ball | red numbers on dial |
| G | phosphor bronze | 150 | 1/16" ball | red numbers on dial |
| H |  | 60 | 1/8" ball | red numbers on dial |

The Rockwell C test is used for determining the hardness of hard steels. A conical diamond indentor is employed, called a Brale, under a 150-kg load. The Rockwell B test applies 100 kg to a $\frac{1}{16}$-in. diameter steel ball and is used on mild steels. Both tests, and the other Rockwell tests in the table, can be executed on the same instrument. A 10-kg minor load is applied to the indentor to take up elastic deformation in the specimen. Then a major load, 150 kg for Rockwell C and 100 kg for Rockwell B, is applied. The depth of indentation caused by the major load is measured and subtracted from a constant number. This is the Rockwell hardness and is registered on the dial on the front of the instrument. Hardness is recorded as $R_C$ 35, $R_B$ 85, etc.

A Rockwell hardness test can be performed in seconds, and therefore this test is preferred for production testing. To obtain some notion of hardness ranges, the maximum hardness of any steel is about $R_C$ 65; a steel is almost unmachinable when the hardness exceeds $R_C$ 35.

The scleroscope test uses a steel weight dropped from a fixed height onto the surface of the specimen. The height of rebound measures the hardness. Although not as widely used, this instrument has the virtue of small size and portability. The explanation for this hardness method is simply that hard materials have high elastic limits and absorb little of the kinetic energy of the dropped weight. Softer materials will have lower elastic limits and thus deform plastically, absorbing more energy.

The hardness of plastics may be measured on the Rockwell scales $R_R$ and $R_M$, which use large ball indentors. The plastics, as their name implies, deform plastically, a basic requirement of the Rockwell method. The M scale uses a $\frac{1}{4}$-in. steel ball for an indentor, preloaded with 10 kilograms, and measures the additional penetration from a load of 100 kg. The hardness is read from the red Rockwell scale of numbers. The M scale is used for the harder plastics. Softer plastics may be tested with the Rockwell R scale, using a $\frac{1}{2}$-in. ball and loads of 10 and 60 kg. The hardness of a plastic measured by the Brinell or Rockwell methods may be influenced by the thickness of the specimen.

Hardness measurements on rubbers cannot be made with the Rockwell or Brinell tester. Rubbers are elastic and not subject to permanent deformation under usual circumstances. The hardness of a rubber must therefore be measured during application of the load from the indentor. The Shore durometer is used for rubbers, though it may also be used for plastics. The Shore A durometer is suited to rubbers and soft plastics, and the D type is suited for harder plastics. The A durometer is shown in Fig. 8.16. Both indentors have the same diameter but different shapes of point. Both durometers have a flat bottom surface which is pressed against the specimen to be tested. The relative movement of the indentor into the specimen is read out on the dial of the instrument immediately after the instrument is seated on the specimen. Rubber may creep, and the reading after 15 seconds may be lower than the initial reading. Very hard rubbers, harder than automobile tires, use the D durometer. An automobile tire has a hardness of A70. A comparison of the two durometer scales is given in Fig. 8.17.

An interesting method of testing for the hardness of plastics is the pencil test. Actually this is a test of scratch resistance, like the Moh's scale of hardness for rocks. The test uses the whole range of pencil hardnesses from 2B, B, HB, F, H, 2H up to the maximum hardness obtainable 9H. Each pencil is

Fig. 8-16 Shore durometer reading hardness of a block of polyurethane tooling rubber.

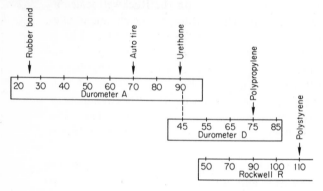

Fig. 8-17 Durometer scales. (From Du Pont, *Mechanical Properties of Adiprene*)

sharpened to a wedge and dressed with abrasive paper. The pencil is drawn across the surface of the plastic as it is held at a right angle to the surface. If the plastic is harder than the pencil carbon, it will not be scratched. If the pencil is harder than the plastic, it will groove the plastic slightly. Pencils are applied in increasing hardness until a hardness is found which first leaves a visible mark in the surface. The hardness of the plastic is reported as the pencil hardness grade.

## 8.11 dimensional stability

The machinist who grinds a part to an accuracy of half a thousandth assumes that this accuracy will be preserved in the part. Such accuracy would not be sought if it could not be maintained. The assumption of dimensional stability similarly dictates that structural metals should not be loaded beyond their elastic limits or endurance limits. Nevertheless dimensional stability is not always easy to maintain, and many material properties work against it.

For dimensional stability, an ideal material would have the following properties:

1. coefficient of thermal expansion—zero
2. modulus of elasticity—infinite
3. no phase changes producing small changes in volume
4. no plasticity or creep
5. high tensile strength
6. absolute corrosion resistance
7. high hardness and wear resistance

The metals and ceramics most nearly approach the above properties, which accounts for the use of these materials in those components that require the best dimensional stability, such as measuring gauges. Of the basic groups of solid materials, the organics have the poorest dimensional stability, because of high expansion, low modulus of elasticity, plasticity, creep, low hardness, and in some cases, dimensional changes with changes in humidity.

Dimensional stability is most critical for the master standards of length used in metrology. The standard inch was formerly defined as 0.0254 times the length of the international meter, the latter measurement being the distance between two gauge marks on a platinum-iridium bar stored in France. A platinum-iridium alloy is an excellent choice for a dimensionally stable material. But a metal bar does not have *absolute* dimensional stability, and it may be destroyed by fire or other accident.

Absolute dimensional stability is provided by photon wavelengths, however. The standard inch is now defined as 41,929.399 wavelengths of light from Krypton-86. Such a master dimension can be reproduced at any time and place as desired.

## 8.12   coefficient of friction

The coefficient of friction is both a processability factor and a serviceability factor. It is a ratio or pure number: the least force required to cause movement of one surface over another, divided by the force between the contacting surfaces. When such a coefficient is quoted, the usual assumption is that the mating surfaces are smooth and not lubricated.

A low coefficient of friction is required for sleeve bearings. In other cases a high coefficient is required, as with vehicle tires, brakes, tracked vehicles, and conveyor belts. The material offering lowest coefficients, as low as 0.04, is the plastic Teflon. Rubbers offer high coefficients, though the harder the rubber the lower the friction coefficient. Urethane rubbers in hard formulations offer coefficients as low as 0.20 and have occasional uses as bearings. Because of the slipperiness of lead, it is used in many bearing alloys and is added to metals to improve their machinability.

## 8.13   angle of repose

Angle of repose refers to the angle with the horizontal that a pile of bulk material will make. If a bulk material such as sand or coffee has a steep angle of repose, then more of it can be piled in a given area, on a truck, or on a conveyor belt. Angle of repose, therefore, is a significant characteristic for material handling purposes. For a high angle of repose, the particles must have a high friction coefficient against each other or must interlock.

## 8.14   machinability

Machinability has been defined as "a complex property of a material that controls the facility with which it can be cut to the size, shape, and surface finish required commercially." A brief definition is "relative ease of machining on machine tools," and a less reverent definition: "the ability of a material to go through a machine shop without irritating the machinist."

Machining with standard cutter bits is possible only within a restricted range of material hardness. Machining of materials with a hardness greater than $R_c$ 35 is difficult and slow and may require special tools and techniques. Harder materials must be ground with abrasive wheels. At the other extreme, soft plastics and rubbers may deform too readily to be machinable or they may require refrigeration in order to be machined. The softest steels are not readily machinable; soft aluminum is not as machinable as its harder alloys.

The relative machinability of machinable materials, however, is subject to a host of variables besides hardness. The temperature at the point of contact between cutting tool and metal chip will be approximately 1500°F when machining steel at production rates (1500°F is a cherry red heat). The properties of the steel at this cutting temperature are therefore among the important variables, chiefly tensile strength and elongation. The velocity of the cut in surface feet per minute and the depth of cut are two more important variables. The angles of the cutting tool are other significant variables, complicated by the fact that these angles must be adjusted to suit various metals; sharper angles are needed for machining aluminum and rubber than for steel. The cutting oil used has an influence on friction and cooling effects. Other variables have more indirect effects. For example, stainless steel has a low thermal conductivity, and during machining, considerable heat builds up in the cutting area. Certain metals, such as soft steels and titanium, tend to gall or weld to the cutting tool. Finally, the type of machining operation influences the machinability of the material; a metal may be easy to broach but difficult to turn.

The human race seems to associate brass with machinability. This relationship is borne out in practice: brass is an ingratiating metal to cut, and where machinability has a major bearing on cost, brass may be selected for the raw material.

Relative machinability ratings are given to metals based on steel alloy AISI B1112, which is discussed in the next chapter. This is a sulfurized low carbon steel and is assigned the arbitrary machinability rating of 100. A metal with a machinability rating of 200 would be twice as machinable as B1112. When machined by methods appropriate to it, including suitable tool angles, presumably the more machinable metal could be machined at twice the speed of B1112. This is only partially true: machinability is too complex and insufficiently understood for such ratings to have dependable accuracy.

A great many steel, copper, and aluminum alloys are blended to provide a high machinability rating. These are called free-machining or freecutting alloys. Copper is added to aluminum to promote hardness and therefore machinability. Lead is added to brasses, the lead inclusions serving to lubricate the tool. Selenium, lead, or sulfur are added to stainless steels, and lead, phosphorus, or sulfur to carbon, low alloy, or tool steels for improved machinability. Phosphorus is added to steels for its hardening effect on soft and stringy steels. Sulfur is added to produce manganese sulfide inclusions in the steel; like lead inclusions, these serve to lubricate the cutting tool. In addition to these alloying methods, steel bars specially prepared as machining stock are often cold-drawn, giving a surface hardening effect with improved machinability.

The selection of free-cutting grades of bar is of course a matter of economics. There is no purpose in paying a premium price for free-cutting quality throughout the cross-section of the bar if only a light skin cut will be made on the bar. Free-cutting grades are also uneconomic except in production machining on automatic lathes, numerically controlled machine tools, or other high-production equipment. As a rule of thumb, 10 percent of the bar must be removed as chips in order to justify the higher cost of free-machining stock.

In the case of beryllium, there may be an unusual limitation on machining speeds. Since finely divided beryllium is exceedingly toxic, machining chips must not be allowed to fly about at random as they are delivered from the cutting tool. The machining area is fitted with an exhaust hood that uses a high air velocity to capture chips and carry them into the exhaust system. The machining speed will be limited by the capacity of the exhaust system to capture the chips, since too high a machining speed will throw the chips beyond the capture area of the exhaust hood. This case serves as one more example of the infinite variety of problems generated by the many materials now in use in industry.

Some of the more difficult alloys to machine are the high-nickel heat-resisting alloys used in aircraft gas turbines, such as the Inconels. These may have machinability ratings as low as 5. Such materials cannot be cut except with heavy and rigid machine tools equipped with large driving motors and sharp cutting tools. The following list is the approximate order of metals from least machinable to most machinable, assuming all metals to be in the soft or annealed condition:

| | |
|---|---|
| *Least machinable* | mild steel |
| titanium | free-cutting mild steel |
| inconel | aluminum alloys |
| nickel | brass |
| stainless steels | zinc alloys |
| tool steels | free-cutting brass |
| copper | magnesium alloys |
| gray cast iron | *Most machinable* |
| high carbon steel | |

## 8.15  formability

Broadly speaking, two methods are available for shaping materials: machining and forming. Machining removes metal or material in the form of chips; forming shapes material in forms or dies. Although machining is possible on either brittle or ductile materials (though it is easier to machine brittle materials, in general), forming is not possible on materials that do not have a considerable degree of plasticity. This statement is subject to some exceptions, since brittle materials can be formed by such methods as casting or compacting of powders. In this section, however, we are concerned with the problem of forming sheet between male and female dies.

To form a sheet in brake dies or drawing dies, the stress in the metal must exceed the yield strength, but it must not exceed the ultimate tensile strength. Should the yield strength of the material be very close to the ultimate strength, then the metal will have only a very limited plastic range. Bending and drawing then must be carefully done, since slight overloading will fracture the material. Those metals are easiest to form which have yield strengths well below their ultimate strengths, such as pure aluminum and most stainless steels. Stainless steel alloy 301 will serve as an example. This metal is used in the hubcaps of automobiles. It has a yield strength of

40,000 psi and ultimate tensile strength of 110,000 psi, a ratio of almost 3 to 1. This ratio explains its suitability for the complex drawing operation involved in forming an automobile hubcap.

The typical stress-strain curve of a drawing-quality metal sheet is shown in Fig. 8.18. Both the elastic and the plastic region present problems in form-

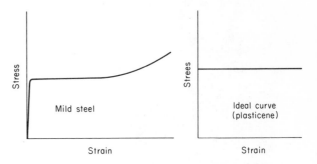

Fig. 8-18 Stress-strain curves. The curve for mild steel shows the elastic range, the cause of springback, and the work-hardening effect of plastic deformation. The ideal curve, which might be that for children's plasticene, has neither elastic deformation nor work-hardening.

ing operations. Let us suppose that the sheet is to be bent 90° in a braking die. As the sheet is bent between the mating dies, it is first stressed to the elastic limit. After bending is completed, and the dies are opened, this elastic strain is recovered. This *springback* in the metal must be compensated for by overbending. The higher the elastic limit of the material, the greater this springback will be. Next consider the plastic region. After yielding, the material can stretch with no rise in stress, but after considerable stretching the stress begins to increase. This is the strain-hardening effect possessed by all metals as the straining operation gradually uses up the dislocations in the space lattices of the metal. Such hardening reduces ductility. In severe drawing such as is required to make a welding gas cylinder or an artillery shell case, this strain-hardening effect will require the drawing to be done in graduated steps in a series of dies, the various stages of the drawing operation being separated by annealing operations (softening by heating) to remove the strain-hardening effects. This is especially true with most of the stainless steels, which though possessing considerable ductility will strain-harden rapidly. The ideal metal for forming would have a stress-strain curve like that of Fig. 8.18 (right side), with no elastic springback and no strain-hardening.

Several methods are available for comparing the formability of metals. The standard tensile test will give data such as yield strength, which affects the amount of springback, modulus of elasticity, which determines the force required for drawing, and elongation, which is a measure of the basic formability. An even simpler test is that of bending. If the material will bend 180° flat on itself, it has unusually good formability. More sophisticated tests are cup tests, such as the Erichsen or Olsen cup test. In these tests the metal sheet is clamped between ring-shaped dies, while a rounded punch is forced into one side of the sheet until fracture is initiated. The height of the dome produced is measured in millimeters and is taken as a measure of the formability of the sheet. The Olsen cup tester is illustrated in Fig. 8.19.

Fig. 8-19   Olsen cup tester for assessing drawability of sheet metal. (Courtesy Tinius Olsen Co.)

## 8.16   weldability and hardenability

Weldability does not mean the ability to be welded, just as machinability did not mean the ability to be machined. It must be presumed that there is a method of welding any material to itself, and even very dissimilar materials can be welded together, such as glass and metal. The term *weldability* means the relative ability of metals, usually steels, to be welded *without cracking*. In general, new structural steel alloys are formulated for good weldability, since this is a critical requirement in their use and assembly.

Weldability, like machinability and formability, can be rated on a rough numerical scale. The rating methods and tests are ingenious but cannot be discussed at this point. Cracking of steels during welding operations results from the heat-treating effects of welding; these will be taken up in a later chapter.

*Hardenability* is a factor of concern in the processing of heat-treatable steels such as tool steels. Steels can be hardened only by rapid cooling, called quenching. If the cooling rate is not fast enough, the steel will not harden. If the hot steel is plunged into cold water, the exterior of the steel, in contact with the cold water, will harden by cooling at a sufficiently rapid rate. But in the interior of a thick section of steel, heat will be lost more slowly. The interior therefore may not harden. Hardenability is measured by the depth in the steel to which hardening occurs.

So far this book has attempted to discuss a rather wide range of the properties of materials, under the groupings of physical, electrical, and mechanical characteristics. A later chapter will deal with an equally important group of characteristics, the economic ones. All material properties are really economic properties, of course. Needless to say, only the more common and most universal of material characteristics have been considered. This wide range of characteristics, any of which may be specified on buyers' purchase orders and supplied by producers within acceptable tolerances, are sufficient evidence of the high degree of sophistication reached in materials technology. The advances of the next few years will produce new requirements and new characteristics. All this explains why universities and other institutes establish departments for Materials Engineering, why engineers are hired for positions as purchasing agents, and why some companies, knowledgeable in materials, succeed when less knowledgeable ones do not.

One of the most perplexing questions to ask is, "Which is the cheapest material to use?" In this age of rapid progress, there is no definitive answer. Consider the following case.

A steel part must be machined, then welded to the end of a steel shaft. The cheapest raw material is mild steel. But mild steel does not machine well. A leaded steel will be slightly more expensive than a mild steel but will machine at half the cost of a mild steel. It is therefore cheaper in terms of raw material plus machining costs. But a leaded steel is tricky to weld, since the lead boils out of the steel during welding. Which steel is now the cheaper, leaded or mild steel? One shop might select mild steel, and another shop leaded steel. Each shop would make its selection in the light of its own difficulties, resources, abilities, and experience, and each would presumably be correct in its selection. Materials selection always involves compromise, and compromises call for more skill and judgment than simple choices between black and white.

## PROBLEMS

1   A steel has a BHN of 240. What is its ultimate tensile strength?

2   Define a micron.

3   Why is purity critical to the proper functioning of a transistor?

4  Which space lattice is not prone to brittle fracture?

5  In order to be sure of the strength of a butt weld, welders will build up the weld so that it is thicker than the adjacent plate. Discuss the implications of this practice for both brittle fracture and for fatigue failure.

6  By and large, which is the best choice in the design of metal fabrications, a butt weld or a fillet weld? Why?

7  What is an endurance limit?

8  What metals or other materials can you name that will creep at room temperature, given a sufficient period of time?

9  Differentiate between abrasion and impact.

10  Is the Rockwell hardness test a suitable one for a perfectly elastic material such as glass? Why or why not?

11  What is the relationship between coefficient of friction and angle of repose?

12  What reasons can you give why Rockwell hardness does not correlate well with machinability ratings?

13  How does sulfur contribute to machinability of metals?

14  What is the meaning of springback?

15  Tool steels have poor weldability. If their thermal conductivity were higher, would their weldability be improved?

16  Rubber is soft, yet resistant to abrasion. Why?

17  Why is the Rockwell hardness test not applied to rubbers?

18  Investigate the pencil hardness test for plastics.

19  Make 500 kg Brinell hardness readings on thick plastics.

Steel is the basic material on which the Western World has built its technical culture. Though steel has been known and used in limited quantities for a few thousand years, it did not come into extensive use until about a hundred years ago, at which time methods were finally established for producing steel in tons rather than in pounds. Since then steel has been the colossus among metals. Although other materials such as aluminum, concrete, and plastics have made some inroads into the markets served by steel, nevertheless far more steel is consumed than the total tonnage of all other metals and plastics combined. Annual consumption of all steels in the United States exceeds 150,000,000 tons, in Canada 12,000,000 tons, and in the whole world half a billion tons.

The remarkable versatility of steel in part explains its wide use. There is a steel for almost every purpose: soft and hard steels, steels of low conductivity and high conductivity, ferromagnetic and paramagnetic steels, weldable and heat-treatable steels, wrought and cast steels, steels to resist heat, cold, corrosion, impact, and abrasion. Notable too is the readiness

*STEEL*

*9*

with which most other metals can be blended with steel to provide an almost infinite variety of useful alloys.

Though by no means the largest group of steel alloys by tonnage, the tool steels deserve special attention. Just as the toolmaker may be considered the most critically important of all the skilled trades in a society that values its high standard of living, the tool steels, which are the toolmaker's raw materials, must be accounted the most critically important group of metals. These are the special steels from which are made the tools, forms, dies, gauges, and machinery that produce finished consumer goods. And just as toolmakers are high-paid craftsmen, so their tool steels are among the most expensive of the steels.

Only limited quantities of elemental iron are used in industry. Instead, iron is alloyed with a small amount of carbon, the resulting alloy being a *steel*. There are always present in steel small percentages of sulfur and phosphorus originating from the iron ore and the coke fuel used to reduce the ore. However, these are not counted as alloying elements, since they are usually not advantageous but detrimental. In addition, almost all steels include small amounts of silicon and manganese, which are added to the steel in the steel mill. Except in larger amounts, these too are not counted as alloying elements. The most commonly used alloying metals in steels are nickel and chromium. A *carbon steel* therefore will contain carbon and minor percentages of sulfur, phos, silicon, and manganese. A *low alloy steel* will contain carbon plus a few percent of another metal or metals such as nickel. A high alloy steel will contain over 5 percent of alloying metals.

Those steels which contain more than 2 percent carbon are usually called irons, and the group of such irons includes gray cast iron, white cast iron, malleable iron, and ductile iron. These are all cast products of the iron foundry and not rolled products from a steel mill.

The total number of steel alloys in use is probably not known, but it is unquestionably in the thousands. New steel alloys are developed every year. In the last ten years particularly steel technology has seen rapid development.

## 9.2 the extraction of iron
## from its ores

Most ores of iron having industrial importance are iron oxides, and most of these are magnetic. Iron ore occurs quite commonly over the earth's surface, particularly in low-grade deposits, but the steel industry operates on such a large scale that only the largest ore bodies are developed. Currently the American steel industry consumes 140,000,000 tons of ore, supplemented by an unknown quantity of scrap steel. Some statistics of the large Hull-Rust open-pit mine in Minnesota dramatically suggest the magnitude of iron ore operations. One billion tons of ore and overburden have so far been excavated from this mine, leaving a hole in the ground measuring $3\frac{1}{2}$ square miles in area with a depth exceeding 500 ft.

Typical iron ores include the following three:

1. *Magnetite*, $Fe_3O_4$, brown in color, strongly magnetic, containing about 65 percent iron. This is a rich ore and therefore less commonly found than lower-grade ores.
2. *Hematite*, $Fe_2O_3$, red in color (hematite, hemorrhage, and hematology have the same derivation), paramagnetic, containing about 50 percent iron.
3. *Taconite*, usually green, strongly magnetic, containing about 30 percent iron and much silica. Such low-grade ores must be beneficiated or upgraded at the mine site by removal of silica.

It is not yet commercially possible in the United States to convert iron ore directly into steel in a single-process operation. Such a method is called "direct reduction." Direct-reduction methods have been the subject of aggressive research since 1950. A few small direct-reduction plants are operating in other countries, and there is some likelihood that in the near future direct-reduction methods will be used for the making of steel in North America. At present, however, the process from iron ore to steel takes two stages:

1. The ore is converted into hot metal (raw iron, or pig iron) in a blast furnace.
2. The hot metal is refined into steel in a steel refining furnace such as an open-hearth furnace.

The blast furnace, Fig. 9.1, is the largest of all the big industrial furnaces, though the rotary kilns used for the manufacture of cement clinker are longer. A blast furnace, of which there are about 200 in the United States, is a vertical shaft furnace about 200 ft high, with a diameter inside the firebrick lining of about 30 ft, and shaped like an oversize Coke bottle. It requires for its refractory lining a supply of refractories by the trainload rather than the carload. High heat duty firebrick is employed, except that carbon brick is used to line the hearth walls below the molten metal line. The furnace itself weighs about 10,000 tons and is supported on steel piling and a 12-ft-thick pad of reinforced concrete about 60 ft in diameter. Refractory concrete is poured on top of the foundation concrete, and about 15 ft of large firebrick is laid on the refractory concrete. The cost of this sort of construction explains why the steel industry is more interested in methods of increasing production from existing blast furnaces than in building new installations. A single blast furnace installation may cost $60,000,000, but its product sells for only 5 cents a pound.

Raw material is charged into the top of the furnace by means of skip cars. Production rate for a blast furnace may run to 3000 tons of metal per

Fig. 9-1   Blast furnaces on Route 30 through Pittsburgh.

24-hr day, and each finished ton requires the charging into the furnace of 3350 lb of iron ore, 570 lb of limestone, and 1350 lb of coke, for a total of about 5300 lb of raw materials. In addition, about 40 tons of cooling water must be passed through the furnace per ton of product, and $3\frac{1}{2}$ tons of combustion air. Nothing can match the awesome continuous appetite of this ugly monster.

The limestone serves as the slagging material to remove sulfur and phosphorus. The coke burns principally to carbon monoxide, which reduces the iron oxide by combining with its oxygen at a temperature of about 3000°F to form carbon dioxide. This is the essence of the chemical reactions of the blast furnace, though a complete understanding of the chemistry is still somewhat beyond our grasp. The carbon found in steel is, of course, derived from the coke originally charged into the blast furnace.

Once the blast furnace is fired, it is never cooled down until the refractory lining must be replaced. This means continuous operation for at least five years. The molten iron, called hot metal or pig iron, is tapped from the furnace periodically throughout the day.

The hot metal thus produced is actually an impure and low-quality cast iron containing about 4 percent carbon. This high carbon content makes the metal unusable as a steel. With so much carbon it has no ductility and cannot be rolled into finished shapes such as plate, sheet, or bar. This is the reason why the blast furnace iron must be further processed: among other considerations, the steel refining furnace must reduce this carbon content.

## 9.3 steel-refining practice

The principal function of a steel-refining furnace is to control the amount of carbon, impurities, and alloying elements in the steel. This is done under a basic slag using lime, since an acid slag using silica does not remove sulfur and phos. The hot metal delivered to the steel-refining furnace contains 4 percent carbon, and the steel furnace will reduce this to less than 1 percent by oxidizing the carbon. At the same time, any alloy metal additions will be made to the molten steel, such as small percentages of manganese or nickel. Certain ingredients such as aluminum are too readily oxidized and would disappear into the slag, but these can be added to the finished steel after it is poured from the furnace into the pouring ladle.

Only three types of steel-refining furnace are important enough to deserve mention here. These are:

1. the open-hearth furnace, invented about 1860
2. the electric arc furnace, invented about 1900
3. the oxygen converter, invented in Austria, first used in North America in 1954

The open-hearth furnace is becoming obsolete, since steel companies are retiring these furnaces in favor of the much faster oxygen converter. At the time of writing, about 10 percent of all steel production comes from electric furnaces, another 50 percent from converters, and the remainder from open-hearths.

The specialty steels such as tool steels and stainless steels are reserved for the electric arc furnace, which also is used in geographical areas which are not served by blast furnaces.

All three refining furnaces may be charged with either hot metal or steel scrap, but each shows some preferences. It is not usual to charge blast furnace iron into an electric furnace. This furnace uses virtually 100 percent scrap. The converter is not charged with more than 25 percent scrap, since it does not have a burner or other external source of heat and cannot therefore melt steel. The open-hearth furnace uses either hot metal or scrap, employing more scrap when scrap steel prices are low and less when prices are high.

The basic oxygen process was first used in Austria in 1952. The furnace (Fig. 9.2) is a pear-shaped vessel open at the top. No source of heat is supplied. The converter is first charged with hot metal. Then a retractable water-cooled lance, actually a pipe, is lowered into the converter above the surface

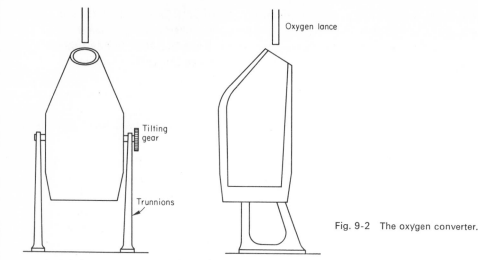

Oxygen lance

Tilting
gear

Trunnions

Fig. 9-2  The oxygen converter.

of the bath. Pure oxygen at about 150 psi is directed from the lance onto the bath. The stream of oxygen burns out the carbon in the hot metal as carbon monoxide in a very short time as compared with other steel-refining methods. This chemical reaction is the only source of heat during the operation. As the temperature of the process increases, it frequently must be held back by additions of scrap steel. The open-hearth and electric furnace require a few hours to melt and refine a heat of steel. The oxygen converter prepares a heat of steel at a furious rate of 30 minutes or less; an early problem with this method was getting an analysis of the steel back from the laboratory by the time the heat was ready to pour. Indeed, the converter works so vigorously that it burns out its firebrick lining in a week.

Typical capacity of a converter is 100 to 200 tons of steel per heat. The older open-hearth furnaces are much larger, more complicated in construction and operation, and more expensive than the converter. There are however still about 900 open-hearth furnaces in the United States. Neglecting any description of its complicated gas flues, the open-hearth furnace is a shallow melting bath with an arch roof, melting 200 to 400 tons or more steel per heat. This furnace is heated by a large gas flame for several hours; this may be supplemented by oxygen lancing.

Lime or limestone is first charged into the open-hearth furnace, then scrap metal. The scrap is melted, and the slag floats. Hot metal is then charged after the beginning of operations. Refining of the steel commences after melting is completed. During the refining stage, the carbon content is lowered and sulfur and phosphorus reduced to lowest possible levels. Common practice is to reduce the carbon below the required level, finally recarburizing by adding hot metal, coke, or coal just before tapping the furnace. If the carbon must be reduced to very low levels, say 0.03 percent for culvert steel or porcelain enamelling steel, then the use of oxygen lancing is necessary.

Since alloy metals melt at lower temperatures than pure metals, as the carbon is burned out of the steel the melting temperature constantly rises. Low carbon steels therefore impose severe temperature conditions on the furnace. Burner heat requirement per ton of steel exceeds 3,000,000 Btu in the open-hearth furnace.

Electric arc-melting furnaces operate on scrap metal. Such furnaces are simply large-scale arc welders, using three-phase electric power brought to the furnace by means of three large graphite electrodes that enter vertically through the dome roof of the furnace (Fig. 9.3). Current travels down one electrode, arcs through the scrap and up another electrode with much noise. While arc furnaces as small as half-ton capacity are available, for tonnage steel the capacity ranges from 30 to 150 tons or more.

Fig. 9-3 Electric arc melting furnace. The three electrodes are partially withdrawn from the furnace while the furnace crew makes up a slag. (Courtesy American Bridge Division, United States Steel Corp.)

Electric transformer capacity for an arc furnace may be of the order of 30,000 kva. As in arc welding, conditions approach short-circuiting. Meltdown begins at about 200 v, and even a modest size furnace may draw 25,000 amps per electrode. While carbon electrodes are much cheaper, graphite is used because of its lower electrical resistance and greater current capacity. Electrode diameters range from about 12 to 24 in. The electrodes are slowly consumed by the heat of the arc at their ends, usually at a rate of about 15 lb of electrodes per ton of steel. A new electrode can be screwed onto the top end of the one in use.

Operation of the furnace begins by raising the electrodes, then swinging the roof of the furnace to one side. A charging bucket then dumps scrap into the furnace, usually filling the furnace to the top. The roof is swung back and the electrodes lowered. Each electrode is automatically positioned at the proper level of about one inch above the charge. Arcing begins with much heavy noise until the charge is melted. Thereafter the furnace is quieter. Melting requires 400 or more kilowatts per ton of steel.

In the refining operation that follows melting, the carbon content of the steel is brought to the desired level. Iron ore or oxygen lancing are used to reduce the carbon; coke or scrap electrodes can be added to increase carbon.

Generally a double-slag method is used in electric refining of steel. This method is perhaps best explained in terms of a heat of stainless steel containing about 20 percent chromium. The chromium is added to the melt in the form of ferrochromium, an alloy of two-thirds chromium and one-third iron. The first slag is a basic (lime) oxidizing slag with the purpose of oxidizing phosphorus, silicon, manganese, and carbon. But chromium is a readily oxidizable metal also, and much chromium will also be removed by such a slag. To recover this chromium, a second slag is made up of lime, silica, fluorspar, and coke. The coke and lime react to form calcium carbide, which yields a reducing slag. This slag reduces and returns to the bath such reducible metals as manganese and chromium.

Besides alloy additions, certain other additives are used for quality control of the steel. All steels must contain a minimum of about 0.3 percent manganese. The manganese combines with any residual sulfur in the steel to form harmless manganese sulfide inclusions of microscopic size. Without manganese, sulfur in steel would cause the steel to crack during rolling-mill operations, sulfurized steels being "hot-short" at rolling temperatures. A second additive is silicon, which is used to deoxidize the steel. Molten steel dissolves oxygen, one of the interstitial elements that promote brittle fracture. Sometimes aluminum is added, either to deoxidize the steel or to produce a fine-grained steel. Some of these additives may be added after the steel is poured from the furnace into the holding ladle, especially if they are readily oxidizing, as aluminum is.

Whether the heat of steel is refined in an open-hearth furnace, a converter, or an electric furnace, the laboratory must provide a chemical analysis of the steel before it is poured. The analysis reports carbon to the nearest hundredth of a percent, plus all other significant constituents of the steel, of which there may be as many as 20, including the interstitials hydrogen, nitrogen, and oxygen. If the heat meets the requirements of the analysis, the steel is tapped into the holding ladle in preparation for pouring into ingots.

Thus steel is born. The instruments of its birth are furnaces—massive, dirty, ugly, and noisy. Yet in these fiery hellholes, whose chemistry we only dimly understand, men can formulate steels with a precision that the druggist in his pharmacy cannot match. Science did not invent steel. Rather, steel is a heritage of Western culture, for it required the work of practical men for over 600 years to develop these steel-making processes.

## 9.4 rolling-mill practice

The molten steel at a temperature of about 2900°F has been poured into the ladle. The ladle contains a pouring nozzle at the bottom. A long steel stopper rod running from the top of the ladle holds a refractory stopper on its bottom end to close the nozzle. The stopper rod can be raised with a lever when the steel is to be poured. The steel stopper rod does not melt in the molten steel, first because it is protected by refractory sleeves, and second because it is made of pure iron, which melts at a higher temperature than steel. The pouring of the steel must be done with care, since splashing metal could seriously injure personnel.

The steel is poured into ingot molds and freezes. The molds are tapered to facilitate stripping of the mold from the solid ingot.

Plain carbon steels are made in three grades: rimmed, semikilled, and killed. A killed steel is deoxidized by ladle additions of silicon or aluminum. These two metals are strong oxide formers and remove the dissolved oxygen in the steel as silicon or aluminum oxide inclusions. Since there is no evolution of dissolved gas when the molten steel cools in the mold, such steels lie quiet in the mold, and the steel man calls them "killed." A semikilled steel is partially deoxidized. A rimmed steel does not receive this deoxidation treatment and is named for the very characteristic rim of parallel crystals that grow inward from the mold surface as the steel freezes. Killed steels are more resistant to brittle fracture than rimmed steels and are commonly rolled into plate and structurals. Rimmed steels are chiefly used for drawing-quality sheet.

The killing process removes only oxygen from the steel. The oxygen remains as ceramic inclusions visible under the microscope. For some steels these inclusions are harmful, one of their effects being a reduction in the fatigue resistance of the steel. Such inclusions, for example, cannot be tolerated in ball and roller bearings. Where the inclusions cannot be tolerated, and when other dissolved gases such as hydrogen must be removed (hydrogen causes cracking in heavy forgings), the steel is degassed by pouring in a vacuum. The improvement in quality gained by vacuum degassing is considerable and degassed steels have been increasingly in demand.

A more recent development in pouring techniques is continuous casting, a technique that will be increasingly integrated with the oxygen converter process. The ingot mold for continuous casting is a water-cooled copper mold open at the bottom. Casting is carried on without interruption as the long frozen ingot is pulled out of the mold by pinch rolls and cut into lengths. Because of their high shrinkage killed steels are not produced by continuous casting.

The steel lies in the ingot mold until frozen. The molds are then stripped. The ingots are reheated in soaking pits until they are at a uniform temperature throughout for hot rolling into finished shapes. During hot rolling the coarse grain structure produced in the slowly cooling ingot is broken up to make a finer-grained steel.

Although hot rolling between roll stands is the principal method of reducing ingots, other methods are used for special products. The ingot may be forged on a large hydraulic press. Seamless tube is extruded from ingots through an extrusion die with hydraulic pressure. In hot rolling, the ingot is passed through a succession of rolls until it is reduced to the desired shape and size. The final shape may be rods for reinforced concrete, bar, structural shapes, rail, strip, sheet, plate, or skelp (plate for welding into pipe). A larger proportion of steel is rolled into sheet than into any other form, and sheet may be obtained in either the hot-rolled or cold-rolled condition. Hot-rolled sheet is cheaper and therefore more commonly used and can be recognized by its black coat of mill scale. Cold-rolled sheet is rolled at lower temperature in the final pass. It has a bright finish but is less ductile and thus not suitable for more severe forming operations. The thinner gauges of sheet are produced only in the cold-rolled condition.

## 9.5 carbon in steel

The most important alloying ingredient in steel is carbon. This element is very potent in its effect on steel characteristics, and few steels contain more than 1 percent carbon. Those steels in which carbon is the only significant alloying element are termed carbon steels. Low carbon steel, also called mild steel, is produced in greater tonnages than all other steels combined. It is cheap, soft, remarkably ductile, and readily welded. It cannot be usefully heat-treated however. This is the material from which ships, tanks, pipe, car bodies, household appliances, bridges, culverts, and building frames are made.

Steels are designated according to their carbon contents in the following classification:

| Designation | Carbon Content | Area of Use |
|---|---|---|
| 1. low carbon steel | 0.03 to 0.30% C | sheet and structural |
| 2. medium carbon steel | 0.35 to 0.55% C | machine parts |
| 3. high carbon steel | 0.60 to 1.5% C | tools and tooling |
| 4. cast iron | over 2% carbon | castings |

The medium carbon steels are used for reinforcing rods for concrete, harrow teeth, shafting, gears, and other components of agricultural implements, automobiles, aircraft, and machinery. High carbon steels are used in knives, files, springs, piano wire, hammers, axes, chisels, and shovels. Very low carbon is necessary in some steels, which are therefore virtually irons. These include many stainless steels, porcelain enamelling sheet, and some types of welding rods.

A small increase in the carbon content of a steel, even as little as a tenth of a percent, has a strong effect on all the properties of the steel. If the carbon content is increased, these are some of the effects:

1. the melting point of the steel is lowered
2. the steel becomes harder
3. the steel has a higher tensile strength
4. the steel is less ductile
5. the steel becomes more wear-resistant
6. the steel becomes less easily machined
7. the steel is more difficult to weld without cracking
8. the steel becomes heat-treatable
9. the steel is more expensive because of the smaller volume of production

There is virtually no change in the modulus of elasticity with carbon content. If a mild steel and a high carbon steel are both loaded to the same stress, both will show the same strain. The higher carbon steel, however, will have a higher yield stress, higher tensile strength, and less elongation at rupture.

Carbon has a powerful effect on the melting point of steels. A pure iron melts at 2800°F. Increasing carbon reduces the melting point until at 4.3 percent carbon the melting point falls to 2065°F. The high melting point of pure iron makes severe demands on the refractory linings of steel melting furnaces and is one reason why pure iron is not in common use. By leaving a little carbon in the steel the melting point is lowered substantially, and so is the cost of the steel.

The effect of carbon on the hardness of steel is shown in Fig. 9.4. The hardness of any steel may be varied by suitable heat treatments; this graph

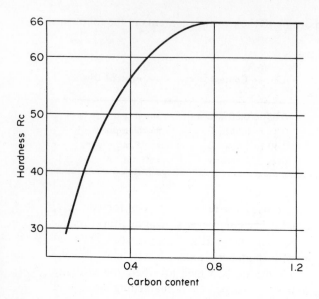

Fig. 9-4 Rockwell C hardness vs carbon content for carbon steels heat-treated to highest possible hardness.

shows the hardest possible condition at any carbon content. The graph indicates that the maximum hardness is reached at 0.8 percent carbon. This raises the question of the purpose served in going to still higher carbon contents. Carbon above 0.8 percent gives increased wear-resistance of the steel and is necessary in such tools as files, knives, wood-cutting tools, and hard-facing welding electrodes.

While carbon dissolves in molten steel, and in steel at red heats, only a few thousandths of a percent of carbon will dissolve in the space lattice of iron at room temperature. In steels cooled to room temperature the carbon is found to be combined with iron as iron carbide, $Fe_3C$, distributed through the steel. If a hard or wear-resistant steel is required, this is obtained by high carbon content to increase the amount of hard ceramic carbide.

An easily welded steel will have a low carbon content, and steels specially formulated for welding, such as tank and ship plate or welding electrodes, will have only one or two tenths of a percent carbon. If, however, the steel must be heat-treatable, a minimum carbon of 0.3 percent is needed. This gives the minimum amount of carbides required for heat-treating purposes.

Figure 9.5 shows the performance of steels of different carbon contents under conditions of the Charpy impact test.

Figure 9.4 is the phase diagram for iron-carbon alloys. Notice that for a pure iron, the transformation temperature separating the fcc iron phase from the lower bcc iron phase is 1666°F. As the amount of carbon in the iron is increased, this transformation temperature falls, reaching 1333°F at

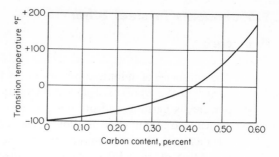

Fig. 9-5 Fifteen ft-lb Charpy transition temperature vs carbon content for plain carbon steel. Few standard steels will match the optimum performance shown in this graph.

0.8 percent carbon. The fcc phase of steel is termed *austenite*, and the bcc phase termed *ferrite*. Since carbon enlarges the austenite range, it is an austenite promoter, at least up to 0.8 percent carbon. The composition 99.2 percent iron, 0.8 percent carbon, though in the solid state, corresponds to a liquid eutectic in that it represents the lowest transformation temperature. It is referred to as the *eutectoid* composition. Steels containing less than 0.8 percent carbon are referred to as hypoeutectoid. Most carbon steels are hypoeutectoid. A carbon steel with 0.8 percent carbon is referred to as a eutectoid steel; greater carbon than this amount produces a hypereutectoid steel. When other alloying elements are added to steel, the eutectoid composition moves from the 0.8 percent position, usually to lesser carbon.

This transformation diagram of Fig. 9.6 is the basis for heat-treating of steels. Most heat treatments, either for hardening or softening the steel, require that the steel be first heated above the transformation temperature into the austenite phase, followed by cooling at some controlled rate—fast cooling for hardening, slow cooling for softening.

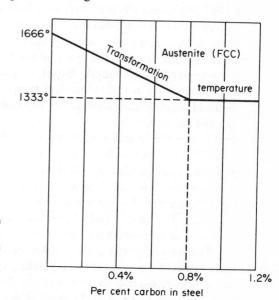

Fig. 9-6 Transformation diagram for carbon steels. Above the transformation temperature line steel is austenitic (bcc).

## 9.6 elements in steel other than carbon

The presence of minute amounts of hydrogen, oxygen, and nitrogen dissolved in the iron space lattice interstitially, even a few thousandths of a percent, is harmful and promotes cracking, flaking, and brittleness. These elements are rarely used as alloying elements, though nitrogen is occasionally used as an austenite promoter in a few stainless steels.

Most of the alloying elements added to steels are metals. It is rather confusing to try to remember the specific effects of each element on the properties of steel. Instead, it is best to begin with the following simpler and more generalized approach to these matters.

By and large, all alloying elements fall into two groups. They either (1) dissolve in the iron space lattice by replacing iron atoms in substitutional solid solutions or (2) form carbides with the carbon in the steel. A few elements such as manganese may show both tendencies, but even so one effect will predominate over the other, so that this simple approach is substantially true.

A second important effect of alloying elements must be borne in mind. One of the effects of carbon was austenite promotion, or the enlargement of the austenite range. Any alloying element must promote either austenite or ferrite. A ferrite promoter will reduce the austenite field and raise the transformation temperature; an austenite promoter such as carbon will do the reverse.

*Silicon* is used in small quantities to produce a killed steel. Larger silicon additions produce electric steels with higher electric resistance for such uses as transformer cores. Silicon dissolves in ferrite, replacing iron atoms in the space lattice. Since this means a change in interatomic distance where a silicon atom is located, the space lattice will be distorted and the metal will show higher tensile strength, higher hardness, and decreased ductility.

*Aluminum* is not added to steel except in small quantities for killing the steel and for producing a fine grain. The fine grain size produced by aluminum inoculation results in a tougher and somewhat stronger steel. Aluminum dissolves in ferrite and is a ferrite promoter, like silicon.

*Nitrogen* dissolves interstitially in the iron space lattice. It is an austenite promoter and is added to certain stainless steels for this purpose.

*Manganese* is added to all steels to make them easier to roll, as previously discussed. Manganese dissolves in ferrite and is a powerful austenite promoter. The addition of 12 percent manganese results in a steel which is austenitic at room temperature.

The ferromagnetic metals *nickel* and *cobalt* both dissolve in the iron space lattice. Cobalt is used in high alloy steels such as the high speed steels and in heat-resisting alloy formulations. Nickel is one of the most important alloy additions to steels and is used in a wide range of steels in amounts as little as 1 percent or as much as 50 percent. Like manganese, nickel is a powerful promoter of austenite; by adding sufficient amounts of either alloy, an austenitic steel at room temperature can be produced. Nickel improves most of the characteristics of steel, notably low-temperature toughness, heat resistance, and corrosion resistance. The most important high nickel austenitic steels are the 300 series stainless steels.

The above alloy additions all dissolve in ferrite. The following alloy metals do not dissolve in ferrite, at least to any notable degree, but tend rather to combine with carbon in the steel to form hard carbides. They are therefore usually associated with higher carbon contents. These metals are to be found chiefly in tool steels and hard-wearing steels. The carbide-formers also share the effect of decreasing the austenite field and raising the transformation temperature. This group includes the metals chromium, molybdenum, tungsten, vanadium, columbium, and titanium. Of these, titanium has the greatest affinity for carbon, though this metal is found in only a few steel alloys.

*Columbium* is added to a number of structural steels in very small amounts to increase the tensile strength. Typically, the addition of 0.02 percent Cb provides an increase of 10,000 to 15,000 psi in tensile strength.

*Vanadium* is not used in large amounts in any steel alloy. It promotes a fine-grained condition and increases strength. It is a frequent constituent of tool steels.

Both *tungsten* and *molybdenum* are used in large amounts, up to 18 percent, in high speed steels, and in lesser quantities in other types of hot-working steels, such as forging die steels. The presence of the carbides of these metals enables such steels to maintain hardness even at red heats.

*Chromium* is the most common of the carbide formers. Low alloy steels may contain as little as 1 percent chromium. In addition to providing hardness and strength, chromium in amounts of 12 percent or more produces the corrosion-resistant steels known as stainless steels. Corrosion resistance, however, is provided by chromium, not by chromium carbide. To avoid the formation of chromium carbides, those stainless steels which are not intended for heat-treatment have extremely low carbon content and are really stainless irons.

Phosphorus, copper, lead, sulfur, and boron are used in certain special steels. Both phos and copper form substitutional solid solutions in the iron lattice and therefore have a strengthening effect. Phosphorus is added to produce a free-machining steel; copper gives corrosion resistance. Lead does not dissolve in steel, and it does not form carbides. It is distributed through free-machining steels in the form of microscopic globules. The only use for sulfur is likewise to produce a free-machining steel containing inclusions of manganese sulfide. Boron is added to steels in amounts of a few thousandths of a percent to improve the hardenability of steels which are to be heat-treated. Too much boron, like too much phosphorus or sulfur, will make the steel impossible to roll.

## 9.7  industrial iron

Commercially pure iron has limited uses in industry. Armco iron is an iron of 0.01 percent carbon, which is used for its corrosion resistance in rolled products such as culverts. Iron, like any other metal, is more resistant to corrosion than its alloys. Iron powder is pressed and sintered in molds to produce finished small parts such as components of office machines.

A special type of iron that has been in use for centuries is *wrought iron*. This is an iron with slag stringers distributed through it, some of which are visible to the naked eye. These stringers are silicate slag and form effective barriers to corrosion of the iron. Though not produced by the methods used in melting steels, wrought iron is rolled and shaped by the usual steel mill equipment, as its name suggests. Carbon content is minimal, rarely as high as 0.035 percent.

Pure iron, like wrought iron, has an ultimate tensile strength less than that of steels, about 36,000 psi, which is the strength of the weldable grades of aluminum. Structural steels and mild steels have strengths from 60,000 psi up. Ductility of pure iron is unusually high, however, and its corrosion resistance is superior to steel's.

## 9.8  cast steels and irons

Although most steels are converted into wrought products, any steel may also be formed by casting; indeed all steels must first be cast into ingots before being rolled. About 2 percent of steel production goes into finished castings. Although this is a small fraction of total steel consumption, steel castings are of critical importance in primary and heavy industry. The rolls used in rolling steel shapes are made of cast steel. Such complex shapes as pump casings and large valve bodies cannot be formed by any other economical method than casting.

While most steel castings are low carbon steels, nickel and other alloy steel castings are used for special purposes. Most heat-resistant and corrosion-resistant castings are made of stainless steel alloys. The high melting temperatures of steels require that steel castings be formed in sand molds, although it is a standard practice to use graphite molds for cast steel railroad wheels.

A steel containing in excess of 2 percent carbon is called an iron. Such steels have almost no ductility and must be shaped by casting.

The commonest type of cast iron is *gray cast iron*, containing about 3 to 3.5 percent carbon and over 1 percent silicon. These are relatively large amounts of both carbon and silicon. With so much carbon, all of the carbon cannot be taken up as iron carbide. A considerable fraction of the carbon separates out as long graphite flakes distributed through the cast iron, as illustrated in Fig. 9.7. These graphite flakes produce the characteristically dark appearance

Fig. 9-7  Micrograph of gray cast iron, showing graphite flakes in a pearlite structure.

in a freshly broken surface of gray cast iron. The reason for the high silicon is that this element promotes such graphitization of the carbon.

Gray cast iron has almost no ductility, and if heated or cooled too rapidly it may crack. The graphite flakes make for some difficulty in welding, for it is not possible to make weld metal bond to graphite. Another of their effects is a lower modulus of elasticity, about $18 \times 10^6$ compared to about $30 \times 10^6$ for steel. A fourth deficiency of gray cast iron is its low tensile strength compared to steel. This is explained again by the graphite flakes, which have the effect of a network of cracks within the metal. The strength of graphite in cast iron is no better than its strength in a lead pencil.

However, gray cast iron is not without its virtues. It is a low-melting steel and easy to cast into complex shapes. It can withstand higher temperatures than steels without warping or oxidizing, even as high as 1000°F, hence its use for doors and other furnace parts. It is readily machinable because of the lubricating effect of graphite. The graphite network also provides a considerable degree of corrosion resistance. Still another advantage of cast iron is vibration-damping. The graphite network rapidly damps out vibrations and "whip" in the metal, so that gray cast iron is frequently selected for crankshafts and machine bases.

Much gray cast iron is melted in foundry cupolas, which are miniature blast furnaces ranging from three to six feet in inside diameter. Cupolas are charged from the top with scrap iron, coke, and limestone. A better quality cast iron can usually be produced in an electric arc furnace, such as is used for melting steel, or in an induction furnace.

Gray cast iron results from relatively slow cooling in the sand mold. Fast cooling by such methods as the use of steel or graphite "chills" (inserts) in the mold produces *white cast iron*. White cast iron is extremely hard and abrasion-resistant, unmachinable, and unweldable.

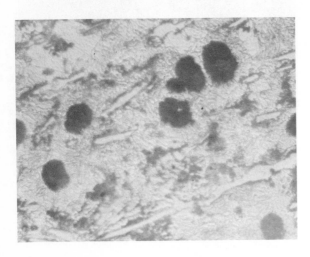

Fig. 9-8 A molybdenum ductile iron.

*Nodular iron*, also called ductile iron, is a special type of gray cast iron. This iron has a microstructure that overcomes many of the limitations of gray cast iron. When the iron is poured from the melting furnace or cupola into the ladle, it is inoculated with a small amount of magnesium. As a result of this treatment, the graphite takes a nodular (almost spherical) form. The result is a cast iron of considerable ductility and high tensile strength. The operation of inoculating the iron is one of the more dramatic routines of industry: a blinding white flare is produced.

*Malleable iron* is a cast iron which by special heat-treatment acquires a degree of ductility, 10 percent or more in a 2-in. gauge length. The graphite in malleable iron is somewhat nodular in shape.

*Ni-Resist* is a 15 percent nickel cast iron for corrosion resistance. For wear resistance, the alloy cast iron *Ni-Hard* is employed.

## 9.9  low alloy steels

Low alloy steels are those steels containing not more than 5 percent of any alloying metal or metals. Such steels are used for applications requiring either extra toughness or higher strength. High alloy steels contain over 5 percent of alloying elements and find their chief applications under conditions of wear, heat, and corrosion.

Until about twenty years ago, the method of building great strength into equipment was simply to make the equipment from heavier sections. This method is in general neither efficient nor economical, and indeed for an increasing number of products is impossible. The high strength required in the frame and skin of a supersonic aircraft cannot be obtained by indefinitely "beefing up" structural sections. If a truck body can be made lighter in weight by one ton through the use of high strength, low alloy steels, then an extra ton of revenue-producing payload can be carried on each truck haul for the life of the truck. Again, consider the case of a high-rise building of 50 stories. If the use of high strength, low alloy steel permits a reduction in the depth of floor beams of 2 in., then the total building height is reduced by over 8 ft. This represents 8 ft less of curtain wall covering the building of steel framing, of fireproofing, ducting, piping, insulation, and installation labor, not to mention lower foundation costs. In the final result, those 2 in. per story represent savings of many thousands of dollars after allowing for the slightly higher cost of the alloy steel.

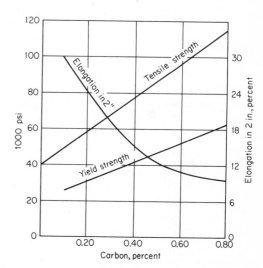

Fig. 9-9 Effect of carbon content on ultimate tensile strength, yield strength, and elongation on annealed hot-rolled steel.

These are the considerations that explain the growing popularity of low alloy steels. Every year sees a number of new steels introduced, and the number now available is perhaps confusingly large.

To sort out the whole field of low and high alloy steels in general, they can be grouped into six classes:

1. high strength, low alloy (HSLA) steels, chiefly used as construction steels in buildings, cranes, construction machinery, etc.
2. American Iron & Steel Institute (AISI) and Society of Automotive Engineers (SAE) steels, chiefly used as machine parts in automobiles, aircraft, and other machinery

3. alloy tool steels for tooling, either low alloy or high alloy
4. stainless steels
5. special heat-resisting steels
6. electric steels, chiefly alloyed with silicon or aluminum

## 9.10 high strength, low alloy steels

The standard structural steel in use for the last 60 years was simply called "structural steel." It was made to ASTM (American Society for Testing and Materials) standard A7. It was a low carbon steel, but the specification did not require the steel to have a definite chemical analysis. The specification laid down requirements for strength and ductility. Yield strength was 33,000 psi, ultimate tensile strength 60,000 psi, and elongation about 20 percent in a 2-in. gauge length, all these being minimum values. The design stress used with this steel was 20,000 psi.

The newer high strength, low alloy steels have made rapid inroads into the applications formerly filled by A7 steel. These are all "lean" steels, that is, very low in alloy content, with ultimate strengths as high as 120,000 psi. By keeping alloy additions lean, the cost increment over A7 steel is held to a minimum, and the steels do not raise any unusual difficulties for the welder, though the higher strength levels require the use of welding rods with commensurate strength. Such HSLA steels are rolled into plate, sheet, bar, and structural shapes.

To ensure toughness and weldability, the HSLA steels for construction purposes have carbon contents of 0.20 percent or less. The alloy formulation of these steels is arranged to give them maximum resistance to atmospheric corrosion, and occasionally such steels have been left unpainted. Of the wide range of these newer construction steels, over 300 brand names, perhaps the best known are "Cor-Ten" and "T-1." These two steels are typical of the HSLA steels as a group. They contain small amounts of the ferrite strengtheners, including manganese, silicon, nickel, and copper, and still lesser amounts of the carbide-formers, such as titanium, vanadium, and columbium.

## 9.11 AISI alloy steels

These are machinery steels which have been catalogued by number by the American Iron and Steel Institute (AISI) and the Society of Automotive Engineers (SAE). An AISI 8620 is the same steel as SAE 8620. The steels have a number system which is in part descriptive of the alloy; although it seems simple in principle, the system is in practice a bit complicated, and only the bare bones of the scheme will be discussed here. For example, consider AISI 2345. The last two digits, in this example xx45, always designate the carbon content, or 0.45 percent. The first two digits designate the alloy. In this example 2xxx indicates a nickel steel, and x3xx indicates approximately 3 percent nickel. A designation such as 10100, with five digits, would indicate 1.00 percent carbon or more. The basic digits in the system are the following:

| | |
|---|---|
| 10xx | plain carbon steels |
| 11xx | sulfurized free-machining carbon steels |
| 12Lxx | leaded free-machining steels |
| 13xx | manganese 1.75% |
| 23xx | nickel 3.50% |
| 25xx | nickel 5.00% |
| 3xxx | nickel-chromium steels (not stainless steels) |
| 40xx | molybdenum 0.20–0.30% |
| 41xx | low chromium-molybdenum |
| 4xxx | nickel 1–4%, chromium to 0.55%, moly 0.20–0.30% |
| 5xxx | up to 1% chromium |
| 6xxx | low chrome-vanadium steels |
| 7xxx | not used |
| 8xxx | low nickel-chromium-molybdenum |
| 9xxx | usually nickel-chromium with low molybdenum |

The letter B in the middle of the number, such as 81B45, signifies a minimum of 0.0005 percent boron for hardenability.

The 2xxx and 3xxx series of AISI-SAE steels are no longer in common use. Most of the other steels of this series have a wide range of uses in automobiles, agricultural equipment, and aircraft. In general these steels are fabricated into small tools and hand tools, and machine components, including those of the family automobile. The following list is a selection of uses:

| | |
|---|---|
| 1010 and 1015 | —steel tubing for general use |
| 1022 | —piston pins of automobiles |
| 1040 | —connecting rods of automobiles |
| 1055 | —gardening tools |
| 1065 | —spring wire |
| 4140 | —socket wrenches |
| 4130 | —small rocket motors |
| 4340 | —highly stressed aircraft parts, such as landing gear |
| 52100 | —ball and roller bearings |
| 8620 | —automobile gears and parts of universal joints |

# 9.12  tooling and tool steels

The subject of tooling appears not to be well understood by personnel outside the field of production technology. Tooling is often confused with hand tools, which are not tooling but are produced by tooling. The word "tooling" embraces the special equipment which is used to shape materials into finished components. Such equipment adapts the capacity of a production machine to the requirements of the part to be produced. A simple example of tooling is the combination of a punch and a die, which if mounted in a large press can produce holes in sheet metal or plastic.

Tooling materials are not consumed in massive tonnages, yet they are probably the most critically important of the industrial materials, since without them almost nothing can be manufactured except by prohibitively expensive hand methods.

Although it is possible for almost any industrial material to be used for tooling purposes, tooling materials are generally restricted to a relatively small group of materials. Dimensional stability is one of the dominant requirements for a tooling material, for without it, identical and interchangeable parts cannot be made. Most of the special tool materials are the select group of steels known as tool steels, but epoxies, polyurethane rubber, silicone rubber, magnesium plate, and cast irons serve useful functions in tooling, and these accessory additions to the toolmaker's repertory will be examined in their appropriate chapters.

The service requirements to be met by a tool steel may be most conveniently discussed in terms of punching 1,000,000 holes in 16-gauge mild steel sheet, not an especially demanding service for tool steels. The operation is performed on a punch press. In the punch press a standard tooling item called a standard die set is mounted. This is a pair of heavy steel plates carefully aligned. The male punch is attached to the upper plate of the die set, the punch plate; the female die containing the hole is attached to the lower plate or die shoe. The rigidity of the die set ensures that the punch will always be aligned with the hole. Misalignment will break the punch.

Suppose the holes to be punched are 1 in. in diameter. To punch a hole, the steel sheet must be sheared all the way round the 1-in. periphery for a shearing length of 3.14 in. The sheet is 16-gauge, or approximately $\frac{1}{16}$ in. thick. The punch must therefore shear approximately $\frac{3}{16}$ sq in. of steel to produce the hole. The mild steel sheet has a shear strength of perhaps 45,000 psi. But there are friction and binding effects to account for, and to take care of these conditions the shear strength of the sheet is often taken to be the ultimate tensile strength, which perhaps is 60,000 psi. The total shearing force required to punch the hole is then

$$\frac{3}{16} \text{sq in.} \times 60{,}000 \text{ psi} = 11{,}000 \text{ lb (approximately)}$$

The compressive strength of the punch material must support this force of 11,000 pounds, plus any bending force that the operation may produce in the punch. Similarly, the die must withstand the punching force of 11,000 pounds through thousands of cycles without any mushrooming of the material around the hole. The punch and the die must not fail in fatigue. Further, when the blank is punched out of the sheet, the sheet will grip the punch and must be stripped off it by a stripper plate. This stripper operation, together with the punching operation, will wear away the surface of the punch, so that the hole size will become smaller after some thousands of punching operations. Such wear must not be excessive, or the punched holes will in time become too small.

The material for both punch and die must be readily machinable during toolmaking operations, but it must also be heat-treatable, for after the toolmaking both punch and die must be heat-treated to exactly the required hardness, toughness, and wear-resistance. Further, these tooling materials must not distort or alter their dimensions to an excessive degree during the phase changes involved in heat treatment.

Fig. 9-10 High-speed steel milling cutter notching a small workpiece. For this application the cutter material requires the property of hot hardness.

This example represents a very simple case of tooling, yet only a special steel can meet such service requirements. More severe standards of performance are demanded of hot forging dies, or dies used to form vinyl plastics (which are corrosive to steels), or wire-drawing or extruding dies subjected to extreme pressures. As tooling practices become ever more exacting, users of tool steels demand ever better performance from these materials. The chief complaints made against tool steels are probably these four:

1. they wear out in service too soon
2. they chip or break during use
3. they change size during heat-treating
4. they distort or crack during hardening

From the previous remarks it should be apparent that progress in materials technology is as dependent on tooling materials as on materials progress in general, for no material becomes significant until it can be reduced to useful shapes by economic methods.

The several classes of tool steels are classified by letter designation:

| | |
|---|---|
| type W | water hardening. High carbon steels |
| O | oil hardening. Lean alloy steels |
| A | air hardening. Medium alloy steels |
| S | shock-resisting steels |
| H | hot-working steels |
| T | high-speed steels containing tungsten |
| M | high-speed steels containing molybdenum |

The simplest of the tool steel alloys are the high carbon water-hardening steels. These, as their name suggests, are hardened by fast quenching with water. These have limited uses in cold-working operations, such as the shearing of leather and paper. Plain carbon steels are capable of great hardness but begin to lose hardness at the very low temperature of 300°F, so that they are not suited even for drill bits for drilling metals. As a result, the general trend has been away from type W steels.

All the other groups of tool steels use for their alloying elements the carbide-forming elements tungsten, molybdenum, chromium, and vanadium. To supply sufficient carbides, the carbon content of such steels must necessarily be high. Carbon therefore may be as little as 0.4 percent or as high as 2.25 percent. The hot-working and high-speed steels, which must operate at elevated temperatures, must be rich in carbides because these are refractory and relatively insensitive to temperature. Because they are very rich in carbides, these steels can be hardened by relatively slow cooling in air. Figure 9.11 is a micrograph of a high-speed steel, showing the dispersion of hard carbides throughout the steel. These are largely tungsten carbides, since the steel contains 18 percent tungsten.

Shock-resisting steels are used in applications where toughness and shock resistance are paramount requirements. They are made into punches, chisels, pneumatic tools, and shear blades. These steels contain carbide formers, silicon to strengthen the ferrite, and sometimes manganese or nickel for increased toughness.

The oil-hardening type O steels are considered the most important group of tool steels because of their wide range of applications. Carbon

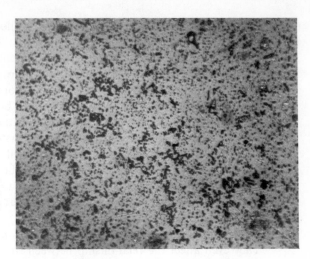

Fig. 9-11 Micrograph of a tungsten high speed steel, showing the
distribution of tungsten carbides
in the annealed state.

content is about 1 percent, with about 0.5 percent chromium and 0.5 percent
or more tungsten.

Type H hot-working steels are employed in forging dies, aluminum
extrusion dies, and die-casting dies for plastics. These are medium carbon
alloy steels, with 5 percent chromium or several percent tungsten.

The high-speed steels earned their odd name in earlier days when they
were substituted for carbon steel cutter bits on machine tools. They permitted
the cutting speed of the machine tool to be greatly increased without losing
their cutting edge. These steels can maintain their hardness to temperatures
as high as 1200°F, which is a dull red heat. Originally the high-speed steels
were type T (tungsten), but in late years molybdenum has tended to displace
tungsten. Molybdenum is not so heavy as tungsten, yet it forms an equal
amount of carbide, and steels of course are sold by weight, thus making
molybdenum cheaper. A very high alloy, high-speed steel may contain 18.5
percent W, 4 percent Cr, 1 percent V, 4 percent Co, and 0.8 percent carbon.
Nowadays these steels can no longer meet the most stringent demands of the
machining industries, which are better served by sintered carbide cutters.

Some of the hot-working tool steels have found occasional use in gas
turbine engines as turbine blading by reason of their hot-hardness and hot-
strength. Another alloy used both as a hot-working tool material and as a
heat-resisting material is *stellite*. It is not a steel, as its name suggests, but a
cobalt alloy. It is very hard and is formed only by casting and grinding.

### 9.13 stainless steels

Heat-resistant alloy steel castings, designated type H, and corrosion-resistant
alloy steel castings, designated type C, may be thought of as types of cast
stainless steel. For improvement of both heat and corrosion resistance,

generally the nickel content of these casting alloys must be increased, up to a maximum of 66 percent nickel. Such cast alloys are used in gas turbine engines and afterburners, discharge ends of cement kilns, fixtures and baskets for furnaces and heat-treating operations, furnace dampers, burner nozzles, centrifugal pumps, and agitators for mixing of chemicals.

Perhaps the most useful remark that can be made about the large group of stainless steels is that a little knowledge of these materials is of little value and is a cause of some regrettable errors in materials selection. The stainless steels are often viewed as the panacea for all steel ills. When an ordinary steel fails in service, a stainless steel may be substituted. Sometimes the substitution is successful, but there are occasions when the stainless steel does not perform as well as the previous material. Pious faith is not desirable in materials selection.

Certainly the stainless steel group has some interesting characteristics. They even affect the relationship between the sexes. Consider the stainless steel kitchen knife. The wife likes it because it doesn't rust when left in the dishpan. The husband dislikes it because it doesn't keep its edge. Stainless steels are poor tool steels, an argument to which the female sex is impervious.

The use of stainless steels has been growing at a more rapid rate than that of most other steels. Annual consumption of stainless steels in the United States is now about a million tons. These are somewhat costly steels, but their advantages and market acceptance outweigh their additional cost in a wide range of applications, such as kitchen knives. The stainless steels are a group of low carbon, high-alloy steels which offer the advantages of attractive appearance, low maintenance, great corrosion resistance against oxidizing chemicals (but not reducing chemicals or chlorides), oxidation resistance, and high strength and toughness at extreme temperatures. However, these are heavy steels: there is a double price premium in the higher cost per pound and the higher specific weight.

Thermal conductivity of the stainless steels is one-third that of carbon steels. Thermal stresses are therefore a more serious consideration when designing with stainless steel. The low thermal conductivity also makes these steels more damaging to machining cutters through inability to remove the heat generated during machining. The nickel-chromium grades have an expansion coefficient about a third higher than that of carbon steels, and they produce more warping during welding operations.

The stainless steels, like the tool steels and construction steels, are not included in the AISI-SAE steel coding system. Instead, the stainless steels have a numbering code of their own which uses three digits instead of four. Neglecting a few 200 and 500 series stainless steels, all stainless steel compositions are assigned numbers in either the 300 or 400 range. The first step in understanding the characteristics of these steels is to differentiate clearly between the 300 and 400 groups. The second step is to break the 400 series into two groups, heat-treatable and not heat-treatable. The third step is to become acquainted with some of the more common grades in these groups.

The stainless steels may be summed up as very low carbon irons with a minimum of 12 percent chromium, although stainless 501 and 502, used in the oil industry where a lesser degree of corrosion resistance suffices, contain 4 to 6 percent chromium. The corrosion and heat resistance of these steels may be considered proportional to the chromium content, which for extreme service conditions may be as high as 35 percent. Unlike the high-speed steels, this heat resistance of the stainless steels does not depend on the presence of carbides but on chromium metal itself uncombined with carbon. Chromium quickly acquires a thin skin of tightly adherent chromium oxide which blocks corrosion or oxidation of the stainless steel.

All stainless steels have considerably higher strengths than low carbon steel. A mechanical shear rated for $\frac{1}{4}$-in. steel plate will not safely cut $\frac{1}{4}$ in. of stainless steel, as is repeatedly proved by numbers of broken shears across the country.

The 400 series are straight chromium steels. The 300 series are chromium-nickel low carbon steels. Many of the 300 series contain approximately 18 percent chromium and 8 percent nickel and used to be referred to as "18-8" stainless. This practice is not recommended.

## 9.14   the 300 series
### stainless steels

The peculiarities of the 300 series may be thus summed up:

1. they are low carbon, none containing more than 0.25% carbon
2. almost all contain 16–20% chromium, 8–13% nickel
3. they are austenitic, face-centered cubic, paramagnetic
4. they are not heat-treatable
5. they work-harden when formed, shaped, or cut
6. they have an expansion coefficient a third higher than carbon steels

Because of work-hardening effects, these steels are machined with slow speeds and deep cuts to get underneath the work-hardened layer.

The standard or basic 300 stainless steel alloy is 304, and all other 300 stainless steels may be considered as modifications of 304 to serve special applications. Type 304 analyzes

There is also about 1 % silicon and 1 % manganese. The silicon gives additional corrosion resistance and oxidation resistance at high temperatures. This is the general-purpose austenitic stainless steel.

The other 300 alloys are modifications of 304. The more important of these modifications are the following:

1. *The economical alloy, 301.* Consumer products require materials of lowest cost. Type 301 is formulated for low cost and is a 304 except that nickel is reduced to 6–8 percent. Type 301 is used in wheel covers and hub caps of automobiles.

2. *The machining alloy 303.* Because of their work-hardening characteristic, the 300's do not machine readily. Type 303 contains 0.15 percent sulfur for improved machinability. Type 303 Se contains about 0.25 percent selenium to serve the same purpose.

3. *The forming alloy, 305.* The 300's, because they work-harden, may crack during severe forming and drawing operations. Type 305 is a 304 modified by increasing the nickel content to 10–13 percent, thus reducing the tendency to work-harden. This steel therefore is used in ashtrays, kitchen sinks, and stainless rivets.

4. *For enhanced corrosion resistance, 310.* Corrosion resistance and heat resistance can be considered proportional to the chromium content. Type 310 contains about 25 percent chromium.

5. *For acid resistance, 316.* To resist the attack of strong acids, especially against pitting corrosion, 316 is used. It is a modified 304 incorporating 2 to 3 percent moly.

6. *For welding rods and wires, 308.* Chromium metal tends to oxidize in the arc. A welding rod of stainless 304 could lose as much as $1\frac{1}{2}$ percent chromium in transfer of the chromium across the arc. The deposited metal then would have less chromium and would be less corrosion resistant than 304. To compensate for chromium loss in the arc, welding rods are made of 308 alloy, a 304 modified by increasing the chromium to 19–21 percent. This is the standard welding wire alloy, although other stainless wire alloys are also available.

7. Other grades of austenitic steel serve many purposes, such as 348, an alloy formulated for the requirements of nuclear reactors. But the most frequently met problems are served by 304 and the few alloy modifications mentioned here. Only the specialist needs to be intimately familiar with other 300 stainless alloys. Stainless 302 though should be mentioned. This was the first of the stainless steel alloys. It is a 304 with 0.15 percent carbon. As explained below, this is excessive carbon, especially if the alloy is to be welded or heated. Type 302 will become obsolete in the future.

## THE AUSTENITIC STAINLESS STEELS

| Type | C | Cr | Ni | Mn | Other |
|------|-----------|-------|-------|-----|----------|
| 301 | 0.15 max | 16–18 | 6–8 | 2 | — |
| 302 | 0.15 max | 17–19 | 8–10 | 2 | — |
| 303 | 0.15 max | 17–19 | 8–10 | 2 | sulfur |
| 304 | 0.08 max | 18–20 | 8–10 | 2 | — |
| 305 | 0.12 max | 17–19 | 10–13 | 2 | — |
| 308 | 0.08 max | 19–21 | 10–12 | 2 | — |
| 316 | 0.08 max | 16–18 | 10–14 | 2 | 2–3% Mo |
| 321 | 0.08 max | 17–19 | 9–12 | 2 | titanium |
| 347 | 0.08 max | 17–19 | 9–13 | 2 | Cb + Ta |

## 9.15 chromium and carbides

Stainless steel alloys are recognized by their low carbon, high chromium analysis. Consider two similar alloys:

$$0.1\% \text{ C} \quad 12\% \text{ Cr}$$
$$1.7\% \text{ C} \quad 12\% \text{ Cr}$$

Here are two high-alloy chromium steels. The second steel also contains a high carbon content, indicating that the intent of this alloy is the formation of large amounts of chromium carbides (for wear resistance in excavator buckets and teeth). Carbides are the distinguishing feature of tool steels: the second alloy is a tool steel and the chromium will be taken up in carbides. The first alloy has a low carbon content. The intent then is not to form carbides in the first case. Since the low carbon limits the formation of carbides, the chromium will exist in the alloy largely as elemental chromium for corrosion resistance. The first alloy is a stainless steel, actually stainless 410.

Chromium therefore serves two purposes in steel alloys: corrosion resistance or carbides. For corrosion resistance there must be low carbon; for carbides, high carbon. Clearly, the intent in austenitic stainless steels is to keep the chromium from forming carbides. This becomes a problem in welding these steels. Chemical activity increases rapidly with temperature, and in the temperature range 800–1600°F there is a rapid combination of chromium and carbon in stainless steels to form chromium carbides. These carbide particles tend to precipitate in the grain boundaries, the effect being called *carbide precipitation*. The loss of chromium as carbide impairs the corrosion resistance of the steel. Carbide precipitation is significant only if corrosion resistance is significant: it would be disastrous in the equipment of a paper mill but of no importance in the kickplate of a hospital door.

Carbide precipitation is more prominent at the edge of a weld (the heat-affected zone) than in the weld itself. There are two methods of preventing this defect. One method is to keep the carbon content to very low levels. Carbide precipitation does not occur in austenitic stainless alloys if the carbon level is 0.05 percent or less. Extra-low carbon alloys, 304L and 316L, are available; they are also numbered 304ELC and 316ELC. A second method is to add to the stainless alloy a metal that has a greater affinity for carbon than does chromium. Of all the metals, titanium has the greatest affinity for carbon. Type 321 stainless contains titanium, so that the carbides precipitated from this alloy are largely titanium carbides, leaving the chromium for corrosion resistance. An alternate to 321 alloy is 347, containing columbium.

## 9.16 the 400 series
### stainless steels

The 400 series stainless steels contain no nickel. The absence of nickel, the austenite promoter, and the presence of chromium, a ferrite promoter, indicate that these steels cannot be austenitic. They are ferromagnetic and do not work-harden to an unusual degree. These steels do not possess quite the high levels of corrosion and heat resistance that characterize the richer 300 series. The 400 series actually includes two groups of steels, *martensitic* and *ferritic*.

Martensite is the hard phase of steel resulting from the fast quenching of carbides, and the word martensitic implies that a steel may be heat-treated. Martensitic stainless steels contain more carbon and during heat-treatment form chromium carbides. They are used for machine parts that must combine some degree of corrosion resistance with high strength levels. The most commonly used martensitic stainless steels are 410 and 416, containing 0.15 percent maximum carbon; others may contain as much as 1 percent carbon. Type 420, with 0.20 percent carbon or higher, is used for stainless cutlery.

The ferritic stainless steels cannot be heat-treated. They are low carbon, high chromium irons containing 16 percent or more chromium, commonly used in sheet form. Type 430 is used as a stainless steel chimney liner.

The 300 series, being austenitic (face-centered cubic), remain notch-tough down to absolute zero. On the other hand, the 400 series have poor impact strength at low temperatures. In the annealed (soft) condition the 300 series are slightly stronger (about 90,000 psi) and have greater ductility than the 400 series.

The two 200 series stainless steels 201 and 202 are interesting in that 0.25 percent nitrogen is substituted for nickel as an austenite promoter. These two steels substitute for 301 and 302.

The maraging steels (pronounced "mar-aging") are a recent discovery but are rapidly finding new applications, especially in the aerospace industry. These are extremely low carbon iron-nickel alloys with a minimum of 18 percent nickel (manganese may be substituted), about 9 percent cobalt, 3 percent molybdenum, and small amounts of titanium and aluminum. Until the introduction of the maraging steels, it was not possible to heat-treat steels to tensile strength levels beyond 225,000 psi, because higher strengths were accompanied by dangerously low ductility. The maraging steels can be heat-treated to 300,000 psi and still retain minimum ductility, which is generally considered to be about 10 percent in a 2-in. gauge length.

Another steel of remarkable characteristics is 12 percent manganese steel, also called Hadfield manganese steel. This contains about 1 percent carbon besides the high manganese. Manganese, like nickel, is an austenite promoter, and this steel contains sufficient manganese to be austenitic at room temperature, like the 300 group of stainless steels. The austenitic steels work-harden more rapidly than other metals, and Hadfield manganese steel has remarkable work-hardening ability. It is used for such applications as excavator teeth, rock crushers, railroad frogs, and hard-surfacing welding rods, where constant battering will continually harden the steel. The work-hardening tendency is so powerful that the steel is virtually unmachinable and must be formed by casting and grinding.

### PROBLEMS

1  It is not impossible to produce stainless steel alloys in a basic oxygen converter. What might be the chief difficulty in this process, however?

2  What is the source of heat in an oxygen converter?

3  Why is the steel industry losing interest in the open-hearth furnace?

4  Explain the need for double-slagging in the melting of some steels.

5  Why is sulfur generally harmful in steels?

6  How does manganese overcome the harmful effect of sulfur in steel?

7  What advantage does vacuum degassing of steel offer?

8  What general characteristics are required of tool steels?

9  What purpose does chromium serve in a tool steel? In a stainless steel?

10  What are the general differences between 300 and 400 stainless steels with respect to:
(a) alloy content
(b) crystal structure

(c) magnetic properties

(d) low-temperature behavior

(e) work-hardening characteristics?

11   Why is a 304 stainless superior to 302 in a weldment? Why is 304L still better in this respect?

12   Account for the use of 308 stainless alloy in manual welding rods.

13   What does the "killing" operation do to molten steel?

14   What influence does the graphite network have on the properties of gray cast iron?

15   What is a high-speed steel?

16   What is meant by "carbide precipitation" in austenitic stainless steels?

17   For what uses is a Hadfield manganese steel suited?

18   Why is hot metal from a blast furnace not usable for metal products?

19   A machine shaft is subject to wear during operation. Why would you select a 1040 alloy instead of a 1020 one for this shaft?

20   Ball bearings are made of AISI 52100. Does this appear to be a good alloy for the purpose? Look up the alloy composition in a handbook.

21   In selecting alloys for machine parts, the author adopts the principle of selecting the lowest carbon content that will do the job. What is your opinion of this principle in machine design?

22   What is the carbon content and the maximum hardness obtainable in the following alloys (Rockwell C): (a) 1040, (b) 1080, (c) 10100, (d) 10120, (e) 52100, (f) 0.8 percent C, 18 percent W, 4 percent Cr, 1 percent V?

23   A drill bit can be made of 1080 or of alloy (f) in the previous question. Both alloys have the same maximum hardness at room temperature. Why would you select the richer and more expensive alloy?

24   Pure metals are more corrosion resistant than alloys. Nevertheless, 300 stainless steels are much more corrosion resistant than pure iron. How do you explain this?

25   List the more important carbide formers used in steel alloys.

26   List the more important alloy additions to steels that are not employed for carbide formation.

27   Which of the following alloys are probably tool steels:
(a) 0.80 percent C, 0.3 percent Si, 0.6 percent Mn
(b) 0.025 percent C, 1.8 percent Mn, 20.6 percent Cr, 9.7 percent Ni
(c) 0.10 percent C, 15 percent Ni, 34 percent Cr
(d) 1.2 percent C, 1.75 percent W, 0.75 percent Cr
(e) 0.7 percent C, 9 percent W?

28   Deposit a welding bead from a mild steel welding rod (6013 or other arc welding rod) on a sheet of 304 or other austenitic stainless steel. Measure the Rockwell C hardness over and around the weld deposit. You will obtain several readings that will be considerably higher than the hardness of either the sheet or the rod. Can you account for this unusual result?

29  If you must butt-weld a mild steel sheet to a sheet of 304 alloy, would you use a mild steel electrode or a stainless electrode?

30  Obtain a small piece of austenitic stainless bar $\frac{3}{4}$ $\phi$. Measure the Rc hardness. Hammer the bar vigorously with a heavy ball peen hammer so that the bar is deformed. Measure the hardness of the deformed area.

31  Heat a bar of stainless 410 in a furnace set at 1800°F (25 minutes for a $\frac{3}{4}$ $\phi$ bar). What is the highest hardness that you can obtain from this heat-treating? Bend the bar in a press to see if it has any ductility. If this steel is to be welded, will the weld deposit and heat-affected zone harden?

32  Select a stainless steel for the following purposes and defend your selection:
(a) head of a golf club
(b) a kitchen knife—Rc 50 required
(c) a surgical scalpel
(d) a stainless pump shaft
(e) an austenitic stainless part that must be tapped $\frac{1}{4}$-20 without breaking the tap
(f) a stainless socket for an incandescent lamp
(g) the hull of a mine sweeper that must sweep magnetic mines
(h) a stainless door hinge

33  Define the following in terms of carbon content: (a) a machinery steel, (b) a tool steel.

34  Why is a continuous chip not produced when machining gray cast iron?

35  Why would you not select a steel with 0.8 percent carbon for a machine shaft?

36  Why do welding rods have a low carbon content?

37  Name a free-machining mild steel and a free-machining austenitic stainless steel.

38  State the two methods used to prevent carbide precipitation in austenitic stainless steels.

39  What is the simplest way to differentiate a 300 from a 400 stainless steel? If you don't know, ask a scrap dealer.

## ABBREVIATIONS FOR THE METALS

| aluminum | Al | iridium | Ir | sulfur | S |
|---|---|---|---|---|---|
| beryllium | Be | lead | Pb | tantalum | Ta |
| cadmium | Cd | magnesium | Mg | tin | Sn |
| carbon | C | manganese | Mn | titanium | Ti |
| chromium | Cr | molybdenum | Mo | uranium | U |
| cobalt | Co | nickel | Ni | vanadium | V |
| columbium | Cb | phosphorus | P | tungsten | W |
| copper | Cu | selenium | Se | zinc | Zn |
| gold | Au | silicon | Si | zirconium | Zr |
| iron | Fe | silver | Ag | | |

## 10.1 metallography

*Heat-treatment* refers to the heating and cooling operations performed on a metal, a plastic, or a ceramic for the purpose of changing its properties. Such property changes arise as a result of microstructural changes in the material. In the case of metals, some of these changes are on a sufficiently large scale to be observed under a microscope, though a few are not.

The microscopic study of metals is called *metallography*, and it has been one of the most important contributors to advances in metallurgy. Metallographic examination of a metal discloses a wealth of information quickly and economically, including such details as carbon content, grain size, identification of the metal, defects, inclusions, and its previous fabrication and heat-treatment history. Metallography is the detective work of the broad field of metallurgy.

Little of significance is disclosed by looking at a piece of unprepared metal under a microscope, however. To disclose the microstructure, the metal sample must first be carefully polished by suitable techniques and then etched in a suitable corrodent. The corrodent attacks the different micro-

# MICROSTRUCTURE
# AND HEAT-TREATMENT
# OF METALS

# 10

constituents in the metal at different rates and also attacks the grain boundaries. As a result the separate grains are disclosed in outline and the different constituents appear in different colors or different shades of black, white, or gray. Thus in Fig. 9.11 the carbides in the steel appear dark because of the etch.

## 10.2 recrystallization and annealing

Above the transformation temperature carbon steel is in the austenite phase. When steel in the ferrite phase is heated into the austenite region, the original ferrite grains must be replaced with austenite grains. The new grains nucleate at grain boundaries or other regions of high energy concentration, and as heating continues, the grains grow to a size visible under the microscope, gradually consuming the original grain structure. If the heating is stopped at an early period, a fine-grained structure results, and except for metals to operate at high temperatures, this type of structure is the most desirable one for both highest strength and highest notch-toughness. Prolonging the heating period will promote larger grains.

Now any metal can be made to recrystallize in the above fashion, even without a phase change, if it has previously been plastically deformed. The recrystallization temperature is defined as the lowest temperature at which grains free of the previous distortion stresses appear in the microstructure of the previously plastically deformed metal. The recrystallization temperature is subject to many influences, the principal ones being these:

1. the severity of the plastic deformation
2. the temperature at which plastic deformation occurs
3. the presence of alloying elements and inclusions

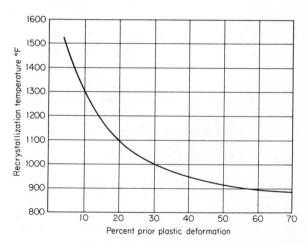

Fig. 10-1  Effect of plastic deformation on the recrystallization temperature of a low carbon steel.

Figure 10.1 shows the effect of prior deformation on the recrystallization temperature of low carbon steel. For zero plastic deformation the recrystallization temperature corresponds to the transformation temperature; increasing deformation lowers the recrystallization temperature.

APPROXIMATE MINIMUM RECRYSTALLIZATION
TEMPERATURES

| lead | below 70°F | nickel | 1110° |
|---|---|---|---|
| zinc | 70° | molybdenum | 1650° |
| aluminum | 300° | tantalum | 1830° |
| copper | 390° | tungsten | 2200° |
| iron | 840° | | |

The above table suggests that minimum recrystallization temperatures are approximately proportional to melting points. It should be noted that the recrystallization temperature imposes some kind of a service temperature ceiling upon a metal. If the metal part is operated above its recrystallization temperature, it will tend to recrystallize into a single-grained metal part, a condition of great weakness and brittleness.

Plastic deformation followed by recrystallization provides a method of obtaining grain refinement in metals and alloys which do not undergo a phase change. The recrystallization serves also as an anneal, which is a relaxation or stress relief for deformation stresses.

## 10.3 annealed carbon steel

Below the transformation temperature, there are two constituents observable in hypoeutectoid (less than 0.8 percent carbon) carbon steel: iron which is called *ferrite* (bcc), and iron carbide, $Fe_3C$, which is called *cementite*. Figure 10.2 shows the appearance of ferrite under the microscope. Cementite contains 6.67 percent carbon. All the carbon in the steel can be considered to be taken up as cementite, since at room temperature only a few thousandths of a percent of carbon is dissolved interstitially in ferrite.

The cementite is found in grains of *pearlite* in a peculiar fashion. Pearlite grains are alternate laminations of ferrite and cementite. Figure 10.3 shows an AISI 1080 plain carbon steel. This steel contains only pearlite grains, the only ferrite present being in the laminations.

The amount of pearlite in a hypoeutectoid carbon steel is proportional to the carbon content. A pure iron can have no pearlite since it has no carbon or iron carbide. The composition of pearlite is 0.8 percent carbon, the

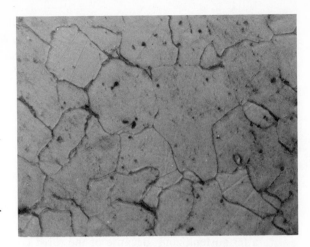

Fig. 10-2　Ferrite.

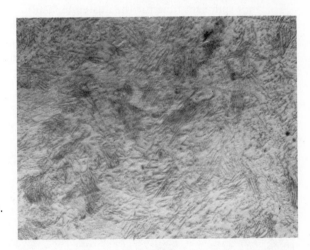

Fig. 10-3　Pearlite.

eutectoid percentage. Therefore a steel with 0.8 percent carbon will show 100 percent pearlite under the microscope, like Fig. 10.3. A steel with 0.4 percent carbon will be 50 percent pearlite and 50 percent ferrite. The metallurgist can thus estimate the carbon content of a carbon steel by looking into the microscope and by estimating the fraction of pearlite present. The method obviously is not applicable to hypereutectoid steels, but since these are less used, they will not be discussed.

We can now turn to the heat-treating processes of hardening and softening steels. Softening is called *annealing*, though annealing often has the additional purpose of stress-relieving. In annealing, a steel is first heated about 50°F above the transformation temperature (also called the critical temperature), held at the temperature a sufficient time to dissolve carbides in the austenite, then allowed to cool slowly in the furnace. The structure of an annealed steel is the mixture of pearlite and ferrite just discussed. This is the softest condition of steel and the condition in which steel usually is supplied to buyers by the steel mill or the warehouse. A brief scrutiny of Fig. 9.6 will reveal that the annealing temperature of a carbon steel can be closely estimated from the carbon content.

## 10.4 hardening of carbon steels

To harden a steel, the metal is heated to about 50° above the transformation temperature, held at that temperature for a sufficient time to dissolve carbides in the austenite, and then quenched rapidly in water or oil. The difference between hardening and annealing therefore is entirely in the cooling rate after austenitizing. Because of the rapid cooling rate of hardening, there is not sufficient time for the austenite solution to dissociate into the usual ferrite and cementite. What comes down with sufficiently rapid cooling is a supersaturated solution of carbon atoms interstitially trapped in a body-centered tetragonal structure of iron. This suddenly frozen solution is called *martensite*. The microstructure of martensite is shown in Fig. 10.4; it resembles a loose pile of straw in appearance.

Fig. 10-4 Martensite.

The transformation from austenite to martensite does not occur at the transformation temperature between ferrite and austenite. Instead, the martensite transformation occurs over a *range of temperature*. Austenite may begin to transform to martensite at 800°F for a low alloy steel. As the temperature continues to fall, more martensite is formed, until at room temperature the structure of the steel may be 99 percent martensite and 1 percent austenite. Complete transformation probably never occurs. With greater alloying additions to the steel, the temperature at which the martensite transformation begins is lowered, and the transformation may be less complete at room temperature. In a high-speed steel, martensite may not begin to appear until a temperature of 600°F is reached, and at room temperature perhaps only 80 percent of the austenite will have transformed. Similarly, an austenitic stainless steel is austenitic by virtue of the fact that the temperature of initial transformation to martensite lies below room temperature.

The temperature at which the transformation to martensite begins is designated $M_s$. One of the many formulas for determining $M_s$ is the following. Percentages of the various elements are substituted for the symbols in the formula.

$$M_s = 930-570C-60Mn-50Cr-30Ni-20Si-20Mo-20W$$

Note that all alloy elements in steel have the effect of lowering the $M_s$ temperature.

The transformation to martensite ceases if cooling is interrupted, and it will proceed further if cooling is resumed. Changing the cooling rate, say from oil-hardening to air-hardening, does not change the $M_s$ temperature.

The hardness of martensite depends on the amount of carbon in the steel, that is, on the amount of carbides, as shown by Fig. 9.4. Thus the maximum hardness obtainable in a steel is built into the steel at the mill and cannot be changed by the fabricator. The fabricator may, however, carburize or case-harden the exterior of the steel by allowing extra carbon to diffuse into the steel.

Certain tool steels will harden at relatively slow rates of cooling. These are called air-hardening steels. Air-hardening characteristics are obtained by using large amounts of carbide-formers such as chromium, tungsten, or molybdenum.

Martensite is hard and brittle, with no ductility, so that there is always danger of cracking because of thermal stresses. Worse still, there is a volume expansion when martensite forms. The part of the steel that is cooling but not transforming is contracting, while the fraction that is transforming is expanding. This makes for dangerous cracking possibilities, either during cooling or soon after the quenching operation. To prevent cracking, tempering is done immediately after hardening.

Tempering, or drawing, is a softening and toughening heat-treatment performed immediately after hardening. Almost always the steel is drawn back to a hardness of $R_C$ 60 or less from the original hardness of martensite, $R_C$ 65. To temper a hardened steel, it is heated to some temperature below the transformation temperature, in the range of 300–1200°F. Since the transformation temperature is not passed, the rate of cooling back to room temperature is not significant. The higher the tempering temperature, the softer and tougher the steel will be. The effect of tempering temperature on the hardness of a carbon steel is shown in Fig. 10.5.

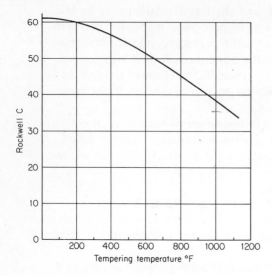

Fig. 10-5 Hardness vs tempering temperature for a waterquenched machinery steel with 0.4 percent carbon and 1.25 percent nickel. (Atlas Steels SPS)

The purpose of tempering may be understood by contrasting it with hardening. In hardening, we obtain the maximum possible hardness in a steel but no toughness or ductility. In tempering, we "sell" some of this hardness in order to buy toughness. If a great deal of toughness in required, then much of the hardness must be given up. If toughness is not especially critical, as in a scriber for marking metal, then most of the hardness can be retained, and we may temper simply for the purpose of relieving the hardening stresses, which may cause later cracking of the steel.

Tempering is a well-known operation because of the beautiful colors that appear on carbon steels at various tempering temperatures. These colors do not always appear on alloy steels. The list below shows how colors serve as a built-in thermometer:

| | |
|---|---|
| straw yellow | 400°F |
| brown | 460 |
| bright purple | 500 |
| dark blue | 550 |
| loss of temper color | 700 |

There are no further temper colors until the temperature of the steel reaches about 1200°F, whien a dull red heat begins to show. This is the start of the red-hot range. The colors result from the formation of transitory oxide films on the steel surface.

Tempering is actually a stress-relieving operation, but it is not called stress-relieving unless the tempering temperature is taken above 1100°F, specifically for the stress-relieving of weldments or machined parts.

Tempering and annealing are both softening operations. However, tempering does not involve a phase change. Further, tempering is performed only after a hardening operation, and its purpose is not to soften but to toughen.

The heat-treating of carbon steels containing 0.9 percent or more carbon follows somewhat different procedures from these and will not be discussed here.

## 10.5  normalizing

Normalizing is a heat treatment somewhat similar to annealing. In annealing, the steel is taken slightly above the critical temperature and then furnace-cooled or otherwise cooled very slowly. In normalizing, the steel is heated about 100°F above the critical temperature and then cooled in air. Hence the cooling rate for normalizing is between the cooling rates for hardening and for annealing. The purpose of normalizing is to produce a harder and stronger steel than annealing can produce and to refine the grain. Certainly the grain size of the steel ought to be finer with the faster cooling rate provided by a normalizing treatment. The microstructure of a normalized steel resembles that produced by annealing, except that the laminar structure of the pearlite grains is no longer visible under the microscope.

## 10.6  summary of the microconstituents of steel

The principal microconstituents of steel have been discussed at some length. The following list briefly summarizes them.

1. Ferrite. Essentially pure iron, body-centered cubic, soft, ductile
2. Cementite. Iron carbide, $Fe_3C$, 6.67% carbon, hard and brittle
3. Pearlite. Laminar ferrite and cementite, 0.8% carbon
4. Martensite. Interstitial carbon atoms in body-centered tetragonal iron. Hard and brittle quenched steel
5. Austenite. Ductile face-centered cubic iron existing above the transformation temperature

A heat-treatable steel must have two characteristics:

1. It must undergo a phase change across a transformation temperature.
2. It must contain carbides, either iron carbide or alloy carbides. Plain carbon steels with less than about 0.3 percent carbon have insufficient carbides to make possible any significant heat-treating effects.

All steels contain fractional percentages of silicon and manganese, but in small amounts these two elements are not important influences on the heat-treating characteristics of steels. Most of the other alloying additions made to heat-treatable steels are carbide-formers and influence the heat treatment procedures.

## 10.7  heat-treatment

## of alloy steels

The basic heat-treating theory of the plain carbon steels is somewhat complex. When this theory is extended to the alloy steels, the basic operations of hardening, annealing, and so on remain, and there are no new phases to consider beyond those already set out. For the alloy steels, however, the heat-treating processes are somewhat modified.

Nickel and manganese lower the transformation temperature of a steel, both elements being austenite promoters. If these elements are present in sufficient amounts, the transformation temperature becomes so low that the steel becomes austenitic at room temperature, as in the case of Hadfield manganese steel and the 300 series stainless steels. The carbide-formers, titanium, chromium, etc., move the transformation temperature to ranges higher than those indicated in Fig. 9.6. Large amounts of tungsten, for example, move the transformation temperature to about 2000°F, as in the case of the high-speed steels, some of which must be heated to 2400°F for heat-treating. Another effect of carbide-forming elements is more advantageous in that the transformation to martensite can be made more slowly: by the addition of over 5 percent of carbide-formers the steel becomes an air-hardening steel. This much less severe type of quench reduces the risk of cracking. A third effect is the lower $M_s$ temperature.

The carbide-forming elements such as tungsten and chromium also have strong effects on the tempering behavior of alloy steels. Figure 10.5 indicates that a carbon steel progressively softens as its tempering temperature increases —as is proven whenever a carbon steel drill bit becomes overheated. The presence of considerable amounts of carbide-formers, however, produces a tempering curve such as the one in Fig. 10.6. The alloy steel may not soften at all until a temperature of 1000°F is reached, or it may actually increase in hardness in the region of 900°F. This highly advantageous effect is called "secondary hardening" and is explained as follows. The addition of large amounts of carbide-forming alloys reduces the start of martensite transformation and is a cause of incomplete transformation of austenite to martensite by the time room temperature is reached. On tempering to a high temperature such as 900°F, a temperature is reached at which the mobility of the consti-

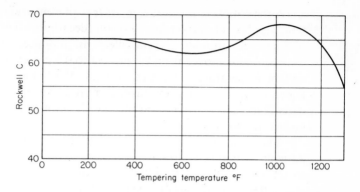

Fig. 10-6 Secondary hardness developed in tempering to 1000°F. The steel is a high-speed steel consisting of 1.30 percent C, 9.5 percent W, 5.2 percent Mo, 3.5 percent Cr, 10 percent Co.

tuents of the steel is improved, and the untransformed austenite completes its transformation to martensite, with increased hardness.

Nickel and manganese enlarge the austenite region; the carbide-formers restrict the austenite field. Chromium is one of the more common alloying additions to steel, and its effect on the austenite region may be considered typical. The effect is shown in Fig. 10.7. The lines marked 0 percent Cr are the boundaries of the austenite field for a plain carbon steel. A steel may be hardened or annealed from any temperature within this austenite area, but only the lowest possible temperatures within the field are used, those just

Fig. 10-7 Effect of chromium on the austenite range.

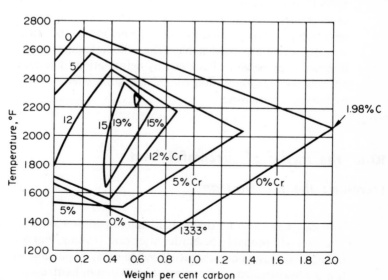

above the critical temperature, in order to prevent grain growth, scale formation, and loss of carbon by oxidation (decarburization). With the addition of 5 percent chromium, several effects are produced. The eutectoid carbon composition is moved from 0.8 percent to 0.5 percent carbon. The transformation temperature is increased to about 1500°F. Finally, the whole austenite region is narrowed. For example, there is no austenite region for 5 percent chromium and 1.4 percent carbon, which means that such a steel could not be heat-treated.

The next boundary shown is for 12 percent chromium. Again, the austenite region is reduced in area, the transformation temperature is raised to the neighborhood of 1600°F, and the austenite region pinches out at about 0.9 percent carbon. Type 410 stainless steel contains 12 percent chromium and 0.15 percent carbon. The diagram shows that this steel can be hardened, since it has an austenite region.

In the boundary for a 15 percent chromium steel it will be noted that a steel containing 15 percent chromium, 0.15 percent carbon, and no other alloying elements would not be martensitic (heat-treatable). The austenite region is being pinched out on both the left-hand and the right-hand side. Finally, the boundaries for a 19 percent chromium steel show that the austenite region has nearly disappeared; if such a steel is to be hardened, the carbon content must be very closely controlled, and so must the heat-treating temperature. Type 430 stainless steel, containing 14–18 percent chromium and a maximum of 0.12 percent carbon, is without an austenite region and the accompanying transformation temperature. This is a ferritic stainless steel and cannot be hardened, at least not by heat treatment.

Because chromium raises the transformation temperature, all the martensitic stainless steels have hardening temperatures in the range of 1700–1850°F. The ferritic stainless steels (type 430, etc.) cannot be hardened because they lack an austenite phase. Only the martensitic stainless steels have two phases separated by a transformation temperature, and only these stainless steels are hardenable. The 300 series austenitic stainless steels, however, may be work-hardened by deformation. The explanation for this effect is the transformation caused by strain effects of a small amount of austenite to martensite.

## 10.8 the tactics of successful heat treating: the I-T diagram

You can never assume that a hardening operation on a steel is successful. The steel will probably be harder after heat treating, but how hard ought it to be? You must know what Rockwell hardness is possible with any steel and test the heat-treated steel against this maximum hardness:

| 0.4% C | Rc 60 |
| 0.8% C | Rc 62–67 |
| 1.0% C | Rc 62–67 |

It is not always possible to obtain a Rockwell of 60 in a 0.4 percent carbon steel: Rc 55 is considered acceptable.

Suppose that the hardened steel shows a Rockwell of 35 after quenching. The heat-treater must decide what has gone wrong. Perhaps there was insufficient residence time in the furnace. Twenty minutes for a $\frac{3}{4} \phi$ is a rough rule, though this time must be varied to suit the alloy, the hardening temperature, and the furnace. If the part is in the furnace too long, or if the furnace is at a temperature more than 100° above the critical temperature, then the surface of the part may be decarburized because of oxidation of the carbon at the surface. The most likely possibility is that the heat-treater was too slow in transferring the piece from the furnace to the quench. Slow quenching softens the steel and fast quenching hardens the steel. Intermediate quenching gives intermediate hardness. This leads to the question of how fast is fast enough for quenching?

To set up the conditions for successful heat treating, a tactical "map" is required that will predict at all temperatures exactly what a steel will do. Such a "map" for a 1080 water-hardening steel and for an oil-hardening steel is given in Fig. 10.8. Such diagrams are called isothermal transformation or I-T diagrams and are available for all commonly used machinery steels.

Consider the I-T diagram for 1080 steel. Note the critical temperature line at 1333° (740°C) marked $A_s$ (start of austenite formation on heating). The austenite region above this critical temperature line is designated A. Note also the $M_s$ line (start of martensite transformation) at about 430°F. Finally note also the "nose" shape of the two phase boundary lines. Between the two phase boundary lines a mixed phase exists, A + F + C (austenite plus ferrite plus cementite); to the right of the I-T diagram the phase is F + C, ferrite plus cementite, which is pearlite.

Using this I-T diagram, consider an annealing operation in which the workpiece is cooled from the critical temperature to room temperature in two hours. The cooling rate of two hours will take the material through the two phase boundaries into the F + C or pearlite region, resulting in the expected anneal.

Hardening a 1080 steel will be more difficult. For full hardening the steel must remain austenitic until it cools to $M_s$, that is, in cooling the steel must not cross the phase boundary lines. This means that the cooling line must remain to the left of the leftward boundary line all the way to $M_s$. But this allows only about 1 second to go from 1333°F past the nose at 1000°F (1000°F is just under a red-hot condition). Since one second is very little time, it is necessary to get the steel out of the furnace and into the quench rapidly. A thick piece of 1080 steel cannot be cooled below 1000°F in 1 second, except

Fig. 10-8 Isothermal transformation diagrams for a 1080 and a 4140 steel.

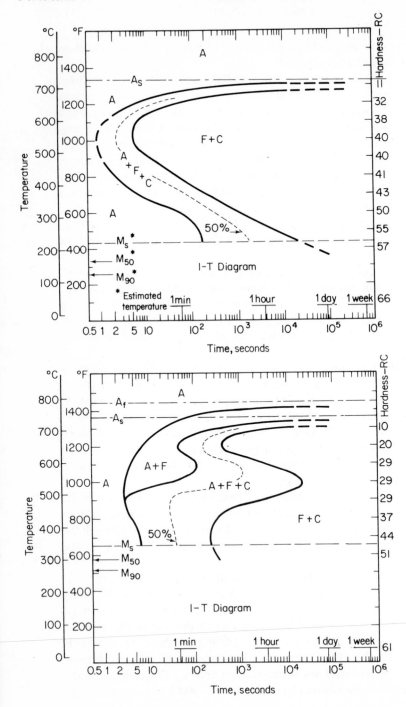

perhaps at its surface. Once the cooling has dropped below 1000°F at the nose, the steel may cool more slowly; from the nose to $M_s$ about 1 minute is allowed.

Suppose that the steel is not quenched quite fast enough. Then in this case the leftward phase boundary is crossed into the region marked A + F + C. The fraction of the steel that is A will harden to martensite, but the fraction that is F + C will be soft. The steel will show an intermediate hardness, probably in the range of $R_C$ 35.

Successful and predictable heat treating requires the use of the I-T diagram for the steel to be heat-treated. The diagram tells the heat-treater how much time he has to get past the dangerous nose, which for any plain carbon steel is at about 1000°F or just below a red heat.

The I-T diagram provides the three temperatures which the heat-treater must know if the heat-treating operation is to be under control:

1. critical temperature ($A_s$), above which the furnace temperature must be set
2. the "nose" temperature, at which the heat-treating operation may fail by insufficiently rapid cooling
3. the $M_s$ temperature, at which transformation begins. This is the temperature below which the steel may crack

Skilled heat-treaters use the I-T diagram to control a number of special quenching methods that ensure hardness without cracking the steel. These special methods are imaginative and interesting, but they cannot be discussed here.

The second I-T diagram is for the low-alloy 4140 steel used in socket wrenches and other hand tools. Over 2 seconds are allowed to pass the nose, so that a slower cooling rate will still give full hardness. This more extended cooling rate allows 4140 to be oil-quenched. Oil gives a slower cooling rate with less distortion and less tendency to crack, and of course the ability to harden deeper. Low-alloy steels and machinery steels are almost always oil-hardening. A high-alloy steel allows still more time to pass the nose, so that such steels can be air-hardened. Stainless 410, with 12 percent chromium, allows over 3 minutes to pass the nose. The difference between a water (W), oil (O), and air (A) hardening steel then is simply a matter of time to the nose. More time is obtained by alloying. If a steel must be hardened to great depth, and not just near its surface, then it must have sufficient alloy content to ensure deep-hardening, since the interior of the steel cools more slowly than the surface.

When a steel alloy is quenched to martensite from austenite, dimensional changes occur. In martensite the iron atoms are not so densely packed as in austenite. The volume increase is of the order of 4 percent. In a plain carbon steel, the quenched part may increase in length as much as 0.002 in. per foot. The plain carbon water-hardening steels show the largest dimensional changes and the most warpage, but the oil-hardening, low-alloy steels are much better in this respect, some of them being called "non-deforming" steels.

## 10.9 heat-treatment of

## nonferrous metals

To reiterate, those steels are heat-treatable which provide the two requirements of carbides and a transformation temperature. Although no nonferrous metal offers the wide heat-treating possibilities of steel (of which only
the basic methods are discussed in these pages), some of the nonferrous
alloys are heat-treatable. While the heat-treating approach is different from
that of steels, there are similarities. A phase change is required for heat
treating, as well as an alloy ingredient which will deposit in different forms or
phases, such as carbon does in steel.

No pure metal can be heat-treated, just as pure iron is not heat-treatable,
although pure metals if work-hardened can be reannealed. Most nonferrous
alloys do not undergo a suitable phase transformation on heating, while
for others the phase transformation is too close to the melting point of the
metal to be of practical use for heat-treating purposes. For such metals and
alloys, heat treating in order to change the hardness, strength, or ductility is
not possible. The only hardening methods available in such cases are cold-
working and alloying, though alloying is not a method available to the
fabricator.

There is only one industrial alloy of importance that can be heat treated
in a manner similar to steel. This is aluminum bronze with a composition of
approximately 10 percent aluminum and 90 percent copper. With furnace
cooling this alloy gives an annealed structure like pearlite, and it may be
quenched to a martensitic structure for strength and hardness.

The more important of the heat-treatable nonferrous alloys, in addition
to aluminum bronze, include the following:

> copper with 2% beryllium
> aluminum with 4% copper
> zinc alloys of aluminum
> aluminum alloys of magnesium
> certain alloys of nickel

The method of hardening these alloys is termed *precipitation hardening* or
*age-hardening*. Like other metals, heat-treatable or not, these alloys may
also be softened by annealing. Two stages are involved in producing the
hardening effect: *solution heat treatment* and *aging*. The method of precipitation hardening will be explained here in terms of the aluminum alloy 2024,
also named 24S, a high strength aluminum containing 4.5 percent copper
plus lesser constituents.

It is a familiar fact that water dissolves more sugar as the temperature
is raised. This is also true of many solid solutions of one metal in another,
though certainly not true of all such solutions. More carbon can be dis-

solved interstitially in body-centered cubic iron at high temperatures (a maximum of 0.025 percent) than at room temperature (0.008 percent). This is also the case with the 4.5 percent copper in 2024 aluminum. Figure 10.9 is a part of the equilibrium diagram for copper alloys of aluminum. At 1018°F aluminum can dissolve a maximum of 5.65 percent copper. On cooling, aluminum can dissolve progressively less copper, as shown by the phase boundary displacing to the left at lower temperatures. The $\kappa$ or kappa phase is a solution of copper in aluminum. The $\theta$ (theta) phase is the intermetallic compound $CuAl_2$. Now suppose that the 2024 alloy is heated to 1000°F into the $\kappa$ region. The copper is held in solution in the solid aluminum.

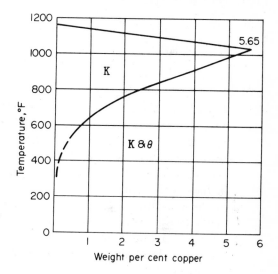

Fig. 10-9 Equilibrium diagram of aluminum alloyed with copper.

To harden this alloy, it is rapidly cooled. This cooling does not produce a martensitic structure, but freezes the solution, holding the copper as a supersaturated solution by not providing time for the copper to precipitate out of solution. This action is called *solution heat treatment*. The metal is not in its hardest condition yet however. A second process called *aging* must follow. Aging occurs over a period of hours or days, during which the supersaturated copper gradually comes out of solution as copper aluminide, $CuAl_2$. By means of precipitation hardening, the ultimate tensile strength of 2024 aluminum may be doubled, from approximately 30,000 psi to approximately 60,000 psi.

These then are the requirements for precipitation-hardening. There must be a decreasing solubility of the alloy element in the base metal as the temperature is lowered, so that a supersaturated condition can be produced by rapid cooling. With time, the solute atoms precipitate out to form a separate phase which is an intermetallic compound, such as $CuAl_2$. The precipitation process may occur naturally over a period of time at room temperature, but sometimes it requires the extra energy provided by a temperature higher than room temperature.

Precipitation-hardening effects are quite pronounced with beryllium copper, which can be heat treated to the remarkable strength level of 220,000 psi.

## 10.10  martensitic precipitation-

## hardenable steels

The martensitic precipitation-hardenable steels are a more recent development and include certain precipitation-hardening (PH) stainless steels and the maraging steels.

The martensitic PH steels have a composition such that martensite forms in the usual manner at a temperature in the range of 100–250°F. This is obtained by using slightly lower nickel and manganese than the austenitic stainless steels. These steels have a lower carbon content than the usual martensitic steels, 0.12 percent maximum. The hardness of martensite depends solely on the carbon content (Fig. 9.4). As a result of the low carbon in these PH steels, a martensite is produced which is soft enough to be machined in the range of $R_C$ 25 to 30 by using a hardening temperature of about 1900°F. The precipitation-hardening process precipitates a complex intermetallic compound usually involving titanium and aluminum, giving final strengths of about 200,000 psi with about 10 percent elongation in 2 in. and $R_C$ 40 to 45.

The maraging steels, mentioned in the previous chapter, are heat-treated similarly to the PH steels, that is, they are first quenched like any hardenable ferrous alloy to produce martensite, then precipitation heat-treated (maraged) like any hardenable nonferrous alloy to produce the intermetallic compound dispersion. Strength levels of about 300,000 psi are possible with maraging steels, at the relatively low hardness level of $R_C$ 50. The martensitic quench is produced simply by air cooling from 1500°F with no risk of cracking and virtually no distortion. Machining is done in the martensitic condition, after which the steel is maraged at the low temperature of 900°F with minimal warping and oxidation.

The following table gives analyses of Stainless W, a PH stainless steel, of 18 percent nickel maraging steel, and of Inconel 718, a precipitation-hardening nickel alloy chiefly used in gas turbines. Certain similarities in these analyses should be noted. All contain nickel, aluminum, and titanium, the metals involved in the intermetallic compound that appears after precipitation-hardening of these alloys.

| Element | Stainless W | 18% maraging steel | Inconel 718 |
|---|---|---|---|
| C | 0.12 max | 0.03 max | 0.04 |
| Mn | 1.00 max | 0.10 | 0.20 |
| Ni | 6–8 | 18.5 | 52.5 |
| Cr | 16–18 | – | 19 |
| Fe | balance | balance | 18 |
| Al | 1.0 max | 0.10 | 0.6 |
| Ti | 1.0 max | 0.40 | 0.8 |
| Co | – | 9.0 | – |

## 10.11 how to read and interpret a steel analysis

The chemical analysis of a steel reveals a great deal of information about the steel, including the following:

(a) $M_s$ temperature.
(b) the relative weldability, machinability, and hardenability of the steel.
(c) the probable hardening temperature.
(d) the quenching medium to be used in hardening.
(e) the uses of the steel.

This kind of metallurgical detective work requires a degree of practice, however. To see how to read a steel analysis, consider a few examples.

1. 0.80% C, 0.30% Si, 0.60% Mn

All steels contain fractional percentages of silicon and up to about 1.65 percent manganese. These two elements therefore disclose little about this steel, but if they dissolve substitutionally in the ferrite lattice, they strengthen the steel (they may instead combine with oxygen and sulfur). The significant element in this analysis is carbon.

(a) This is a high carbon steel, and therefore it is presumably a tool steel.
(b) Being a plain carbon steel, it is water-hardening.
(c) Its critical temperature is 1333°F; therefore it should be hardened from about 1400°F.
(d) If annealed, its microstructure will be 100 percent pearlite.
(e) It has sufficient carbon to be hardened to $R_c$ 65.
(f) Since the carbon is high, this steel will tend to crack if welded.

This steel might be used for cold chisel stock or other cold-work hand tools such as scribers. It is a W-1 tool steel.

2. 0.90% C, 0.50% Cr, 0.50% W, 0.2% V, with small amounts of silicon and manganese

(a) The high carbon content, with small amounts of chromium, tungsten, and vanadium, all carbide-formers, indicates hard carbides when the steel is hardened. This must be a tool steel.
(b) Since it is almost a plain carbon steel, it is restricted to cold-working tools, such as punches and dies.
(c) The addition of small amounts of alloying elements signifies an oil-hardening tool steel.
(d) It will harden to the maximum steel hardness of $R_c$ 65.
(e) Being little different from a plain carbon steel, presumably it would be quenched from about 1400°F.

This is an 0–1 cold-work steel.

3. 0.73% C, 18.0% W, 4.0% Cr, 1.1% V, with silicon and manganese

(a) Obviously a very high alloy steel. The alloy ingredients are all carbide-formers. This is a hot-working tool steel for such purposes as cutter bits for machining or for forging dies.
(b) Because of the high alloy content, this steel will exhibit secondary hardening when tempered and should not lose hardness below about 1000°F.
(c) With such a high tungsten content, it probably must be hardened from about 2200°F or higher.
(d) Carbide-forming alloy ingredients in excess of 5 percent usually indicate an air-hardening steel.

This is a T-1 high speed steel.

4. 2.25% C, 12.0% Cr, 1.0% V, 1.0% W, minor amounts of other elements

(a) This steel obviously is compounded to give large amounts (14 percent) of hard chromium carbides. It will therefore be a wear-resistant steel.
(b) It must be air-hardening because of the high chromium content.
(c) If tempered, it should remain hard at fairly high temperatures.
(d) Although it has almost as much carbon as a cast iron, the carbon must be in the form of chromium carbide rather than graphite.
(e) It must be hardened from a high temperature.

5. 0.08% C, 12.0% Cr

(a) This has the same chromium content as the previous steel, but there is little carbon. Chromium carbides cannot be formed in any significant amount. This cannot be a tool steel.

(b) The very low carbon suggests a stainless steel.

(c) This is a ferritic stainless steel.

6. 0.15% C max, 12.0% Cr

This is a martensitic stainless steel.

7. 0.15% C max, 13.0% Cr, 0.15% S

(a) A martensitic stainless steel.

(b) The high sulfur content indicates a free-machining grade.

8. 0.10% C, 0.35% Mn, 0.15% Si, 0.25% Cu, 0.35% Ni, 0.10% Cr

(a) A very low carbon steel, with an almost complete absence of carbide-formers. This cannot be a tool steel. Neither can it be a stainless steel.

(b) Such elements as copper, nickel, and silicon form substitutional solid solutions in the iron lattice, their effect being to strengthen the steel.

(c) This is a high strength, low alloy, construction steel.

9. 0.13% C max, 9% Ni

(a) Low carbon with high nickel. Nickel does not form carbides, but it does lend toughness to a steel, especially at low temperatures.

(b) This steel is a construction steel used for cryogenic purposes.

10. 0.30% C, 1.0% Cr, 0.20% Mo

(a) There is sufficient carbon for heat-treating purposes, though the carbon is not high.

(b) There is not enough carbon for tool steel use. Also, the alloy content is small.

(c) This is a high strength machinery steel for automotive parts, agricultural equipment, etc. (actually SAE-AISI 4130).

Now try to interpret the following steel analyses. You must be on guard for certain clues, for example, sulfur means a free-machining alloy. Small amounts of silicon and manganese have no special significance.

The following procedure is suggested:

1. *Carbon analysis.* Separate these 20 steels into low carbon, medium carbon, and high carbon classes.

2. *Rockwell hardness.* For carbon contents of 0.4 percent or greater, estimate the maximum hardness obtainable by heat treating.

3. *Furnace temperature setting for hardening.* For the plain carbon steels and the alloy types containing about 1 percent alloy additions, estimate the furnace temperature setting for hardening. Using Fig. 9.6, make the furnace temperature about 75° above the critical temperature. Do not

attempt to estimate the furnace temperature for a high alloy steel or for steels with less than 0.2 percent carbon.

4. $M_s$ *temperature*. Determine $M_s$ for steels nos. 1 and 12. Decide those steels that are probably austenitic at room temperature because of their high nickel content of 9 percent or more; include no. 14 in this list because of its high Mn. After the carbon analysis, look at the alloy analysis.

5. Separate the steels into plain carbon, low alloy, and high alloy.

6. Divide the alloy steels into two types: primarily carbide-formers and primarily non-carbide formers.

7. Finally, decide a probable area of use for each alloy.

1. 0.2% C, 0.55% Ni, 0.50% Cr, 0.20% Mo
2. 0.3% C, 3.0% Cr, 9.0% W, 0.3% V
3. 1.25% C
4. 0.75% C, 17.0% Cr
5. 0.18% C, 0.30% Si, 0.90% Ni, 0.60% Cu
6. 0.30% C, 3.5% Ni
7. 1.20% C, 1.75% W, 0.75% Cr
8. 0.50% C, 1.40% Mo, 5.0% Cr, 1.0% V, 1.5% Ni
9. 0.80% C, 18.0% W, 4.0% Cr, 2.0% V, 8.0% Co
10. 0.15% C max, 0.15% S min, 18% Cr, 8–10% Ni
11. 0.70% C, 0.25% Mo, 0.75% Cr, 1.50% Ni
12. 0.10–0.15% C, 0.50% Mn, 0.03% Si
13. 0.025% C (max), 1.8% Mn, 20.6% Cr, 9.7% Ni
14. 0.6% C, 13.0% Mn, 4.0% Ni
15. 3.0% C, 15.5% Cr, 0.7% Mo
16. 0.025% C max, 0.4% Si, 19.5% Cr, 9.0% Ni, 2.3% Mo, 0.7% Cb
17. 0.06% C, 17.2% Cr, 0.90% Mo
18. 0.11% C, 11.75% Cr
19. 0.19% C, 1.1% Mn, 0.30% Si, 0.4% Cu, 0.5% Cr
20. 0.10% C, 15.0% Ni, 34% Cr

## PROBLEMS

1   What is the difference between tempering and annealing?

2   Microscopic examination of a plain carbon steel indicates that approximately 30 percent of the microstructure is pearlite. What is the approximate carbon content of the steel? (The carbon content of the steel is less than 0.8 percent.)

3   What purpose is served by tempering a steel?

4   A weld is laid down in two passes. Welding conditions are the same for both passes. Microscopic examination discloses that the first pass has a finer grain than the second pass. Can you explain why?

5   Explain the difference in appearance under the microscope between martensite and pearlite.

6   What are the two constituents of pearlite?

7   How is it possible that the martensite formed in a maraging steel can be easily machined?

8   Using Fig. 9.6, estimate the temperature at which you would set a hardening furnace for the purpose of hardening the following plain carbon steels: (a) AISI 1020, (b) AISI 1040, (c) AISI 10100.

9   Calculate the $M_s$ temperature of the following steels:
(a) stainless 304: 0.08% C, 18% Cr, 8% Ni, 2.0% Mn, 1.0% Si
(b) type O-1 tool steel: 0.90% C, 0.50% W, 0.50% Cr, 1.0% Mn
(c) type T-1 high-speed steel: 0.70% C, 18.0% W, 4.0% Cr, 1.0%V

10  Give a metallurgical explanation of the secondary hardening effect found in some high alloy tool steels.

11  What is the usefulness of secondary hardening?

12  What similarities are there between the hardening of a steel and of a hardenable aluminum?

13  Explain how a maraging steel is fully heat-treated.

14  Define the terms cementite, ferrite, austenite, and martensite.

15  A 1080 steel can be full-hardened to a depth of only $\frac{1}{4}$ in., and a 4140 will full-harden to a depth of $\frac{1}{2}$ in. Explain this difference in hardenability in terms of time past the nose of the I-T diagrams.

16  A 1080 steel is quenched to 800°F and then held at 800°F in a salt bath for 1 hour. Consult the I-T diagram. Does this procedure harden or soften the steel?

17  Explain the heat-treating significance of the three temperatures of an I-T diagram: $A_s$, nose, and $M_s$.

18  Suppose that the $M_s$ of an alloy steel is close to room temperature. What effect would this characteristic have on the amount of austenite that does not transform to martensite?

19  (a) The following hardening procedure, called interrupt quenching, would reduce warping and cracking in a 1080 steel: quench in water to 800°F and then in air to room temperature. If the quenched piece is not very heavy, do you estimate that the piece can be full-hardened by this method? Try this method on a $\frac{3}{4}$-in. $\phi$ bar of 1080. Why won't this method succeed for a 4-in. $\phi$ bar of 1080? For a 4-in. bar, what quench might be substituted for air?

(b) For interrupt quenching of 4140, would you select air or oil? Consult the I-T diagram.

20  Cold chisels are made of 1080 steel hardened to $R_C$ 65 and then drawn back (tempered) to $R_C$ 55. Why?

21  It requires an hour in the furnace to austenitize a 1040 steel shaft 2 in. in diameter. Why would you not speed up the operation by raising the furnace temperature from the usual 1550–1575°F to 1800°F?

**22** The following is a table of tool steel alloys. Make up a graph with a vertical scale for furnace hardening temperature and a horizontal scale for total alloy and carbon content. Plot each alloy as a point. The points will not lie on a straight line. Draw a trend line through the few highest points and another trend line through the lowest points. The two lines thus enclose a "scatter band" which discloses the general trend. Is there a rough relationship between total alloy content and hardening temperature? Is it linear?

| Analysis | Furnace temperature |
| --- | --- |
| 0.8 C, 0.25 Mn, 0.20 Si | 1450°F |
| 1.05 C, 0.20 Mn, 0.20 Si, | 1450 |
| 0.40 C, 1.10 Mn, 0.20 Si, 0.15 Mo | 1525 |
| 0.40 C, 0.75 Mn, 0.60 Cr, 1.25 Ni, 0.15 Mo | 1550 |
| 0.20 C, 0.80 Mn, 0.25 Si, 0.55 Ni, 0.50 Cr, 0.20 Mo | 1600 |
| 0.30 C, 0.80 Mn, 0.25 Si, 1.65 Cr, 0.40 Mo | 1550 |
| 0.15 C, 1.0 Si, 1.0 Mn, 12 Cr | 1800 |
| 0.60 C, 0.50 Mo, 17 Cr, 1 Si, 1 Mn | 1850 |
| 1.60 C, 0.3 Si, 0.25 Mn, 12.5 W, 4.75 Cr, 5 V, 5.5 Co | 2250 |
| 0.78 C, 0.25 Mn, 0.30 Si, 18.5 W, 4.25 Cr, 1.9 V, 8 Co, 0.85 Mo | 2375 |
| 1.05 C, 0.30 Si, 0.25 Mn, 6.25 W, 4 Cr, 6 Mo, 2.5 V | 2200 |
| 0.85 C, 0.25 Mn, 0.30 Si, 6.5 V, 4 Cr, 1.9 V, 5.0 Mo | 2225 |
| 0.75 C, 0.25 Mn, 0.3 Si, 18 W, 4 Cr, 1.1 V | 2300 |
| 0.35 C, 0.3 Mn, 0.3 Si, 9.5 W, 3.25 Cr, 0.4 V | 2100 |
| 0.45 C, 0.75 Mn, 1 Si, 5 Cr, 3.75 W, 1 Mo, 0.5 V, 0.5 Co | 1850 |
| 0.35 C, 0.4 Mn, 1 Si, 1.2 W, 5 Cr, 0.3 V, 1.4 Mo | 1800 |
| 2.25 C, 0.3 Mn, 0.25 Si, 12 Cr, 0.25 V, 0.8 Mo | 1800 |
| 0.9 C, 1.2 Mn, 0.3 Si, 0.5 Cr, 0.5 W, 0.2 V | 1450 |
| 0.6 C, 0.75 Mn, 0.3 Cr, 0.2 Mo, 2 Si | 1600 |
| 0.45 C, 0.25 Mn, 0.3 Si, 2 W, 1.5 Cr, 0.25 V | 1700 |

The light metals are those in the Periodic Table up to and including titanium: beryllium, magnesium, aluminum, titanium, and calcium, potassium, sodium, and lithium. Only the first four are employed as structural metals. The others are too soft and reactive, but because of their chemical reactivity they may be used to extract metals from their ores in the manner that carbon is used with iron ores. Sodium and potassium are also used as molten metal liquid coolants in the sodium-cooled type of nuclear reactor.

## 11.1 the extraction of aluminum

## and beryllium

Though both are light metals, aluminum and beryllium seem to be totally unlike. A comparison of the two is instructive and illustrative of the infinite range of problems with which materials science must contend, particularly in extractive processes.

The most abundant metal in the earth's crust by a wide margin is

# THE LIGHT
# METALS

# 11

aluminum. About 8 percent of any clay is aluminum, so that every excavation is a virtual aluminum mine. For a nonferrous metal, though not for iron, an ore with a concentration of 8 percent metal is a rich ore indeed. Unfortunately, while techniques for extracting aluminum from clay are available, clay is not the cheapest source of aluminum. Extraction costs are cheapest for bauxite ore, $Al_2O_3 \cdot 3H_2O$, a hydrated alumina which is rather fortuitously distributed over the earth's surface. There is, for example, no bauxite in mineral-rich Canada, although this country is a major world supplier of aluminum by virtue of its water-power resources.

Although there is no prospect of a shortage of aluminum, short of removing the first few miles of the earth's crust, the prospect is less favorable for beryllium. This is a comparatively rare metal, like tin, and, also like tin, its reserves will some day be exhausted. About 5 parts per million of the earth's crust is beryllium. The principal beryllium ore, and the only commercial source at present, is beryl; other ores are not important and are found in quantities too small to be mined. Beryl is a beryllium-aluminum silicate, $3BeO \cdot Al_2O_3 \cdot 6SiO_2$, containing about 5 percent beryllium, and is almost always found in association with lithium ores in pegmatite formations. Because processes for concentrating beryl have not been introduced, the only mining process is that of hand-picking large pieces of beryl with cheap labor, unless the beryl is separated as a by-product from lithium minerals.

Both beryl and bauxite are much more stable compounds than iron oxide, and coke will not reduce these minerals to metal in a blast furnace process of the type used for iron. Both metals are more expensive than iron, in part because of smelting costs.

Alumina must be reduced in an electrolytic cell. Such a cell is sketched in Fig. 11.1. Since bauxite is a nonconductor of electricity, electrolytic reduction requires the addition to the cell of an ionizable material. Cryolite, a rare fluoride of aluminum, is placed in the cell and is melted by passing current through it. The bauxite is then added. The anodes of the cell, and the lining of the cell, which is also the cathode, are made of carbon manufactured from petroleum coke, an oil refinery product. The oxygen of the ore reacts with the anodes and is released as carbon dioxide. The operation

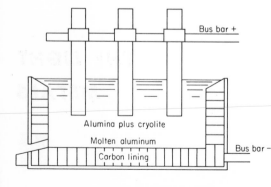

Fig. 11-1  Electrolytic cell for extraction of aluminum from bauxite.

makes heavy demands on the anodes and electric power: a pound of aluminum product requires about 10 kwhr of electric power and over half a pound of anodes. Aluminum smelters therefore are not located at sources of bauxite, but on tidewater near sources of hydroelectric power.

Aluminum then is more difficult to smelt from its oxide ore than iron is. Beryllium is even more difficult to extract. The beryl is converted to beryllium oxide, an oxide even more stable than alumina. The relative stability of these oxides is indicated by their melting points:

| | |
|---|---|
| iron oxide | 2800°F |
| alumina | 3720°F |
| beryllia | 4570°F |

The converted beryllia is converted again to beryllium fluoride and finally reduced to beryllium particles by magnesium in the following reaction. The reaction temperature is above the melting point of beryllium (2370°F).

$$BeF_2 + Mg \rightarrow MgF_2 + Be$$

This reaction is possible because beryllium fluoride is less stable chemically than beryllium oxide. The beryllium product is in particle size and contaminated with over 1 percent magnesium. The alloying of magnesium with beryllium has adverse effects on the properties of beryllium, especially its ductility. Vacuum melting reduces the magnesium content to less than 0.1 percent and produces ingots of beryllium.

This method of reducing beryllium from its fluoride by means of magnesium is used in similar processes to reduce titanium and zirconium from their ores. These metals are reduced from their chlorides, however, in what is known as the Kroll process.

The beryllium, after vacuum melting, has a low level of magnesium, and being in ingot form, would appear ready for rolling or forming to final shape, as other metals such as iron and aluminum would be at this stage. This is not the case. Every operation with beryllium appears to be afflicted with "bugs," and the melting process, in solving or partly solving the magnesium contamination problem, has raised other problems.

First, beryllium, like titanium, is a reactive metal. Both must be melted under vacuum or inert gases, since oxygen, nitrogen, or hydrogen, even in minute quantities, embrittle these metals to the point of uselessness. Their reactiveness is not confined to gases, however. Both metals when molten are a kind of universal solvent and will react with any crucible material. If melted in a graphite crucible, considerable beryllium carbide is formed. If melted in an oxide crucible, even beryllium oxide, the metal will be contaminated with beryllium oxide. It is true that almost all metals will react in some degree with their melting refractories. If aluminum is exposed to refractory walls for a long period, it will reduce silica to metallic silicon, which will alloy with the aluminum. But these metal-crucible reactions are only serious with the reactive metals, chiefly beryllium and titanium. The other reactive metals such as tantalum have melting points so high that they are not extracted by melting.

Although no refractory is entirely satisfactory for melting beryllium, beryllia gives the best performance.

In keeping with the innate genius of beryllium for doing things wrong, still another melting problem arises. This metal freezes in large crystals of a size that must be seen to be believed. In a mold of square cross-section, the ingot may grow only four crystals, one from each side of the mold. Magnesium, cadmium, and zinc, also in Group II of the Periodic Table, tend also to form unduly large crystals but not to the extent that beryllium does. Large crystals are inadmissible in metals because they produce a condition of brittleness, especially in beryllium. This large crystal growth could be controlled in part by a rapid quench, but such quenches invariably split the ingot. There is no phase change or recrystallization that will permit heat-treating to produce small crystals.

There is only one possible solution to the problem of crystal size. The ingot is reduced to small particles, ground to about 200 mesh, then sintered with heat and pressure in dies to produce the required shape.

## 11.2  properties of aluminum

## and beryllium

With problems such as these to contend with, and a production cost of $60 or more per lb, one might well ask what properties beryllium offers that warrant the effort of producing it. For comparison, consider first aluminum.

Pure aluminum is a soft metal without pronounced work-hardening properties and with remarkable ductility. It is not strong unless alloyed. Ultimate tensile strength is only about 10,000 psi for the pure metal, but alloying strengths of 75,000 psi are possible. The modulus of elasticity of aluminum is $10 \times 10^6$, one-third that of steel, and its weight also is one-third that of steel. Because of its uses in aircraft, space vehicles, trains, trucks, ships, conveyors, and shipping containers, aluminum has been termed "the transportation metal." Easy to machine, form, and weld, it is also easy to melt, though it tends to dissolve hydrogen with consequent porosity in castings and fusion welds. The melting point is only 1220°F for pure aluminum, less for alloys. Since it is a face-centered cubic metal, it is not subject to brittle fracture.

By comparison, beryllium is brittle. There appears to be no inherent reason why it should be. It is a hexagonal close-packed metal. So are magnesium, titanium, and zirconium. These metals are ductile and can be rolled into sheet. The same operation should be possible with beryllium. After the expenditure of millions of dollars and 15 years on research, beryllium is still brittle, and rolling is impossible. Best ductility is given by extruded beryllium, but at best elongation is usually only a few percent.

However, beryllium is reasonably strong. Maximum strength is of the order of 100,000 psi, higher than aluminum alloys. The rigidity of beryllium surpasses that of steel, its modulus of elasticity being about $40 \times 10^6$. The very high modulus, high strength, and unusually light weight, two-thirds the specific weight of aluminum, make this an attractive metal for aerospace applications if the problem of brittleness can be solved.

Few if any aluminum compounds are toxic, but all compounds of beryllium are dangerously toxic. Although methods of industrial hygiene are now well understood and applied to the processing of beryllium, there were in earlier times a number of cases of berylliosis in industry. There is at least one authenticated case of a beryllium worker's wife afflicted with berylliosis, though the husband was not. She washed his plant clothing regularly, thus contracting the disease that passed the man by. This incident seems strangely in keeping with the peculiar "personality" of beryllium.

Aluminum can be alloyed with a great many metals. Beryllium is not especially receptive to alloying. Alloying cannot solve the brittleness problem, for an alloy is ordinarily less ductile than the corresponding pure metal.

## 11.3 properties of the light metals

The light metals, like all the light elements, share two other common characteristics in addition to low specific weight:

1. High specific heat, 0.22 Btu per lb for aluminum, 0.25 for magnesium, and a very high 0.52 for beryllium.
2. Low absorption for gamma radiation and for neutrons (with the exception of titanium).

The light metals therefore are used as moderators and as fuel rods in nuclear reactors. The uranium fuel particles cannot be exposed because they corrode and are pyrophoric; they must be canned. Zirconium, zirconium alloys, magnesium, beryllium, aluminum, and austenitic stainless steels are the preferred canning materials. Zirconium and stainless steels have suitable nuclear properties but are not light metals.

Beryllium, when used as a moderator, offers the possibility of some neutron multiplication, since it releases one neutron from its nucleus at a much lower energy requirement than any other metal. This characteristic is used in beryllium neutron-generating devices, which release small amounts of neutrons for analytical work.

The low gamma-ray absorption of the light metals indicates that they are easy to radiograph using either X-ray or gamma-ray techniques. X-ray photography of these metals requires only very low voltages, which for aircraft X-ray inspection may be as low as 50 kv. Figure 11.2 compares the relative X-radiation absorption of beryllium with that of aluminum. Below 100 kv beryllium is much more transparent. Hence the use of beryllium as an X-ray "window" in an X-ray tube.

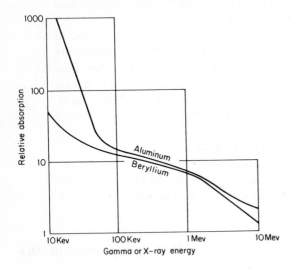

Fig. 11-2 Relative absorption of aluminum and beryllium for high-energy radiation. Both being light metals they are transparent to high-energy radiation above 1,000,000 electron-volts. Note that the differences in absorptivity between different materials become small above 1 Mev; for good contrast in X-ray radiography the lowest practical voltage is used.

## 11.4 aluminum wrought alloys

Since pure metals have better corrosion resistance and better electrical conductivity than alloys, commercially pure aluminum has for its principal uses those applications where corrosion resistance and electrical and thermal conductivity are primary requirements. Because it has a lighter weight than copper, aluminum has tended to make inroads into the wire markets served almost exclusively by copper heretofore, although aluminum has an electrical resistance about 50 percent higher than copper. The thermal conductivity of aluminum is about half that of copper, 1540 Btu/sq ft-in.-°F. It has, however, the significant disadvantages of low strength and has twice the thermal expansion of steel. The high thermal expansion makes for difficulties in the welding and casting of this metal.

Commercially pure grades of annealed aluminum have yield strengths in the range of about 4500 psi, ultimate strength 11,500 psi, elongation 40 percent or more, and low fatigue resistance. It may be cold-worked to strength levels as high as 23,000 psi but with considerable loss of ductility.

In commercial alloys, aluminum is alloyed with silicon, magnesium, manganese, copper, zinc, and chromium. Copper and zinc alloys are heat-treatable and can develop strengths in the range of 65,000 to 75,000 psi. The other types of alloys are hardened by cold-working. The magnesium alloys, which are usually selected for welding, develop ultimate tensile strengths of about 40,000 psi. While only small percentages of alloying elements are used in wrought aluminum alloys, the casting alloys may contain as much as 17 percent silicon, 11 percent copper, or 10 percent magnesium. The hard alloys containing copper are the best machining aluminums, but they have limited ductility for forming and bending.

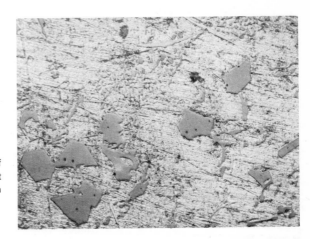

Fig. 11-3 A casting alloy of aluminum containing 13 percent silicon. Individual grains of silicon may be seen.

The aluminum alloys are classified into two groups: the casting alloys and the wrought alloys. The wrought formulations are those used for rolling, extruding, drawing, and forging.

For the wrought alloys, two numbering systems are in use in Canada and the United States. In the United States, the four-digit system is more common. In Canada, a two-digit system is used, the number being followed by the letter S to designate a wrought alloy. The following table compares the two systems:

| | | |
|---|---|---|
| 99.6% purity aluminum | 1S | 1060 |
| 99.0% purity aluminum | 2S | 1100 |
| manganese alloys | 3S to 9S | 3xxx |
| copper alloys | 10S to 29S | 2xxx |
| silicon alloys | 30S to 39S | 4xxx |
| magnesium alloys | 50S to 69S | 5xxx |
| magnesium-silicon alloys | 50S to 69S | 6xxx |
| zinc alloys | 70S to 79S | 7xxx |

Some of the more important alloys of aluminum, with their principal alloy additions, are these:

1. 3S, 3003, with 1.0–1.5% Mn. The manganese addition provides 50% more tensile strength.
2. 24S, 2024, with 3.8–4.9% Cu, 1.5% Mg. A high strength, heat-treatable aluminum commonly used in aircraft work and parts requiring high strength or hardness. Corrosion resistance is poor. This alloy is not usually welded, since welding heat anneals it, causing loss of strength.
3. 28S, 2011, with 5–6% Cu. A free-machining alloy supplied in cold-drawn bar form.
4. 33S, 4043, with 4.5–6.0% silicon. Welding grade aluminum, available as welding wire.
5. 50S, 6063, with approximately 0.5% each of magnesium and silicon. This is an extrusion alloy, familiar as aluminum windows, doors, grilles, and trim.
6. C54S, 5154, with 3.1–3.9% Mg. Welding grade aluminum, chiefly supplied in sheet, plate, and shapes.
7. 65S, 6061, with about 1% Mg, 0.5% Si. Structural aluminum. The geodesic dome of Fig. 11.4 is made of 6061 aluminum.

Fig. 11-4  The geodesic dome over the headquarters of the American Society for Metals in East Cleveland, Ohio. The dome is made of aluminum 6061 tubing.

The following summary is an aid to remembering the important aluminum alloys:

1. for castings—silicon alloys
2. for maching—copper alloys, such as 2011

3. for welding—magnesium or magnesium-silicon alloys
4. for forming—pure aluminum or manganese alloys such as 3003
5. for extrusion—6063 alloy
6. for strength—copper alloys, such as 2024

The temper designation follows the alloy designation, 65S-T6, for example. A "temper" refers to a condition produced in the metal by mechanical working or heat-treatment; the aluminum tempers are produced either by cold-working or age-hardening or both. Only the zinc, copper, and magnesium-silicon alloys can be age-hardened.

The wrought alloys that are not heat-treatable may be supplied either in the "O" (annealed) temper or the "F" (as-fabricated) temper. Mechanical properties are not guaranteed in the "F" temper. Other tempers include:

H1—work-hardened only
H2—work-hardened and partially annealed
H3—work-hardened and stabilized

There may also be a second digit in the temper number, for example, H12. This second digit indicates the degree of work-hardening:

1—eighth hard
2—quarter hard
4—half hard
6—three-quarter hard
8—full hard
9—extra hard

Thus 6061-H12 indicates a magnesium-silicon alloy work-hardened to quarter hard condition.

The heat-treatable alloys such as 2024 (24S) may also be supplied heat-treated for hardness and strength; this temper is indicated by T:

T4—solution heat-treated
T6—solution heat-treated and artificially aged

## 11.5 aluminum casting alloys

Aluminum casting alloys are formulated for sand casting, permanent mold casting, and die casting. These alloys are designated by a three-digit number, such as 108, 356, etc. As was the case with stainless steel alloy numbers, there is little system in these numbers, and the problem is complicated further by the fact that casting alloys have more than one numbering system. Perhaps the most used system is that of the Aluminum Association. In this system, silicon casting alloys have numbers up to 99, silicon-copper casting alloys 100 to 199, magnesium casting alloys 200 to 299, and silicon-manganese 300 to 399. Copper alloys are preferred for their hardness when machinability is a factor; silicon alloys are useful when fluidity to follow intricate shapes is desired. The high magnesium alloys require care in melting to prevent excessive oxidation loss of the magnesium content.

Aluminum is prone to cracking and tearing when hot, that is, it is "hot short." Most aluminum casting alloys are not notably strong in tension, only a few exceeding 40,000 psi at room temperature. This low strength, which drops rapidly at higher temperatures, coupled with a high coefficient of expansion, indicates a tendency toward hot shortness. Aluminum also has a strong tendency to pick up hydrogen in the molten condition, and this is a cause of porosity in castings and welds. Moreover, aluminum castings are not pressure-tight, and to obtain pressure-tightness they must be impregnated with sodium silicate, epoxy, or other material.

## 11.6  use of aluminum

By and large, magnesium alloys of aluminum are selected for weldability, copper alloys for machinability, and silicon alloys for castability. For corrosion resistance or electrical conductivity, the commercially pure material is best.

Aluminum has some special welding problems. The high coefficient of expansion promotes distortion during welding operations. Because of the metal's very high thermal conductivity and low electrical resistance, it is difficult to resistance-weld (spot-weld) aluminum, and currents as high as 100,000 amps may be necessary for such operations. Another preoccupation in welding aluminum is the removal of the refractory oxide skin on the metal.

Under conditions of repeated stress, aluminum parts must be carefully designed against fatigue failure, especially in welded aluminum structures. This metal does not show a true endurance limit. Figure 11.5 shows typical fatigue curves for aluminum alloys.

Aluminum has a high reflectivity for infrared and visible radiation. The metal is frequently deposited on reflectors by a process of evaporation and deposition under vacuum, as in the manufacture of sealed-beam headlights. The high reflectivity also accounts for the use of aluminum foil in insulation materials, the foil serving to reflect radiant heat.

In the form of finely divided flakes, aluminum serves as an excellent paint pigment, providing an impervious coating with excellent hiding power on either wood or metal.

The increased use of aluminum in automobiles deserves mention. As this book is written, aggressive efforts are being made to use aluminum 3003

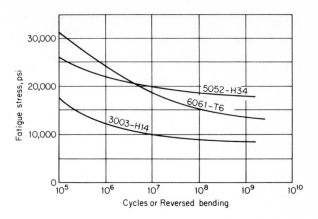

Fig. 11-5 Typical fatigue curves of aluminum alloys. The absence of an endurance limit is a characteristic of these curves.

and 6063 for automobile radiators, and one would assume that the efforts will bear fruit. The cast aluminum engine block has been successfully used; such blocks are usually high silicon alloys such as 356 (9 percent silicon, 3.5 percent copper). Considerable weight saving is possible by substituting aluminum for cast iron in diesel engines. Aluminum pistons are of course standard in automobiles.

As a final salute to aluminum and its alloys, we may note that this metal and United States Steel's T-1 low alloy steel made possible the design of a bulldozer for the U.S. Army that is light enough to be carried by helicopter or dropped by parachute. This flying bulldozer disassembles into three large pieces, which are individually dropped from the air, then quickly assembled on the ground. This remarkable creature weighs only 20,000 lb.

## 11.7 magnesium

Pure magnesium weighs two-thirds as much as aluminum, or 1 oz per cu in. The $E$-value is also two-thirds that of aluminum, $6.5 \times 10^6$. The melting point of magnesium is 1202°F, about the same as that of aluminum. The thermal conductivity is high, but not so high as in aluminum. Coefficient of expansion is 0.000016 in./in., a very high value. Like most of the light structural metals, magnesium is hcp. While aluminum alloys become too weak for use above 300°F, magnesium alloys can be used at twice this temperature. Figure 11.6 is a micrograph of pure magnesium.

Magnesium is an active metal. Its first extensive use was as incendiary bombs for strategic air warfare. Magnesium chips are readily ignited, though there is little hazard in gas welding of magnesium components. Such a metal

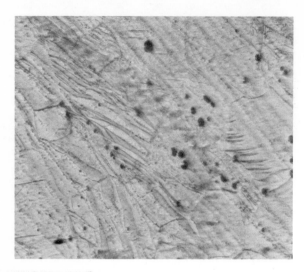

Fig. 11-6  Micrograph of magnesium.

of course has poor corrosion resistance. Its chemical activity, however, is turned to advantage in the use of magnesium anodes for the galvanic protection of hot water tanks, buried pipelines and other structures subject to galvanic attack. It is used also to reduce zirconium, titanium, uranium, and beryllium from their fluorides and chlorides in primary extraction processes. The use of magnesium in the production of ductile cast iron was noted in Chapter 9.

Most magnesium, however, is used for equipment where light weight is a principal requirement, as in aircraft, space vehicles, ladders, portable power tools, luggage, or the dockboards used on warehouse shipping and receiving docks (Fig. 11.7). The large stiffening rings of moon rockets may be as large as 33 ft in diameter; these have been produced in magnesium alloys. Currently the lightest commercial alloy is a magnesium alloy containing 14 percent lithium and 1.25 percent aluminum.

Magnesium has recently taken a place among the tooling materials. The major cost of such tooling devices as fixtures and drill jigs lies not in the material but the manufacturing cost. Magnesium's light weight, weldability, and ease of machining may greatly reduce the cost of manufacture. In use, the light weight of such tooling reduces operator fatigue in handling the fixtures, and it is especially convenient for women operators. Magnesium tooling plate is usually alloy AZ31B.

Alloy designations for magnesium follow a simple system. Two letters represent the chief ingredients, followed by two numbers giving respective percentages to the nearest whole number. A serial letter follows the digits. Thus AZ92A is an Aluminum-Zinc alloy with 9 percent Al and 2 percent Zn. These designations apply both to wrought and cast alloys. The alloy letter code is the following:

Fig. 11-7   A magnesium dockboard providing access to a boxcar from a loading dock for lift trucks.

| | | | |
|---|---|---|---|
| A | aluminum | M | manganese |
| E | rare earths | Z | zinc |
| H | thorium | T | tin |
| K | zirconium | | |

Most of the magnesium alloys are of the AZ type. Rare earth and thorium additions give improved strength at temperatures above 500°F. The amount of thorium used is usually about 3 percent and does not produce a radiation hazard. Zirconium additions of about 0.5–0.7 percent provide grain refinement. Alloy AZ92A is used for normal temperature applications, including sand casting and welding wire. The alloys of highest strength may have ultimate tensile strengths as high as 55,000 psi in extruded articles. Elongation, however, is limited in all magnesium alloys.

## 11.8   titanium

This is the heaviest of the hcp light metals. Only two minerals are of commercial importance: ilmenite, an iron-titanium oxide, and rutile, $TiO_2$. The industrial uses of rutile were outlined in an earlier chapter.

Titanium is a highly reactive metal above 1000°F. Like beryllium,

Fig. 11-8 Micrograph of pure titanium sheet.

molten titanium will attack any refractory oxide: the practice of welding metals as they lie on a firebrick is not possible with titanium.

Titanium can be identified by its light weight, 60 percent of steel, its bluish or silvery tinge, and its blue grinding sparks.

Titanium first began to compete in the metals market when higher strengths at higher temperatures were demanded in high-speed aircraft early in the 1950's. A switch in emphasis from aircraft to missiles almost ruined the market for titanium. However, production now is at a rate of about 15,000,000 lb annually, and the metal is steadily making inroads into the civilian market, where its light weight, strength, and remarkable corrosion resistance at room temperature offer possibilities that compensate for its high price of over $5 per lb. This reactive metal is corrosion-resistant by virtue of an adherent oxide skin such as protects aluminum and the stainless steels.

Titanium has an $E$-value of $15.5 \times 10^6$, half that of steel, and higher than the $E$-values of aluminum and magnesium. It can be alloyed and heat-treated for strengths exceeding 200,000 psi. Pure titanium melts at 3035°F, above the melting point of iron. Thermal expansion and thermal conductivity are quite low and fatigue strength is excellent.

Figure 11.9 shows a stress-strain curve for commercially pure titanium sheet. Both the strength and the elongation, 27.5 percent in 2 in., are excellent.

Titanium undergoes a phase change to bcc when heated above 1620°F. The low temperature hcp phase is termed alpha, the bcc phase beta. It is strengthened and embrittled by minute amounts of carbon, oxygen, and nitrogen. Certain alloying additions such as chromium, molybdenum, vanadium, and iron are beta stabilizers, causing a lowering of the transition temperature. Alloying additions which raise the transition temperature are called alpha stabilizers. These are aluminum, carbon, oxygen, and nitrogen.

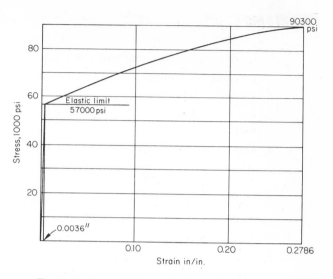

Fig. 11-9 Stress-strain curve for a commercially pure titanium sheet, aircraft quality.

These elements form solid solutions and are employed in nonheat-treatable alloys. Aluminum is a common alpha stabilizer in amounts up to 8 percent, and it is also found in the solution heat-treatable beta alloys.

## 11.9 beryllium

The use of beryllium is still severely limited by its poor ductility. About 80 percent of beryllium production goes into copper alloys, which are discussed in the following chapter. The beryllium coppers are precipitation-hardening alloys, indicative of beryllium's noteworthy capacity as a precipitation-hardening alloy element. Though beryllium nickel has only limited commercial use, it can be heat-treated to a tensile strength of 270,000 psi, with an elongation of 9 percent in 2 in., a combination that almost no steel can match except for the newer maraging steels. This ability to produce hardenable ferrous and nonferrous alloys is interesting when compared with two metallurgical characteristics of beryllium: the beryllium atom is too small to form substitutional solid solutions with other commercial metals, and it is too large for interstitial solid solutions. The hardenable alloys of copper and nickel are possible because of substitutional solid solution at higher temperatures.

A final oddity of beryllium is its low Poisson's ratio. For most metals Poisson's ratio is about 0.3. For beryllium it is about 0.1.

## PROBLEMS

1 Explain why powder metallurgy methods must be employed with beryllium.

2 If beryllium were both nontoxic and ductile, what advantages would it offer in the market for metals?

3 List the general characteristics of the light metals other than low specific weight.

4 Suggest two reasons why aluminum is not easy to spot-weld.

5 What alloying ingredient is added to aluminum for (a) high strength, (b) weldability, (c) machinability?

6 For the metals magnesium, aluminum, titanium, and iron, is there a simple relationship between specific weight and modulus of elasticity?

7 What is the meaning of the magnesium alloy designation AZ62?

8 Exhaust valves of large diesel engines are sodium-filled. Why? What characteristics of sodium make it an excellent choice for this purpose (besides cost: sodium is a cheap metal)? Consult the *Metals Handbook* or other source for data on sodium.

The large group of nonferrous metals and alloys has a very wide range of applications.

The radioactive metals *uranium*, *thorium*, and *plutonium* are chiefly used as nuclear fuels. Thorium however has limited applications an as alloy element in high-temperature magnesium alloys and in tungsten alloys for welding and electronic applications. Uranium was used as an alloy element in steels for improved fatigue strength, but it is no longer employed for this purpose. *Zirconium* also finds its major applications in the nuclear power field, but it is also a useful alloying element for a range of other metals.

*Nickel* and *lead* are as versatile in their applications as steel, though not consumed in the same large tonnages. The uses of *copper* stem chiefly from its high conductivity for heat and electric current and from its remarkable capacity for alloy formation. *Cadmium* and *tin* are more limited in their uses, which are largely in bearing metals, plating, soldering alloys, and copper alloys. *Zinc* has much the same applications as these two metals, and in addition it is well adapted to the die-casting process. *Cobalt* is an alloying element in ferrous and nonferrous alloys and a binder or matrix material in carbide tools. *Manganese* is used not in pure form but as an alloying element.

The most important industrial application of *silver* is its use in the

# METALS
# WITH SPECIAL PROPERTIES
# AND APPLICATIONS

## 12

silver-copper brazing alloys, which can braze most metals together. *Gold* is an occasional alloying element in such brazing formulations. Silver, gold, and the *platinum* group of metals are also used as contacts in instruments and electrical switching devices. *Mercury* is a fluid electrical conductor.

The *refractory metals*, with melting points above 3600°F, are most notable for their high strength, hardness, and stiffness, especially at high temperatures. Finally the *rare earths* include 14 metals very similar in properties, which have not yet fully revealed their industrial possibilities.

## 12.1 nickel

A review of the properties and applications of nickel quickly discloses that this is a most remarkable metal and perhaps the closest approach to an ideal metal. A catalogue of its physical properties makes it appear as a twin brother to steel, however—both mild steel and nickel have approximately the same modulus of elasticity, tensile strength, ductility, endurance limit, coefficient of thermal expansion, thermal conductivity, and electrical conductivity. Also, both metals are ferromagnetic. The melting point of nickel is 2651°F, in the same range as the low carbon steels. If this were the complete personality of nickel, it would be a little-used metal, for it would offer the same properties as steel, but at a higher price, for nickel is not an easy metal to extract from its ores.

Most of the world's nickel is supplied by Canada and Russia from sulfide ores, chiefly pentlandite, an iron-nickel sulfide. The ore is always a copper-nickel ore. Since copper and nickel are similar both chemically and physically, it is difficult to separate the two metals during smelting. This difficulty accounts for the higher price of nickel, about ten times that of steel. In addition to copper and iron, the platinum group of metals, silver, gold, cobalt, selenium, and tellurium, are associated with nickel in the ore and are recovered. Even the sulfur may be recovered as sulfuric acid. The percent of nickel in pentlandite, amid this surfeit of low-grade materials, is only about 1 percent.

The end uses of nickel have the following approximate distribution: about 80 percent of nickel is used in steel alloys such as low alloy and stainless steels, about 10 percent is used in nickel plating, and 10 percent for nickel and nickel alloy uses. Most of the AISI-SAE low alloy steels contain small amounts of nickel. Nickel therefore is an integral element in our complex steel technology, and indeed, steel without nickel is difficult to envisage.

Nickel and steel are alike in the effect of sulfur on these metals. As with steel, sulfur in nickel forms nickel sulfide, which like iron sulfide, melts at a low temperature. Nickel sulfide in a nickel ingot is the last component to

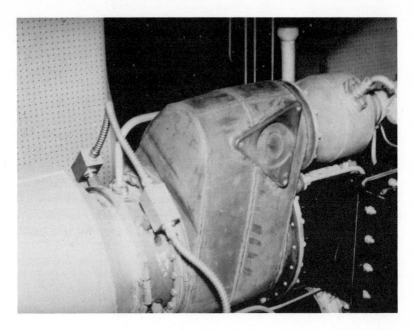

Fig. 12-1   A high nickel alloy combustion chamber of a 60-hp gas turbine engine.

melt and therefore deposits in the grain boundaries. When the ingot is rolled, this low-strength intergranular material causes cracking and crumbling. The problem is cured as it was in the case of steels: manganese is added to combine with the sulfur as manganese sulfide, which deposits as inclusions instead of a grain boundary network. Magnesium may also be used to remove sulfur in nickel. Nickel is severely corroded by sulfur compounds, such as those in crude petroleum or sulfur gases. Since aircraft gas turbine engines are made of high-nickel alloys, the engine fuels for such power plants may contain no more than $\frac{1}{2}$ percent sulfur. Similarly, the stainless steels used in refining crude petroleum must be low-nickel steels such as type 410, since petroleum may contain as much as 4 percent sulfur.

While steels and iron are bcc, nickel is fcc. Nickel therefore has no tendency toward brittle fracture at low temperatures (see Fig. 8.4). Indeed, most alloys for cryogenic applications contain considerable amounts of nickel for toughness. At the other end of the temperature spectrum, nickel has high creep strength and a notable resistance to oxidation and general corrosion. Nickel therefore is the usual constituent in cryogenic alloys and in the superalloys needed in gas turbine and rocket engines. No other metal is equally suited to both high- and low-temperature applications.

Nickel is perhaps the most important metal for electronic and low-current electrical applications. Some of its many electrical uses have been mentioned before—magnetostriction, for example, or permanent magnets. A wide variety of nickel alloys are employed in vacuum electronic tubes, notably for cathodes. Nickel in sheet or other form is used as anodes in low-power tubes and in photocells. A great many thermocouple materials are also high nickel alloys.

Of the nonelectronic commercial grades of nickel, A nickel (Nickel 205) is the purest, with 99 percent nickel. It is used in the chemical industries for evaporators, kettles, and other processing equipment. Tensile strength is 55,000 psi, yield strength 15,000 psi, hardness $R_B64$. The great interval between yield and ultimate strength is of course indicative of good ductility.

Nickel with about 4 percent cobalt is used for magnetostrictive transducers for such purposes as ultrasonic cleaning. Nickel with 4 percent manganese has improved corrosion resistance at higher temperatures, particularly against sulfur compounds. This material is used for spark plug electrodes.


## 12.2 nickel alloys


Nickel, like copper and iron, is capable of producing a wide range of alloys peculiarly adapted to the demands of special applications, and as with these other two metals, these alloys are available in heat-treatable and nonheat-treatable types. Most of the nickel alloys to be discussed have been developed either for high-temperature use or for resistance to specific chemicals.

Nickel and copper are solid-solution soluble in each other in all proportions. The equilibrium diagram for these solutions is given in Fig. 2.17. Thus an infinite number of nickel-copper and copper-nickel alloys is possible. The names of these alloys, like many of the other copper alloys, is somewhat confusing. If the percent nickel exceeds the percent copper, the alloy will likely be termed a *monel*. If the nickel content is in the range of 12 to 30 percent, the alloy will be termed a *cupronickel*. Nickel coinage contains 75 percent nickel and 25 percent copper, but coinage nickel is not called monel.

The *monels* may be considered to be two-thirds nickel and one-third copper. Monels are more resistant to reducing chemicals than nickel and more resistant to oxidizing chemicals such as chlorides and nitrates, than copper, hence they have better corrosion resistance than either constituent. Resistance to attack by both freshwater and saltwater is outstanding, and the monels are therefore used as piping and pumps for brines and seawater, steam condensers, ships' propellers, and shafting.

R Monel (Monel R-405) contains 66 percent Ni, 31.5 percent Cu, and 0.05 percent S. The sulfur addition is made, as in the case of sulfurized steels, for machinability. This monel then is selected for parts that must be machined. Other monels may contain from 44.5 to 84 percent nickel. The standard monel is Monel 400 containing 66 percent Ni, 31.5 percent Cu, and about 1 percent each of iron and manganese.

The presence of half a percent or more of aluminum or titanium or both in a high nickel alloy indicates a precipitation-hardening alloy. The nature of the compound precipitated is a little uncertain; it may be an intermetallic nickel-aluminum and nickel titanium, $Ni_3(Al, Ti)$.

Permanickel, 98.6 percent Ni, 0.50 percent Ti, and Duranickel with 4.5 percent Al, 0.50 percent Ti, are thus heat-treatable nickels. The 500 Monels, K-500 and 501 both contain 2.80 percent Al and 0.50 percent Ti. Inconel superalloys are all heat-treatable and contain even more aluminum and titanium.

The demand for metals with high strength and high creep and oxidation resistance at high temperatures has so far been met not so much by the refractory metals, which oxidize too readily, but by the nickel superalloys. These are complex alloys usually based on nickel and used at temperatures remarkably close to their melting points. They are expensive metals to develop; the overall research costs of extending their temperature range probably approximate a million dollars per degree F. Strength at high temperatures is contributed by the precipitated $Ni_3(Al, Ti)$ phase, which cannot be found under the microscope. Prominent among the superalloys are the Inconels, which are complex nickel-chromium-iron alloys. Inconel X-750 is employed for gas turbine blading and afterburner combustion chambers and has been used in high-stress applications at temperatures of 1700°F or higher. The fuselage of the famous Bell X-15 rocket aircraft was also fabricated of Inconel X-750. Figures 12.2 and 12.3 compare the microstructure of Inconel X-750 with that of pure nickel. Ultimate strengths of the superalloys at room temperature may exceed 200,000 psi after hardening. Their machinability ratings are low, in the range of about 20 compared to B1112 as 100.

Fig. 12-2 Pure nickel, showing twinned structure.

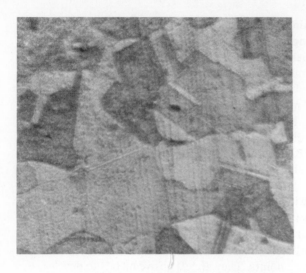

Fig. 12-3   A hardenable high nickel Inconel superalloy.

Analyses of a selection of the superalloys are given in the table following. Some of these alloys contain molybdenum and cobalt, two elements that contribute strength at high temperature. Cobalt forms solid solutions with nickel, while molybdenum forms carbides.

The Hastelloys are nickel-chromium-irons or nickel-molybdenum-irons, developed for great corrosion resistance to chemicals, especially acids, but also used as superalloys. Hastelloy A (see table) is austenitic and thus not heat-treatable.

The alloys constantan, alumel, and chromel P in the table are the commonly used thermocouple materials for temperature control of industrial furnaces. Other chromel alloys are used as electric heating resistances for such applications as household appliances. A very wide range of other high nickel alloys are available with high specific resistance for electric heating, with values for ohms/mil-foot ranging from 10 to 800. For small wire-wound resistors as used in electronic circuitry, the standard alloy is Nichrome, 73 percent Ni, 20 percent Cr, but for precision miniature resistors Karma is preferred, 80 percent Ni, 20 percent Cr, with a specific resistance of 800 ohms per circular mil-foot. Both materials have extremely low temperature coefficients of resistance.

Nilvar and Invar, 36 percent nickel with iron, have the lowest coefficients of thermal expansion of all known materials, 0.000001 in./in. By variation of the nickel percentage in iron-nickel alloys it is possible to tailor an expansion coefficient to any practical requirement. The wire leads of vacuum tubes, for example, can be made to match the thermal expansion of the glass tube.

Elinvar is a high nickel alloy with a constant $E$-value over a considerable temperature range. Such an alloy is invaluable for hair springs and balance wheels in watches and for other instrument components. Finally, the Alnico permanent-magnet nickel alloys should be mentioned again. These are nickel-chromium-aluminum-cobalt alloys.

Beryllium nickel was mentioned in connection with beryllium in the previous chapter. This alloy contains 1.95 percent Be, 0.50 percent Ti, and is marketed by the Beryllium Corporation as Berylco Nickel 440. It is hardenable to remarkable strength, fatigue resistance, and wear resistance, with a maximum $R_c$ of about 50. Its uses are in the field of instrumentation.

## NICKEL ALLOYS

| | Ni | Cr | Fe | Al | Ti | Mn | C | S | Cu | Other |
|---|---|---|---|---|---|---|---|---|---|---|
| A Nickel | 99.5 | | 0.15 | | | 0.25 | 0.06 | 0.005 | 0.05 | |
| Permanickel | 98.6 | | 0.10 | | 0.50 | 0.10 | 0.25 | 0.005 | 0.02 | 0.35 Mo |
| Duranickel | 94.0 | | 0.15 | 4.50 | 0.50 | 0.25 | 0.15 | 0.005 | 0.05 | |
| Monel 400 | 66.0 | | 1.35 | | | 0.90 | 0.12 | 0.005 | 31.5 | |
| Monel R-405 | 66.0 | | 1.35 | | | 0.90 | 0.18 | 0.050 | 31.5 | |
| Inconel 600 | 76.0 | 15.8 | 7.20 | | | 0.20 | 0.04 | 0.007 | 0.10 | |
| Inconel 700 | 46.0 | 15.0 | 0.70 | 3.00 | 2.20 | 0.10 | 0.12 | 0.007 | 0.05 | 28.5 Co, 3.75 Mo |
| Inconel X-750 | 73.0 | 15.0 | 6.75 | 0.80 | 2.50 | 0.70 | 0.04 | 0.007 | 0.05 | 0.85 Cb |
| Incoloy 800 | 32.0 | 20.5 | 46.0 | 0.30 | 0.30 | 0.04 | 0.04 | 0.007 | 0.30 | |
| Rene 41 | balance | 19 | | 2.5 | 1.65 | 3 | 0.30 | 0.10 | | 11 Co, 9.75 Mo |
| Hastelloy A | 53 | | 22 | | | 2 | | | | 22 Mo |
| Hastelloy C | balance | 16 | 5 | | | | | | | 16 Mo, 4W |
| Constantan | 45 | | | | | | | | 55 | |
| Alumel | 95.3 | | 0.1 | 1.6 | | 1.75 | | | | 1.2 Si |
| Chromel P | 90 | 9.5 | 0.2 | | | | | | | 0.4 Si |
| Inco-Weld A | 71 | 16.4 | 6.6 | | 3.2 | 2.3 | 0.03 | 0.007 | 0.04 | |

Special nickel alloys of importance in the fabrication of metals are those formulated for welding and brazing. Such nickel alloys are especially useful for joining metals other than nickel. Cast irons are commonly welded and repaired with high nickel alloys, usually Ni-Rod, 95 percent nickel, or Ni-Rod 55, 53 percent nickel. Inco-Weld A (see table) resembles an Inconel in composition but was developed for the welding of dissimilar metals, such as ferritic stainless steel to high nickel alloys. It is used also in the welding of cryogenic steels such as 9 percent nickel steel and has even welded zirconium alloys. Brazed joints that must resist high temperatures require nickel-base brazing alloys, usually nickel-chromium. These brazing alloys have solidus temperatures close to 2000°F.

In sum, the nickel alloys embrace a remarkably wide range of specialty uses. Nickel is critically important, therefore, in meeting the specialty requirements of our steel technology.

## 12.3 copper

Copper is similar to nickel in its ability to alloy with a wide range of metals, and like nickel it is face-centered cubic. The ores of copper, too, are usually sulfides.

Copper is noted for its high conductivity for both heat and electric current. Much copper production therefore goes into busbars and other types of electrical conductors. The thermal conductivity of copper, about 2700 Btu/°F/in., is almost ten times that of steel. Copper is the preferred material for chills, continuous-casting molds, and other equipment that must remove heat. The high thermal conductivity, however, is a disadvantage in the fabrication of copper. The pure metal is almost unweldable because of the difficulty of retaining sufficient welding heat in the area of the weld. It is rarely resistance-welded.

Pure copper has an $E$-value of $16 \times 10^6$ psi, a yield strength of 9000 psi and ultimate strength of 32,000 psi. Elongation is about 50 percent in 2 in. The wide difference between yield and ultimate strength indicates that copper has considerable work-hardening capacity. The metal is too soft to be readily machinable.

The melting point of pure copper is 1981°F. However, copper cannot be used at temperatures very much above room temperature because of loss of strength and oxidation. The oxide formed at high temperatures is black, but after long exposure to atmospheric conditions copper turns green (so does nickel). The copper industry has devoted much effort to surface treatments that will preserve the original surface appearance of copper and has recently developed treatments that may prove suitable for that purpose.

The current-carrying grades of copper must be relatively pure, since alloying additions, even in minute amounts, greatly increase the resistivity. Silver, gold, and cadmium, however, have lesser effects on resistance. Electrolytic tough pitch copper (ETP grade) containing 0.04 percent oxygen is used for conductors. Oxygen-free high conductivity copper (OFHC grade) is the best grade available. A micrograph of oxygen in furnace copper is shown in Fig. 2.22. The inclusions are $Cu_2O$. Such copper is not welded, since release of the oxygen produces cracks in the copper. To remove oxygen, phosphorus is added to molten copper as small phosphorus-copper shot.

Though much copper is used for tubing, pipe, and heat transfer equipment, it is not used in the food-processing industries in contact with foods because many copper compounds are toxic.

The copper alloys are numerous indeed, but their nomenclature is often inconvenient and confusing. Zinc, tin, aluminum, and silicon all serve as solid-solution hardeners in these alloys. By suitable alloying, a wide range of metallic colors, from copper to white, is available. In general, *brasses* are alloys of copper and zinc, *bronzes* are alloys of copper and elements other than zinc.

The brasses are, of all metals, the most ingratiating ones to machine, so that the word "brass" is almost synonymous with machinability in the minds of production personnel.

The portion of the copper-zinc equilibrium diagram that embraces the brasses is given in Fig. 12.4. Note that the diagram indicates the brasses to

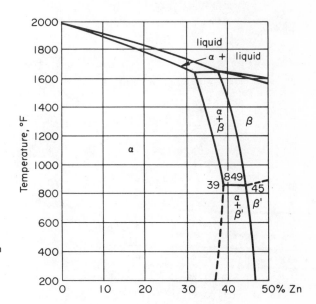

Fig. 12-4   Copper-zinc equilibrium diagram.

be nonheat-treatable: no phase boundary line indicates decreasing solubility of one material in another. The alpha brasses contain up to 36 percent zinc, ranging in color from red to yellow as the zinc content is increased. The alpha plus beta brasses contain 38 to 46 percent zinc. Since copper is fcc, alpha brasses are fcc. The beta phase is bcc. Zinc is hcp. The alpha brasses are good cold-working materials, while the alpha plus beta brasses are not especially ductile at room temperature but are excellent hot-working alloys.

A familiar alpha brass is the yellow *cartridge brass*, 70 percent copper, which is used in a wide range of products from electric light sockets to flashlight casings to rifle shells. Figure 12.5 is a micrograph of this brass.

The red brasses have less alloy content and better corrosion resistance than the yellow brasses. Red brasses are used for plumbing pipe. Alloys such as muntz metal and admiralty metal are formulated for corrosion resistance to freshwater and seawater and find their uses in marine hardware and condenser tubing for thermal power plants.

Fig. 12-5   Micrograph of a leaded cartridge brass. The twinning commonly found in wrought fcc metals can be found here.

## TABLE OF BRASS ALLOYS

|  | Cu | Zn | Sn | Pb |
|---|---|---|---|---|
| 1. electrolytic tough pitch copper | 99.92 | | | |
| 2. gilding brass | 95 | 5 | | |
| 3. commercial bronze | 90 | 10 | | |
| 4. red brass | 85 | 15 | | |
| 5. low brass | 80 | 20 | | |
| 6. admiralty metal | 71 | 28 | 1 | |
| 7. cartridge brass | 70 | 30 | | |
| 8. yellow brass | 65 | 35 | | |
| 9. free machining brass | 61.5 | 35.5 | | 3 |
| 10. muntz metal | 60 | 40 | | |
| 11. naval brass | 60 | 39.25 | 0.75 | |

The usual effect of alloying is increased strength and reduced ductility as compared with the pure metal. Copper-zinc alloys are an exception. The remarkable combination of increased ductility with increased strength in copper-zinc alloys is shown in the following table. Note that the ductility decreasse when the beta phase appears. The table gives comparative values only for one grain size and degree of working and does not indicate the maximum strengths of these alloys. A standard $\frac{3}{4}$-in. diameter bar of yellow brass breaks at about 75,000 psi.

## PROPERTIES OF COPPER-ZINC ALLOYS

| Zn, % | UTS, psi | Elongation % in 2 in. |
|---|---|---|
| 0 | 32,000 | 46 |
| 5 | 36,000 | 49 |
| 10 | 41,000 | 52 |
| 15 | 42,000 | 56 |
| 20 | 43,000 | 59 |
| 25 | 45,000 | 62 |
| 30 | 46,000 | 65 |
| 35 | 46,000 | 60 |
| 40 | 54,000 | 45 |

The brasses present difficulties in welding because of the loss of zinc, which boils at the low temperature of 1665°F. The brasses therefore are gas-welded or brazed rather than arc-welded.

### 12.5 bronzes

The *phosphor bronzes* are actually tin bronzes containing 1 to 11 percent tin. Phosphorus is used in small amounts as a deoxidizer. These bronzes are used in sheet and bar form for their high tensile strength, toughness, formability, fatigue strength, and corrosion resistance. Tensile strength may be as high as 100,000 psi. Phosphor bronzes are frequently specified for instrument parts such as diaphragms, bellows, springs, and electrical contacts.

Cast bronze plumbing fittings are common in domestic and commercial plumbing lines strung in copper pipe. A commonly used alloy for casting these fittings is a tin bronze analyzing 81 percent copper, 3 percent tin, 7 percent lead, 9 percent zinc. The tin and zinc provide castability and the lead gives machinability.

*Silicon bronzes* are more familiarly known by their trade names, such as Everdur and Herculoy. These also are high strength alloys, with strengths as high as 110,000 psi.

The *cupronickels* contain up to 30 percent nickel. Since copper and nickel are completely soluble in each other, these are single-phase alloys like the monels, incapable of heat treatment. They are chiefly employed for tubing which must have resistance to fresh water and seawater attack.

*Aluminum bronzes* contain from 4 to 11 percent aluminum. The alloys containing over 8 percent aluminum are heat-treatable to give microstructures similar to steel. Slow cooling gives a lamellar soft structure like pearlite; fast cooling produces a martensitic structure. The aluminum bronzes are used for a wide range of articles, including electrical hardware, marine hardware, propellers, tubing, and pumps. Tensile strength may exceed 100,000 psi after heat treatment, but ductility is then minimal.

*Beryllium copper* is another heat-treatable bronze, giving the highest strength of all the copper alloys, together with remarkable fatigue resistance, formability, corrosion and wear resistance. Its uses cover a wide range, including instrument parts, springs, diaphragms, dies, firing pins, electrical hardware, and because of its hardness, nonsparking hand tools such as wrenches. A maximum hardness of about $R_C$ 42 is possible. A range of beryllium bronzes is available, from 0.4 percent to 1.9 percent beryllium. Cobalt plus nickel is included in the alloy for grain refinement.

*Zirconium copper* and *chromium copper* are heat-treatable. They are used chiefly for instrument work and electrical hardware such as resistance welding tips and switch parts.

Copper and many of these copper alloys are supplied as cold-drawn leaded bars for ease of machining, and in this form they can often be machined at the maximum speed of a lathe. In addition to lead, tellurium and sulfur may be added for machinability.

The copper brazing alloys include, besides pure copper, high zinc alloys and high phosphorus (5–6 percent P) alloys containing some silver. These will braze most of the copper alloys.

## BRONZE ALLOYS

| | Cu | Sn | Si | Zn | Ni + Co | Other |
|---|---|---|---|---|---|---|
| phosphor bronze | 90+ | 1.25–10 | | | | |
| silicon bronze | 98 | | 1.5–3 | | | |
| architectural bronze | 57 | 3 | | 40 | | |
| cupronickel | 70 | | | | 30 | |
| aluminum bronze | 90+ | | | | | 4–11 Al |
| nickel silver | 65 | | | 17 | 18 | |
| beryllium bronze | 97.9 | | | | 0.2 | 1.9 Be |
| chromium copper | 99.1 | | | | | 0.9 Cr |
| zirconium copper | 99.85 | | | | | 0.15 Zr |

## 12.6 zinc

Zinc is used for an anodic coating on steels, applied usually by hot-dip galvanizing or electrogalvanizing. In sheet form, zinc is most familiar as dry cell casings. Other important uses include die castings and the alloying of nonferrous metals.

Zinc melts at 787°F and boils at 1665°F at atmospheric pressure. Its low boiling point indicates its unsuitability for use under vacuum conditions in industry or outer space. When melted or welded, special techniques are required to reduce zinc losses by evaporation, even in brasses.

Die castings require the use of high purity zinc. The presence of more than trace amounts of cadmium, lead, or tin promotes corrosion in zinc die castings, though die casting alloys may contain a few percent of aluminum or copper. Perhaps the most familiar die castings are the carburetors and fuel pumps of automobiles.

## 12.7 cadmium

*Cadmium* is used for the plating of steel hardware and fasteners and as an alloy in bearing metals. The uses of *tin* are similar to those of cadmium. Since cadmium compounds are toxic, this metal is not permitted in food-processing equipment. Because of its high neutron capture cross-section, cadmium is used in the control of nuclear reactions.

## 12.8 cobalt

Cobalt is a hard, ferromagnetic metal melting at 2719°F. The uses of cobalt largely arise from its strength at high temperature; it is used therefore in superalloys, heat-resistant castings, hot-working steels, and as a binder in sintered carbide cutting tools. It is also found in Alnico and other permanent magnet alloys. The best known of the cobalt alloys are the *stellites*, used for machining cutters and hard-surfacing, which contain also chromium, tungsten, and carbon.

Advances in isotope techniques have made Cobalt-60 one of the more important industrial (and medical) metals. This radioactive cobalt contains an extra neutron, which makes the atomic nucleus unstable. Probably more Co-60 is used in industry and medicine than the total of all other radioisotopes. It is produced in the nuclear reactors at Chalk River and Oak Ridge. Aside from medical therapy, Co-60 is used as a source of deep-penetration (over 1 Mev) gamma radiation for the examination of welds, forgings, and castings.

It is not suitable for the lighter metals with lower absorptions, such as aluminum. Like all radioisotopes, it is sold by the curie (Sec. 2.11) and not by weight.

## 12.9 lead

Pure lead melts at 621°F. This versatile metal has applications that derive from its special properties of high atomic weight and density, softness, ductility, low strength, low melting point, corrosion resistance, and ability to lubricate. Its fatigue strength is poor, so that it cannot be used under conditions of vibration. It will creep even at room temperatures. Still another disadvantage is the toxicity of its compounds. The chief uses of lead include the following:

1. tank linings for corrosion protection
2. storage batteries of the lead-acid type
3. pipe and drainage fittings such as closet bends for toilet bowls
4. machinability additive to free-cutting steels and brasses
5. biological shielding against gamma radiation (but not against neutrons)
6. lead sheathing of electric cable
7. low-melting solders
8. bearing metals
9. fusible alloys and low-melting alloys for various purposes (see Fig. 12.6)
10. type metal for the printing trade
11. terne plate (lead-tin coated steel)
12. tetraethyl lead anti-knock ingredient in gasolines
13. lead compounds in paints

In general, lead does not form useful alloys except with other metals with low melting points. In the free-machining leaded alloys the lead is not an alloy ingredient but an inclusion. For corrosion resistance and for gamma-ray shielding, pure lead gives best performance.

To increase the low strength of lead, two types of alloys are in use: tellurium lead and antimonial lead. The addition of less than 0.1 percent tellurium improves the work-hardening capacity of lead. Pure lead has a tensile strength of only 2500 psi, but antimony improves this strength. A maximum of 11 percent antimony may be used, resulting in an increase in strength of about three times. Alloying, however, reduces the already low melting temperature of lead.

The solders are alloys of lead and tin, the amount of tin used ranging from 5 to 50 percent. The equilibrium diagram for these solders is given in Fig. 2.18. Tin additions are necessary to enable the lead to wet the metal to be joined. The lead-base babbitt bearing alloys are somewhat similar to solders, containing about 10 percent each of tin and antimony. Terne plate,

Fig. 12-6 Experimental aircraft panel under buckling test. The top and bottom of the panel are supported by cast cerrobend, a low-melting-point alloy.

used for gasoline tanks on automobiles and other mobile equipment, is steel sheet coated with a lead solder alloy of about 20 percent tin.

The thickness of lead is usually referred to by weight in pounds per square foot. One square foot of lead $\frac{1}{64}$ in. thick weighs 1 lb. Thus an 8-lb sheet lead is $\frac{8}{64}$ in. thick.

## 12.10  silver and gold

These precious metals have a limited range of uses in industry. Silver is the more useful. Both metals have a tensile strength of about 18,000 psi, though neither is used for structural purposes. Silver has slightly better electrical conductivity than copper; gold has lower conductivity. Both metals are fcc and ductile.

*Gold* has occasional uses as a reflecting surface, as attachments to transistors, and a few brazing alloys contain gold in amounts up to 40 percent.

*Silver* is plated onto low-voltage electrical contacts, but not for its conductivity. When arcing occurs, the contact surface oxidizes. Silver is used because the oxidized silver readily reduces back to silver. Silver is also occasionally used as a corrosion-resistant lining in chemical process vessels. However, it is readily corroded by sulfur.

Silver may be hardened by a few percent of copper, the copper acting as a solid-solution strengthener which permits precipitation heat-treatment. The most important of the silver-copper alloys are the silver solders, brazing alloys which contain considerable silver with copper and zinc. These have solidus temperatures between 1000 and 1500°F.

The photosensitivity of silver salts is the reason for their use in photographic emulsions. When development of photographic film is done on a large scale, silver can be recovered from the emulsion.

Both silver and gold and their brazing alloys are sold by the troy ounce, twelve ounces to the pound. An ounce of silver is heavier than an ounce of lead (avoir.), but a pound of lead is heavier than a pound of silver.

## 12.11 the platinum group

This group of six metals is extracted from nickel ores. Platinum is probably the most useful member of the group. Platinum, palladium, iridium, and rhodium are face-centered cubic. Osmium and ruthenium are body-centered cubic. Osmium and iridium, having twice the specific weight of lead, are the heaviest of all metals, but platinum is almost as heavy. The other platinum metals are slightly heavier than lead. All six of the platinum group have high melting points, above 3000°F.

In addition to being the heaviest metal, osmium has the highest modulus of elasticity, $80 \times 10^6$. This, however, is no virtue because of the remarkable brittleness of osmium. The others of the platinum group also have high $E$-values.

The platinum group collectively shows a high degree of corrosion and oxidation resistance compared to other metals. Iridium is considered the most corrosion-resistant of all the metals.

*Platinum* is the most ductile of the group. Its uses include laboratory apparatus such as small crucibles and dishes, thermocouples and other instrument components, spinning and wire-drawing dies, components of high-power vacuum tubes, glass-working equipment, and electrical contacts. Platinum alloys readily with many metals and is usually solid-solution hardened by ruthenium, iridium, or rhodium.

*Palladium* is used principally in telephone relay contacts. A billion interruptions of current are expected before the relay wears out. Palladium offers for such service freedom from oxidation, sulfidation, and spark erosion.

Iridium is chiefly used as the radioisotope Iridium-192. Together with Cobalt-60, it satisfies most of the requirements of industrial radiography. Cobalt-60 is used for penetrating heavy sections of heavy metals, and Iridium-192 is used for lighter sections in both heavy and light metals.

The rare earths are a large group of 14 metals, actually about a quarter of all the metals. Most of this group are hexagonal close-packed. They are blessed with such exotic names as praseodymium, dysprosium, neodymium, etc., which may or may not explain why they presently lead a disoriented life on the fringes of the society of respectable metals.

Mixtures of the rare earths have been called "mischmetal" since earlier times and are added to stainless steels to provide ease of rolling. Their use as an alloying element in magnesium has been noted. An yttrium alloy of aluminum has been developed.

General Motors Corporation has used a radioisotope of samarium for gammagraphing engine castings. The oxides of cerium, praseodymium, and lanthanum have been considered for the thermoelectric generation of power, and in electronics gadolinium ferrites have been studied. Most, but not all, the rare earths have remarkably high neutron absorption cross-sections and therefore have possibilities for control of nuclear reactors. A few of the rare earths have been employed as doping materials in lasers. By and large, however, interest in the rare earths has developed only in recent years, and it is difficult to predict what uses will be found for these materials. The developmental uses outlined above suggest that they may well settle into the more glamorous areas of engineering.

## 12.13   zirconium

Zirconium is another reactive metal: extraction of the metal from the ore therefore proceeds in the usual method of converting the ore to zirconium chloride and then reducing the chloride with magnesium. Like titanium, molten zirconium reacts with crucible materials, reducing their oxides to metal.

Zirconium melts at 3366°F. It is not characterized by good hot strength, however, and oxidizes readily above 1000°F. Fine zirconium powder or machining chips may even ignite spontaneously, though the metal is not hazardous in massive form.

Crystal structure of zirconium is hcp at room temperature, with a bcc phase above 1590°. In sheet form, zirconium somewhat resembles stainless steel, though it has a somewhat lower specific weight. It has a low thermal conductivity, like stainless steel, but a much lower coefficient of expansion— 0.000004 in./in. Like stainless steel its corrosion resistance is remarkable, though not as good at high temperatures. Small amounts of carbon, oxygen, nitrogen, and hydrogen greatly reduce the corrosion resistance, and they strengthen and embrittle the metal.

Most zirconium is consumed in the manufacture of fuel rods for nuclear reactors. For such an application the amount of impurities of high neutron absorption must be held to very low levels. Zirconium ores invariably contain over 2 percent hafnium, which is a neutron absorber; this hafnium must be removed during processing of the metal. The content of boron and cadmium is held to a maximum of 0.5 parts per million for the same reason.

Fig. 12-7  Bundles of fuel rod sections for a research nuclear reactor. The darker rods are zirconium, the lighter rods are sintered aluminum powder (SAP). (Courtesy Whiteshell Nuclear Research Establishment)

The only alloys of zirconium are zirconium-2.5 percent niobium (i.e., columbium) and the Zircaloys. Zircaloy-2 contains 1.2–1.7 percent tin with small amounts of iron and nickel. Both tin and niobium have low neutron absorption and offset the detrimental influence of impurities such as nitrogen. Zircaloy-2 has a tensile strength of 66,000 psi and elongation of 20 percent in 2 in.

## 12.14  the refractory metals

The refractory metals are by definition those metals with melting points above 3600°F. Such a definition should include osmium, iridium, and ruthenium and should exclude chromium. The first three, however, fall into the group of platinum metals. The refractory metals include only chromium, columbium, hafnium, molybdenum, rhenium, tantalum, tungsten, and vanadium.

The refractory metals have melting points as high as 6150°F, somewhat beyond the useful temperature range of refractories and melting furnaces, and therefore they cannot be melted and cast like most metals. These metals must be prepared by sintering them in powder metallurgy processes similar to those in use with beryllium and zirconium.

It would appear that such metals could be used at temperatures beyond the reach of lower-melting metals. The refractory metals have remarkable strength at unusually high temperatures, but they oxidize catastrophically at rather low temperatures. With respect to oxidation properties, these metals are far inferior to the platinum metals or even nickel.

It is ironical and frustrating that low-melting metals, such as magnesium, aluminum, and calcium, should have highly refractory oxides, but the refractory metals usually have oxides that melt below the melting point of the metal. The following table presents the melting points of the refractory metals and their oxides:

## MELTING POINTS OF REFRACTORY METALS AND OXIDES

| chromium | 2822°F | chromia | 4230°F | $(Cr_2O_3)$ |
|---|---|---|---|---|
| hafnium | 3866 | hafnia | 5140 | $(HfO_2)$ |
| molybdenum | 4730 | molybdenum oxide | 1465 | $(MoO_3)$ |
| niobium | 4380 | niobia | 2500 | $(Nb_2O_5)$ |
| rhenium | 5730 | rhenia | 565 | $(Re_2O_7)$ |
| tantalum | 5425 | tantalum oxide | 2500 | $(Ta_2O_5)$ |
| tungsten | 6152 | tungsten oxide | 1830 | $(WO_3)$ |
| vanadium | 3450 | vanadia | 1270 | $(V_2O_5)$ |

The lowest-melting oxide is reported in the table. Note that the melting point of the metals tends to increase in alphabetical order. The actual relationship of the melting points is disclosed by arranging these metals as they appear in the Periodic Table:

| | | |
|---|---|---|
| Ti | V | Cr |
| Zr | Nb | Mo |
| Hf | Ta | W |

The melting point increases toward the bottom of each of the three columns.

Relative oxidation rates at higher temperatures are shown in Fig. 12.8.

Many of the refractory metals are familiar carbide-formers in steel technology. Comparative properties of the pure metals are listed in the table below.

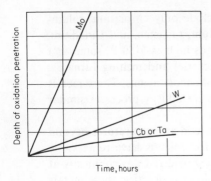

Fig. 12-8 Relative oxidation rates of refractory metals at temperatures in the range of 2000°F. Oxidation loss of tungsten is approximately 0.10 in. per 15 hrs.

## PROPERTIES OF THE REFRACTORY METALS

| Metal | Structure | E(70°) | E(2000°) | Ductility | UTS(70°) | UTS(2000°) |
|-------|-----------|--------|----------|-----------|----------|------------|
| Hf | hcp | $20 \times 10^6$ | | Fair | 60,000 | |
| V | bcc | $20 \times 10^6$ | | Good | 70,000 | 5,000 |
| Nb | bcc | $16 \times 10^6$ | $12 \times 10^6$ | Good | 40,000 | 10,000 |
| Ta | bcc | $27 \times 10^6$ | $24 \times 10^6$ | Excellent | 100,000 | 20,000 |
| Cr | bcc | $42 \times 10^6$ | $30 \times 10^6$ | None | 70,000 | 10,000 |
| Mo | bcc | $46 \times 10^6$ | $40 \times 10^6$ | Poor | 80,000 | 25,000 |
| W | bcc | $60 \times 10^6$ | $50 \times 10^6$ | None | 100,000 | 35,000 |
| Re | hcp | $66 \times 10^6$ | $50 \times 10^6$ | Good | 160,000 | 75,000 |

A review of the properties tabulated above shows why niobium (columbium) and tantalum are the preferred refractory metals at the present time. This preference is based on their combination of good ductility and high-melting point of the oxide, combined with minimum cost. Hafnium has a favorable spectrum of properties, but its cost greatly exceeds that of columbium and tantalum.

Because the properties of these metals are strongly affected by such factors as trace elements, grain size, processing procedures, and even the method of measuring the properties, the values listed above must be viewed as only approximate. This is even true for *E*-values. The table above applies to the "pure" metals, but this means rather the purest available. The presence of 0.01 percent oxygen in a refractory metal might, for example, double the ultimate tensile strength.

Since all of the refractory group except two are body-centered cubic, they show a ductile-to-brittle transition temperature. This transition temperature is well above room temperature for tungsten, molybdenum, and chromium.

*Hafnium* is present in all zirconium ores, and this is the only source from which it is extracted. But while zirconium is commercially available, hafnium is not. It has had limited review for aerospace programs, but its use is limited to rectifiers and a few electronic applications.

*Rhenium* is probably the metal with the most pronounced work-hardening characteristics. This is a high-strength material with good ductility, high-melting point, and high *E*-value, a splendid combination spoiled by its strong oxidizing tendency. This metal has been reviewed for aerospace applications with little success, and future applications for it, as for all the refractory metals, seem likely to lie in electronics, for filaments in high-power vacuum tubes and for structural components under vacuum where its oxidation deficiencies will not be a handicap. It is also used in rhenium-tungsten high-temperature thermocouples. When alloyed with molybdenum and tungsten, it greatly improves the room-temperature ductility of these metals.

The chief uses of *chromium* lie in chrome-plating and in steel alloys, to which it contributes hardness, strength at high temperature, and corrosion resistance. Though a relatively abundant metal, its brittleness may relegate it to the role of an alloying element in the foreseeable future. The research effort devoted to solving the problems of chromium is much less aggressive than that expended on the more promising metals columbium and tantalum.

*Vanadium* has many ores, and is present in crude oils up to a maximum of 0.06 percent by weight. It is added to steels in small amounts up to 1 percent for the purpose of grain refinement. Some promising high strength alloys of vanadium have reached the development stage, but are not yet in use.

### 12.15  columbium and tantalum

These two metals are distinguished by excellent ductility, even at low temperature, tantalum probably having the best ductility of the body-centered cubic metals. It is not certain that tantalum has a ductile-to-brittle transition temperature. Both metals occur together in ores, and as was the case for zirconium and hafnium, they must be separated for nuclear use, since tantalum has a high neutron absorption and that of columbium is low.

Until interest developed in the refractory metals after World War II, columbium was used only as an alloy addition in stabilized stainless steels such as 347 and 316Cb. Tantalum, which has a very low work-hardening capacity, was used for piping, heat transfer coils, and similar applications in chemical process industries, its corrosion resistance being almost identical to that of glass. It has also been used as a melting crucible material for melting reactive metals under vacuum. While columbium is not yet employed in electrical or electronic devices, tantalum is used for anode and grid elements in medium- to high-power tubes, in capacitors, and in tantalum-foil rectifiers. Tantalum oxide has a rectifying effect on alternating current.

Columbium shows promise for high-temperature applications. Alloy development work has produced alloys capable of short-time exposure to temperatures of 2500°F in air, though the objective of these programs is a temperature of 3000°F. Compared to columbium, tantalum has the disadvantages of high specific weight and less abundance in ores. Typical commercial alloys of these metals, all formulated for hot strength, creep resistance, oxidation resistance, and ease of fabrication, are these:

| | Columbium alloys: | |
|---|---|---|
| | FS82 | 33% Ta, 1% Zr |
| | FS85 | 28% Ta, 1% Zr, 11% W |
| | Cb-6 | 10% W, 8% Ti |
| | Tantalum alloys: | |
| | Ta-10W | 10% W |
| | Ta-8W-2Hf | 8% W, 2% Hf |
| | Ta-12.5W | 12.5% W |

Several other columbium alloys are commercially available, but there are only these three tantalum alloys. The Ta-10W alloy has a tensile strength of 50,000 psi at 2500°F, almost the melting temperature of stainless steels. The columbium alloys are used in the nose components and leading edges of hypersonic aerospace vehicles.

## 12.16 molybdenum

## and tungsten

These two metals and tantalum have had industrial uses for many decades. Thomas Edison experimented with tungsten wire in his 98-percent-perspiration, 2-percent-inspiration invention of the incandescent lamp. Molybdenum and tungsten are among the chief carbide-formers in high-speed steels, molybdenum tending to replace tungsten in this application because of its lower weight. Though both metals have unusual strength, hardness, and stiffness, even in pure form, and especially at high temperatures, they both have the serious defects of great brittleness at room temperature and pronounced oxidation at relatively low temperatures. Alloying to reduce oxidation of these metals has not been as successful as in the case of columbium.

Room temperature applications for these metals are numerous. Tungsten has the following applications:

1. electrodes for inert-gas welding (TIG welding)
2. lamp and electron tube filaments
3. anodes for X-ray tubes and electron tubes
4. electric resistance elements

5. wire-drawing dies
6. rocket nozzles

Tungsten is too hard to machine, except by grinding. Molybdenum is machinable. It is also cheaper than tungsten. Its applications include:

1. electron tube anodes
2. grids for high-power electron tubes
3. dies for die-casting, extrusion, and upsetting of metals
4. components in processing molten glass
5. corrosion-resistant valves and piping
6. supports for tungsten filaments in light bulbs
7. resistance-heating elements

The addition of alloying elements to tungsten has little advantage, and tungsten therefore is usually employed in relatively pure form. Molybdenum has a few commercial alloys, chiefly two:

| | |
|---|---|
| TZM | 0.5% Ti, 0.08% Zr |
| Mo-0.5 Ti | 0.5% Ti |

Mo-50W, with 50 percent tungsten, is used only as wire in electron tubes. TZM has outstanding strength up to 3000°F and has a range of uses for aerospace vehicles and metal-forming dies.

The refractory metals will come into wider use when their price is reduced somewhat and when methods of purifying metals are improved. Their shortcomings can be overcome, but probably only to a limited degree. Although they are popularly viewed as exotic metals, their future areas of application will probably be those in which they are presently established:

1. electricity and electronics
2. chemical processing
3. metal fabrication
4. aerospace

## PROBLEMS

1 Compare iron and nickel as to (a) *E*-value, (b) sulfur control during melting operations, (c) melting point, (d) crystal structure.

2 Given the analysis of Monel R-405 (see Table of Nickel Alloys), how do you know that this alloy is not heat-treatable?

3 Distinguish between brasses and bronzes.

4 Give the crystal structure of copper, alpha brass, beta brass, and zinc.

5 Why is zinc gas-welded but rarely arc-welded?

6   What is the thickness of a sheet of 6-lb lead?

7   Summarize the uses of the platinum metals.

8   Why must hafnium be removed from nuclear-grade zirconium?

9   What advantages do the refractory metals offer?

10  List the outstanding characteristics of nickel.

11  Why are solders not made of pure tin or pure lead?

12  Name the two radioisotopes principally used in industrial radiography.

13  What is the principal use of zirconium?

14  Define the term "refractory metal."

15  The oxides of the refractory metals are not always refractory. Name some metals with refractory oxides.

16  Why are nickel alloys and not refractory metals used in gas turbine engines?

17  Name the three broad market areas that nickel serves.

18  Examine the kitchen tools and appliances of your own home. List the uses of nickel that you find in their components.

19  Why is nickel used for spark-plug electrodes?

# ORGANIC MATERIALS

*Part* **4**

## 13.1  a little history

The technical culture of the Arctic Eskimos, now rapidly being forgotten, was at least as gifted as ours in its use of materials. Except for one inland tribe, the Eskimo lived in a country without resources other than rock and snow. The land provided him with insufficient food and materials for survival, so he turned to the resources of the sea for materials. Except for snow as a winter shelter material and stone for tools and a small cooking hearth, the Eskimo used the organic materials of sea animals to survive. Even his fuel had to be animal oils.

Most early civilizations, however, used wood as an industrial material. It was employed principally as a construction material, and still is. Wood and other organic materials, such as canvas, hemp, manila fiber and oakum, made possible the sailing ship, the invention that more than any other raised man's standard of living, for it made worldwide trading possible. Until 125 years ago, the sailing ship was man's most powerful machine. The great sailing "clippers" of the golden age of sail might carry as much as 10,000 square yards of canvas, which propelled them at top speeds of about 25 miles per hour. Such windpower is equivalent to 1000–1500 horsepower.

# THE ORGANIC
# MATERIALS

# 13

The peculiar influence of organic materials on historical development does not end with the sailing ship, the archery that turned the tide in medieval battle, or the wooden palisades of early North American forts. Radar made it possible for the Royal Air Force Fighter Command to win the most crucial battle of World War II, the famous Battle of Britain, but what is less well known is that polyethylene insulation made radar possible. Two years after the Battle of Britain, Japanese forces overran the sources of natural rubber that were to supply the vehicles of American armies, but again new synthetic rubbers saved the day.

## 13.2  general characteristics

The organic materials have a powerful influence on history and on our daily lives. Even the human body is a complex of organic materials, and so is our food.

Wood, paper, asphalt, coal, petroleum, natural gas, natural rubber, foods, wool, and cotton are natural organic materials. All of these materials must be manufactured, converted, and altered in some degree before being put to use, but still they retain their basic characteristics through the manufacturing cycle. The use of natural organic materials has tended to decline throughout this century, with the exception of paper, a material that is critical to a civilization that is organized by documents and documentation. The natural organic materials are often superseded by synthetic materials, which are inventions of the chemist. Notable examples of synthetics are polyethylene, synthetic rubbers, silicones, Teflon, vinyl (polyvinyl chloride), and the many synthetic fibers. Most rubber goods are made of synthetic rubber and most textiles and carpets are made of synthetic fibers. Even paper is replaced by mylar film in drafting offices. Vinyl siding on houses often replaces wood. Four percent of the weight of wood plywood sheets is the synthetic adhesive that binds the plies together. Much synthetic stone (polyester plastic with crushed stone) replaces natural stone. Finally, even natural rubber can be made synthetically, giving us synthetic natural rubber. The chemist's ability to create new synthetic organics exceeds the capacity of markets to absorb them.

The organic materials are the innumerable substances based on carbon chemistry. Parallel to them is the variety of ceramic materials based on silicon, which is in the same column of the Periodic Table. If we compare the structural organic materials, which are wood, rubber, and the plastics, with the structural metals and ceramics, we find a number of differences.

1. The organics are soft and have lower strengths.
2. Moduli of elasticity are lower. For hardwoods $1.6 \times 10^6$, for fiberglass-reinforced plastics $1 \times 10^6$, for other strong plastics less than $1 \times 10^6$.

With the exception of wood and reinforced plastic composites, the organics therefore are not well adapted to sustaining high stresses.

3. Ductility of organics is generally good. All, however, including wood, tend to creep even at room temperature.
4. Temperature limits for these materials are usually low, generally below 500°F. Many are combustible.
5. Thermal and electrical conductivity are both very low. Many organics are excellent electrical insulators, especially vinyl and polyethylene and the synthetic rubbers.
6. Specific weight of any organic material is low.
7. Because these materials contain much hydrogen, specific heats are high, in the range of 0.3 to 0.4 Btu/lb-°F.
8. Many organic materials are degraded by the ultraviolet radiation in sunlight.

Most of the industrial organic materials are *petrochemicals*. A petrochemical is, roughly speaking, an organic chemical derived from petroleum or natural gas. The petrochemicals, like other organic compounds, are chemicals containing carbon and hydrogen (hydrocarbons) but frequently also contain oxygen and nitrogen. An exact explanation of what constitutes a petrochemical, however, is apt to produce confusion. Some petrochemicals aren't petrochemicals, and some are petrochemicals sometimes. Consider sulfur, which is produced from oil and gas. Sulfur is not a petrochemical, because it is not a chemical at all, but an elemental substance. But carbon black, also an elemental substance, is a petrochemical, because a processing operation is required to make it from oil or gas. Benzene, xylene, and toluene can be made from crude oil: these are petrochemicals. But benzene, xylene, and toluene can be distilled from coal: these are not petrochemicals. Ammonia, $NH_3$, which contains no carbon, is a petrochemical. Most rubbers are petrochemicals, but natural rubber is not.

## 13.3 petroleum
## and natural gas

The organic natural resources deposited in the earth's crust may be solid, liquid, or gas—coal, petroleum, or natural gas. Each of these three materials is a mixture of large numbers of hydrocarbons. Many specific hydrocarbons, such as the gas methane, may be found in all three types of deposits. In particular, some petroleum hydrocarbons are always found in natural gas deposits, and natural gas is always found in petroleum. The basic difference between natural gas and petroleum is simply that petroleum has a greater content of liquid hydrocarbons than natural gas, while natural gas has a greater content of gaseous hydrocarbons.

The word "petroleum" means rock oil, signifying oil found in sedimentary rock deposits. Rock oil became an industrial material in the 1860's, the decade that gave us cheap steel, paper from wood, and the decimation of the whale population of the oceans. Whale oil was the principal illuminant at that time, and as it became less available because whales became scarce, the less popular coal oil had to be substituted. Coal oil was kerosene made from coal. Kerosene is a constituent of petroleum, and speculation about the potential use of rock oil for illumination produced the first oil industry north of Pittsburgh and in southern Ontario about 1860. The kerosene fraction was distilled from the petroleum and sold, while the gasoline, asphalt, and natural gas fractions, having no market, were discarded. But the turn of the century brought two inventions that needed the oil industry for engine fuel and for paving: the automobile and the aircraft. A truly gigantic market for gasoline and asphalt was born. This was an ideal market configuration for petroleum, for the "light ends" or light constituents of the crude petroleum could be used in gasoline, and the heavy ends in asphalt.

Later another very important group of products was developed from petroleum and gas. These were the petrochemicals. Though only a few percent of all petroleum production goes into petrochemicals, this group of products serves a spectrum of applications that is simply too vast for brief summarization and extends far beyond the more familiar plastics and rubbers. The history of petrochemicals may be said to date from 1916, when a method of making propyl alcohol from petroleum was discovered. The synthetic plastics and rubbers joined the list of petrochemicals in limited production in the 1930's, barely in time to serve the critical needs of the first war that was fought with materials technology, World War II.

## 13.4  petroleum deposits

Petroleum and natural gas are products of the organic remains of plant and animal life, generally deposited in marine sediments. It is believed that the geological formation of petroleum is not dependent on high pressure or temperature. Vanadium and nickel occur in limited amounts in crude oil in compounds called porphyrins, which are the geological remains of plant chlorophyll. Sulfur is always present in sedimentary oils and natural gas. It reacts slowly with the oil constituents to form hydrogen sulfide. If sulfur compounds are present in considerable amounts, the deposits are referred to as "sour crudes" or "sour gas." Much sulfur production is derived from these deposits. Finally, most petroleum deposits also contain both salt and water.

The analysis of all crude oils appears to correspond approximately with the following:

| | |
|---|---|
| carbon | 83.9–86.8% |
| hydrogen | 11.4–14.0% |
| sulfur | 0.06–2.0% |
| nitrogen | 0.11–1.70% |

However the mixture of chemical compounds may vary greatly from one well to another. Some deposits contain much gas and light oil compounds; others are heavy and asphaltic. The lighter crudes are preferred because of the higher content of light gasoline fractions. Most of the nitrogen and sulfur is found in the heavier constituents of the crude, which are therefore more likely to be sour.

### TYPICAL CRUDES FROM VARIOUS OIL FIELDS

| Area | Specific Gravity | Gasoline | Kerosene | Sulfur | Asphalt |
|---|---|---|---|---|---|
| Pennsylvania | 0.80–0.90 | 30–40% | 15% | 0.1% | 4% |
| Gulf Coast | 0.86 | 5 | 8 | 0.15 | 4 |
| California | 0.96 | 11 | 10 | 3.8 | 50 |
| Leduc, Alberta | 0.82 | 35 | 10 | 0.4 | 7 |
| Venezuela | 0.95 | 7 | 19 | 2.2 | 55 |
| Russia (Urals) | 0.95 | 25 | 0 | 5.0 | 45 |
| Kuwait | 0.87 | 27 | 17 | 2.5 | 25 |

The weight of crude oils in this table is expressed as specific gravity referred to water as 1.0. The oil industry uses also the API (American Petroleum Institute) gravity scale. This API gravity (the units are called degrees) is related to specific gravity by the following formula:

$$\text{API gravity, degrees} = \frac{141.5}{\text{sp gr}} - 131.5$$

Thus water has an API gravity of 10. Gasoline has an API gravity of 52 to 70, SAE #10 lube oil 22 to 33. Crude oil API gravities may range from 4 (unpumpable at 70°F) to 85.

Whether the well is an oil well or a gas well, the mixture that reaches the wellhead consists of gas, oil, and water. These components must be separated. The line from the Christmas tree (i.e., the group of wellhead valves, which doesn't resemble a Christmas tree) runs to a gas and oil separator. The separator is a vertical cylinder which allows the gas to rise out of the oil for collection. The top line out of the separator removes the gas; the bottom line removes the oil. If water is also present, it is allowed to separate out in a field storage tank by gravity.

Raw natural gas is collected in a gas conservation plant for processing. Hydrogen sulfide is removed by contact with ethanolamine and the useful gas constituents separated by distillation processes in three parts. One part is dry gas, methane with some ethane, nitrogen, and carbon dioxide, which is sold as natural gas for heating. Liquified petroleum gas or LPG is the second part, consisting of propane or butane, and is sold as bottled fuel. Natural gasoline or casinghead gasoline is removed as the third part and converted to motor gasoline at an oil refinery.

Almost all oil and gas is transported either by tanker or "big-inch" pipelines from the producing field to the market areas.

## 13.5 carbon bonding

To understand the characteristics of the organic materials, the structure of these carbon compounds must be briefly discussed. These materials are built from carbon atoms with associated other elements such as oxygen and hydrogen. The simplest ones are the hydrocarbons, containing only carbon and hydrogen.

Carbon provides four chemical bonds to other atoms, or in the language of the chemist, it has a valence of four. Hydrogen, chlorine, and fluorine, all of which occur in plastics and rubbers, have a valence of one, or offer only one chemical bond. Since carbon bonds four ways and these gases only one, the compounds of Fig. 13.1 are only a few of the possibilities for

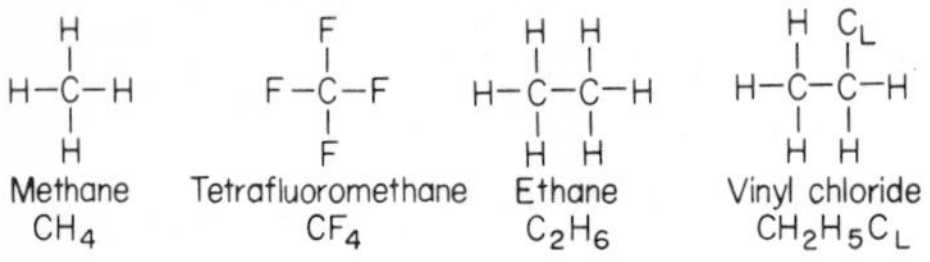

Fig. 13-1   Construction of organic compounds.

constructing compounds from one or two carbon atoms. In Fig. 13.1 the chemical bonding is shown by short lines connecting two atoms; it is more convenient to omit these however and to designate the compound as $CH_4$, $CF_4$, $C_2H_6$, etc.

Oxygen and sulfur have a valence of two. Combinations such as shown in Fig. 13.2 are possible with these elements.

A discussion of the system of naming these chemical compounds is not given here, although the reader will see some system in those names given in the two figures.

The quadruple valence of carbon makes possible the formation of long

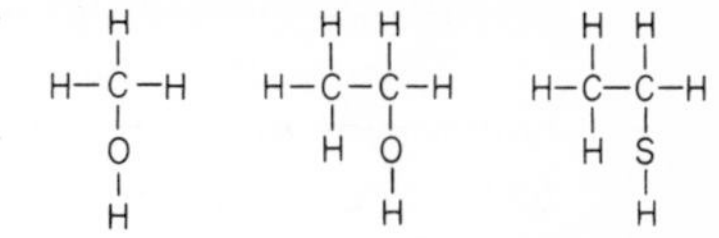

Fig. 13-2  Organic compounds with oxygen and sulfur.

chains of carbon atoms with attached side elements such as hydrogen or chlorine. Some of these long chains are produced by *polymerization*. Polymerization is a process in which two or more molecules of the same substance join to give a larger molecule, such that the molecular weight of the polymer is some multiple of the unit substance or monomer. Thus two molecules of ethylene, $C_2H_4$, can be joined to produce the dimer $C_4H_8$ or $(C_2H_4)_2$. Polyethylene, as its name suggests, is a polymer built from ethylene as monomer. The plastics, rubbers, paints, adhesives, wood, proteins, fats, carbohydrates, and vitamins are only some of the many polymer substances. These are to be discussed; meanwhile any critical reader must be warned that many of the polymerizations to be discussed here will not exactly conform to the definition of polymerization just given. An exact definition would be laborious and obscure and would contribute nothing to an understanding of the polymer materials.

## 13.6  hydrocarbon components of petroleum and gas

The number of hydrocarbon chemicals in petroleum may well be almost infinite. The lighter gaseous and liquid hydrocarbons can be identified and separated from petroleum without difficulty, but the great multitude of compounds are in the heavier fractions. The individual heavy hydrocarbons cannot be identified, nor can they be separated. Instead, they are divided into groups of "cuts" at the oil refinery. The problem may be understood by an illustration. The common gasoline hydrocarbon, octane, which gave its name to the octane number of gasoline, contains 8 carbon atoms and 18 hydrogen atoms. These 26 atoms can be rearranged to form 17 other hydrocarbons. By reducing the number of hydrogen atoms attached to the 8 carbon atoms, still more hydrocarbons are possible. Twenty-five carbon atoms offer the possibility of about 40 million different hydrocarbons, and 25 is by no means the maximum number of carbon atoms in any petroleum molecule.

Most of the hydrocarbons in petroleum can be classified into four groups: paraffin, olefin, naphthene, and aromatic.

The *paraffins* have the general formula $C_nH_{2n+2}$. Methane, the chief constituent of natural gas and the lightest of the paraffins, has the formula $CH_4$. The names of specific paraffins end in -ane: ethane, hexane, heptane, etc. The paraffins are saturated hydrocarbons, the term "saturated" meaning that they contain the maximum possible number of hydrogen atoms in each case. For this reason they are rather inactive chemically. The carbon atoms in paraffins tend to form long chains in polymer fashion. Those which form a straight chain without branches are termed normal or *n*-paraffins; those with branches are isoparaffins. See Fig. 13.3.

```
                                                      H
                                                    H C H
       H        H H       H H H      H H H H       H | H
     H C H    H C C H   H C C C H  H C C C C H   H C C C H
       H        H H       H H H      H H H H       H H H
    Methane    Ethane    Propane   Normal butane   Isobutane


     H H H H H           H H H H H H        Fig. 13-3   Paraffins.
   H C C C C C H       H C C C C C C H
     H H H H H           H H H H H H
   Normal pentane       Normal hexane
```

Paraffins with four or fewer carbon atoms are gases at ordinary temperatures, those with 5 to 15 carbon atoms are liquids, those with over 15 are waxes, asphalts, etc. This general principle is true for plastics; those with only a relatively few carbon atoms in the polymer chain are liquids, while the solid plastics and rubbers are made of very long chains of carbon atoms.

Figure. 13.3 identifies some of the paraffin series. The normal or *n*-paraffins have straight chains without branches. These have low octane numbers and are undesirable in gasolines. Their good burning characteristics however make them preferred constituents of diesel and gas turbine fuels and fuel oils. Gas turbine fuels are largely kerosenes, the original commodity extracted from petroleum a hundred years ago. The isoparaffins, of which isobutane in Fig. 13.3 is an example, are branched. These are the preferred constituents of gasolines, including aviation gasoline, because of their high octane numbers.

The uses of the lighter paraffins may be briefly summarized as follows:

1. methane—chief constituent of natural gas
2. ethane —converted to other organic materials such as polyethylene
3. butane —fuel for refinery boilers, or added to winter gasoline, cigarette lighter fluid, or converted into petrochemicals
4. pentane, hexane, heptane, octane, etc.—raw materials for gasoline

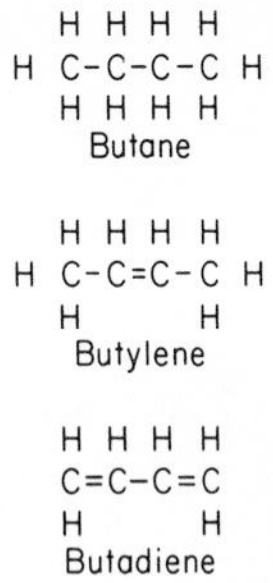

```
        H H H H
     H  C-C-C-C  H
        H H H H
         Butane


        H H H H
     H  C-C=C-C  H
        H      H
         Butylene


        H H H H
        C=C-C=C
        H      H
         Butadiene
```

Fig. 13-4  Olefins.

The *olefins*, Fig. 13.4, are unsaturated. If two hydrogen atoms can be added to an olefin to saturate it, it is a mono-olefin, $C_nH_{2n}$. The mono-olefins have names that correspond to those of the paraffins to which they are related, with the ending -ene or -ylene. If four hydrogen atoms can be added for saturation, the olefin is a diolefin, $C_nH_{2n-2}$, with a name ending in -diene (pronounced dye-een). Olefins are chemically reactive and serve as the principal raw materials for the petrochemical industry in the manufacture of plastics, rubbers, and other materials. Actually olefins are not common in crude petroleum but are formed in the course of oil refinery operations. Gasolines contain considerable amounts of high-octane mono-olefins. The diolefins are undesirable in gasolines because they form gums. The olefins have the straight-chain structure of the paraffins.

*Naphthenes* are saturated ring-shaped hydrocarbons with the general formula $C_nH_{2n}$. The names of the naphthenes begin with cyclo-. The molecule of cyclohexane is shown in Fig. 13.5.

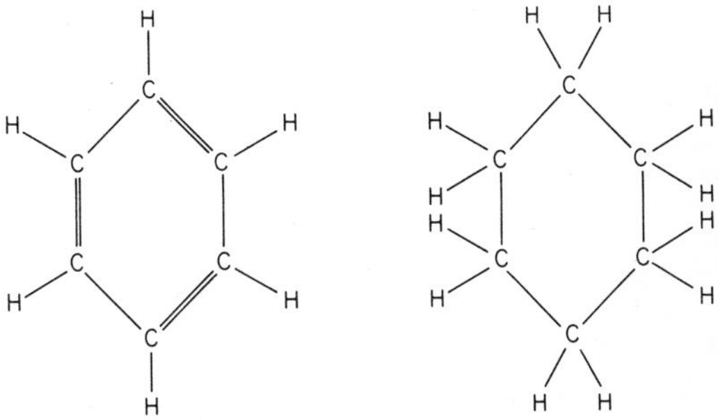

Fig. 13-5  Benzene, an aromatic, and cyclohexane, a naphthene.

*Aromatics* are found in coal and to a lesser extent in crude oil, especially in California crudes. These are ring-shaped compounds like the naphthenes, but are unsaturated. Figure 13.5 includes a sketch of the structure of benzene. Benzene, toluene, and xylene, often referred to in the industry as BTX, are the most important of the aromatics. Aromatics are chemically active, and like the olefins are the starting-points for the synthesis of a wide range of organic chemicals, including explosives, drugs, solvents, dyes, and poly-

*the organic materials | 307*

styrene. They are desirable components of gasolines, but not of diesel and gas turbine fuels because they tend to smoke. The aromatics are excellent solvents, though they present a fire hazard.

## 13.7 the refining of petroleum

Crude petroleum may sometimes be burned as bunker fuel in ships' boilers, but in general it cannot be used as a commercial product. It is therefore pumped by pipeline to oil refineries for conversion into fuels, lube oils, asphalts, petroleum coke, and chemical intermediates, the latter being further converted in chemical plants.

Crude petroleum represents a mixture of a wide range of oil and gas components which differ from well to well. These must be converted into products of fixed characteristics and standard quality, regardless of varia-

Fig. 13-6 Chemical compounds in natural gas and petroleum.

| Formula | | Name of chemical | Condition | Use |
|---|---|---|---|---|
| $H-\overset{\underset{\displaystyle H}{\mid}}{\underset{\underset{\displaystyle H}{\mid}}{C}}-H$ | 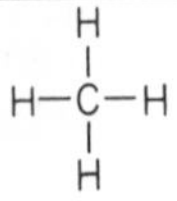 $CH_4$ | Methane | Gas | Heating fuel |
| $H-\overset{\mid}{\underset{\mid}{C}}-\overset{\mid}{\underset{\mid}{C}}-H$ (H H / H H) | 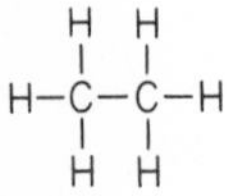 $C_2H_6$ | Ethane | Gas | Converted into plastics |
| $H-\overset{\mid}{\underset{\mid}{C}}-\overset{\mid}{\underset{\mid}{C}}-\overset{\mid}{\underset{\mid}{C}}-H$ | 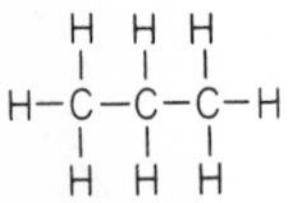 $C_3H_8$ | Propane | Gas | Heating fuel |
| $H-\overset{\mid}{\underset{\mid}{C}}-\overset{\mid}{\underset{\mid}{C}}-\overset{\mid}{\underset{\mid}{C}}-\overset{\mid}{\underset{\mid}{C}}-H$ | 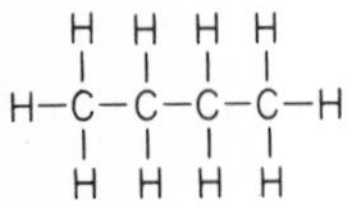 $C_4H_{10}$ | Butane | Gas | Heating fuel or converted into rubbers |
| | $C_5H_{12}$ $C_6H_{14}$ $C_7H_{16}$ $C_8H_{18}$ $C_9$, etc | Pentane Hexane Heptane Octane | Liquid | Converted into engine fuels (gasoline) |
| | $C_{100}$, etc | Asphalts and tars | Solid | Road and roofing asphalts |

tion in characteristics of the crude oil delivered from the pipeline. The petroleum product in greatest demand and of greatest profit is gasoline, though the gasoline fractions in the crude may be small and the heavy asphaltic fractions large.

The first refinery operation is the separation of the crude into various "cuts," using distillation methods. For flexibility of operation, two other basic refinery processes are needed. First, there must be a process for "cracking" or breaking up the larger molecules into shorter gasoline molecules. Second, there must be a method of polymerizing small gas molecules into longer molecules. The first process is called cracking, the second polymerization. These processes will provide the basic operational flexibility to meet changes in market demand for petroleum products and variations in the type of crude oil received from the pipeline. Other refinery processes will be incorporated with these either to improve product quality or to produce special products such as greases, coke, or chemical intermediates.

**Distillation.** Distillation is the first step in reducing the multitude of hydrocarbons in petroleum to usable products. After the crude oil is separated into distilled fractions, these fractions can be refined or converted. The separation of liquid mixtures by fractional distillation is possible because of differences in boiling points, a difference put to use by Allied soldiers in Italy in 1943, who did not like Chianti wine and therefore fractionally distilled it into alcohol and water. The several fractions into which petroleum is usually converted and their boiling ranges are given below.

| Fraction | Boiling Range, °F | Ultimate Use |
|---|---|---|
| methane | $-259.5$ | natural gas |
| ethane | $-128$ | ethylene |
| propane | $-44$ | liquified petroleum gas |
| butane | $+11$ to $31$ | motor gasoline, rubber, etc. |
| light naphtha | $30-300$ | motor gasoline |
| heavy naphtha | $300-400$ | motor and gas turbine fuels |
| kerosene, $C_{11}-C_{15}$ | $400-500$ | gas turbine fuel |
| stove oil | $400-550$ | gas turbine and oil fuel |
| light gas oil, $C_{15}-$ | $400-600$ | furnace and diesel fuels |
| heavy gas oil, $C_{20}-$ | $600-800$ | gasoline and fuel oil |
| vacuum gas oil, $C_{30}-$ | $800-1100$ | gasoline, fuel oil, lube oil |
| pitch | $1100+$ | asphalt |

The fractions to and including heavy gas oil are separated by distillation at atmospheric pressure. The heavier fractions must be distilled under vacuum conditions, using a pressure of about 50–100 torr (1 torr = 1 mm mercury pressure). At atmospheric pressure distillation temperatures would be so high that there would be damage to the product, including depolymerization and excessive formation of coke. The individual hydrocarbon chemicals may be separated from one another by distillation only if they are very light ends from $C_1$ to $C_4$. In the heavier fractions there are too many compounds with identical boiling points.

Petrochemicals are manufactured from light ends such as ethane and butane, while gasoline is made from both the lighter fractions and the heavier gas oils.

**Catalytic Cracking.**   Cracking is the process by which fractions boiling in the gasoline or light naphtha range are produced from heavier fractions. The larger molecules are divided into fragments of lower molecular weight basically by means of heat, but with the assistance of pressure and catalysts. A catalyst is a material which promotes a chemical reaction without being consumed in the reaction: its effect has been compared by chemists to that of moonlight on young love. Without the catalyst, excessive amounts of methane and ethane are produced by cracking, that is, the cracking occurs too close to the ends of the chain of carbon atoms.

Petroleum hydrocarbons are cracked at temperatures above 650°F. Higher temperatures are required to crack larger molecules. The resulting cracked product is a mixture with much the same boiling range as the crude from which it is made, containing fractions from gases to tar and coke. But the cracked material will contain perhaps twice as much material boiling in the gasoline range, that is, pentane, $C_5$, and heavier. The cracked material has no fractions suitable for kerosene, lubricating oils, waxes, or asphalts. These products therefore are not cracked; gasoline, diesel fuel, and furnace oil are cracked. The cracked hydrocarbons are more olefinic than the original crude hydrocarbons. If a $C_{12}$ (dodecane) paraffin is cracked into two parts, one part will probably be a hexane paraffin and the other a hexane olefin. Besides olefins, the cracked material will contain aromatics. All these nonparaffinic components improve the octane rating of the cracked product. Straight-run distillate gasolines made of the pentane, hexane, heptane, octane, etc. of crude oil cannot be burned in the modern automobile, though they were used in the past.

A special type of cracking process used to upgrade the quality of gasolines is called *reforming*. Reforming produces greater amounts of naphthenics and aromatics, converting hexane, for example, into benzene. The products of the reforming process may also be converted into petrochemicals.

**Polymerization.**   Cracking and reforming processes produce large amounts of gaseous hydrocarbons. These light ends may be put through a *poly unit* to produce polymer gasoline. This operation is of course the reverse of cracking.

*Lubricating oils* must meet a wide range of service conditions, and consequently a wide variety of these oils must be produced. Vegetable oils and synthetic oils fill certain special needs, but by and large petroleum oils are the only materials with the necessary properties and low cost for lubrication applications.

The primary requirements for a lube oil are lubricity and resistance to oxidation. Special requirements also apply however. In particular, the viscosity must match the application. A low viscosity is needed for a sewing-machine oil, while high viscosity is necessary for the more severe demands of a worm gear speed reducer. In the case of a fluid power hydraulic system, a low-pressure system will be filled with a low-viscosity oil, perhaps SAE #10, to reduce the horsepower requirements of the pump. But a high-pressure hydraulic system will require a higher viscosity to reduce leakage through the valves of the system. Refrigeration equipment requires an oil suited to low-temperature operation and wax-free, while a steam turbine requires a high-temperature oil which contains antioxidants.

Cutting and drawing lubricants are formulated for thread cutting, drilling, and other machining operations on metals, and for such forming operations as the shaping of sheet metal in dies. Two severe requirements are imposed on such lubricants:

1. The maintenance of a lubricating film between the tool or die and the work, under pressures that may often exceed 100,000 psi.
2. Dissipation of the heat of machining or forming.

Many of these forming lubricants are water-soluble oils or water emulsions.

*Greases* are lubricating oils thickened by fatty acid soaps. The soap holds the oil to the surface requiring lubrication, and it is dispersed in the oil in the form of fibers too small to be distinguished under an optical microscope. Soaps are produced by chemical combination of a metal hydroxide with a fat or a fatty acid. The chief hydroxides used in grease manufacture are those of calcium, sodium, lithium, and barium. The type of fat will influence the properties of the grease, but not to the degree that the metal hydroxide will. Calcium greases are limited to temperatures below 200°F while sodium greases may be used above this temperature. Barium and lithium greases are suitable for both high and low temperatures and are usually designated as multi-purpose greases.

*Petroleum waxes* are used in the packaging industry for waxed paper and waxed containers. A heavily waxed paper may contain as much as 35 percent wax by weight.

*Petroleum asphalts* are used in asphalt concretes for paving and for roofing, paints, varnishes, and insulating compositions.

*Insulating oils* serve for insulation purposes in transformers, X-ray tubes, and oil-filled cables. Such oils must have a high flash point and low vapor pressure, with high dielectric strength.

*Petroleum coke* is the residue from the destructive distillation of petroleum and may be found in car engines that "carbon up." Petroleum coke is not pure carbon but contains up to 16 percent of volatile petroleum compounds and is oily in appearance. It is sometimes used as a fuel in the Eastern States and Minnesota, though it requires special furnace designs and has a very high ignition temperature of over 1000°F. Its chief use, however, is in the manufacture of carbon electrodes for aluminum refining and other such metallurgical processes.

For blast furnace operation, coke is made by distilling coal in coke ovens. Coke is made more cheaply than petroleum coke. But coal contains much sulfur, which is present in coke-oven coke, since no process is in use to remove sulfur from coal or coke. On the other hand it is standard practice to desulfurize petroleum, and petroleum coke thus has almost no sulfur content. The harmful effects of sulfur in metals have been discussed in earlier chapters. Arc-melting electrodes must therefore be made of petroleum coke.

## 13.9  solvents

The industrial solvents used for cleaning metals and paint stripping are petrochemicals with a small molecule that enables them to penetrate into surface coatings. Three types of solvent are in use: paraffins, aromatics, and chlorinated solvents, listed in the order of increasing solvent power. The paraffin solvents are simply members of the paraffin family, such as diesel fuel or gasoline. These have rather limited solvency. The aromatic solvents include such chemicals as benzene and xylene. The chlorinated solvents such as trichlorethylene ($C_2HCl_3$), methyl chloride ($Ch_3Cl$), or carbon tetrachloride ($CCl_4$) are paraffins with the hydrogen atoms partially or wholly substituted by chlorine atoms. The following table compares the characteristics of these three groups of solvents.

### SOLVENT COMPARISON TABLE

|  | Paraffins | Aromatics | Chlorinated |
|---|---|---|---|
| cost | low | medium | highest |
| solvency | limited | good | best |
| flammability | high | very high | none |
| toxicity | low | high | high |
| paint removal | poor | better | excellent |

A fuel is any material which can be made to oxidize at a sufficiently rapid rate, with release of adequate heat of oxidation. The possible number of fuels then is virtually unlimited. Though most fuels have been, and continue to be, hydrocarbons, many inorganic fuels have been tried. Though no longer used, boron hydride, a ceramic material, was a solid rocket propellant, and quite literally thousands of fuels have been tested as potential rocket fuels. One of the finest of solid fuels is the set of tires on an automobile. Scrap tires are actually used as a fuel in municipal incinerators to assist in the burning of wet garbage and refuse (which are used as fuels in some cities). Other types of rubber are used as solid rocket propellants. In World War II the strategic air forces of Germany, Britain, and the United States used magnesium as a fuel in small incendiary bombs designed to ignite the cities of their adversaries. Sewer gas is burned to heat sewage treatment plants. Phosphorus serves as a fuel in matches and mortar bombs. Explosives are fuels designed to create destructive pressures as a result of their combustion. Uranium on the other hand is employed as a nonoxidizing radioactive fuel and plutonium as a radioactive explosive. These random examples should serve to indicate the scope of fuels and fuel technology.

Mankind first used fuels to keep warm and for heat in the processing of meat for food. Later explosive fuels transformed methods of warfare. Today improvements in fuels and fuel technology of rockets are the means of projecting exploratory vehicles into the empty reaches of space.

Neglecting the more peripheral fuels, such as explosives, radioactive and metallic fuels and rocket fuels, the two most important uses of standard fuels are space heating and steam production for power or process use. Fuels for these two purposes may be solid, liquid, or gas.

The commoner solid fuels include:

1. peat, a primitive fuel
2. cordwood
3. bagasse, waste sugar cane
4. hog fuel, which is wood chopped or "hogged" in a wood chipper or hog. Hog fuel is burned at pulp mills for process steam generation
5. sawdust, a poor fuel because of its high moisture content
6. municipal garbage and refuse
7. lignite, or brown coal
8. bituminous or soft coal
9. anthracite, or hard coal
10. petroleum coke

Most fuel coal is consumed in large steam generating plants for the production of electric power in huge amounts. Such a plant may burn a few hundred tons of coal every hour.

The usual liquid fuels are fewer in number than the solid fuels. The liquid fuels are commonly produced by oil refineries.

1. light fuel oils of low viscosity, #1 and #2 fuel oils chiefly for domestic use
2. heavy fuel oils of high viscosity, #4, #5, and #6 (#6 is also called bunker oil or bunker C), which are sold at low prices in large quantities for consumption in heating boilers and ships. These heavy fuels require special burners to atomize them
3. internal combustion engine fuels:
    (a) gasolines for spark-ignition engines equipped with spark plugs
    (b) diesel fuels for compression-ignition (diesel) engines
    (c) kerosene blends for gas turbine engines
    The rocket engine is also an internal combustion engine, but it is not restricted to liquid fuels.
4. black liquor, a heavy black gunk which is a residue from wood pulp operations, requiring a special type of furnace called a recovery furnace

Finally, the commoner gaseous fuels are the following:

1. acetylene, for oxyacetylene welding and thermal cutting of metals
2. liquid hydrogen, a rocket fuel
3. blast furnace waste gas
4. coal gas from coking operations at steel mills
5. refinery waste gas from oil refining operations
6. liquid petroleum gas, LPG, chiefly propane, sometimes butane
7. natural gas, which is chiefly methane

## 13.11 the proximate analysis

## of fuels

The utilization of fuels cannot be understood except in terms of their proximate analysis. Fuels may be analyzed for their chemical elements or compounds, but such an analysis has limited usefulness to fuel technology. Instead, the proximate analysis, though relatively unscientific, provides immediate insights into the method of employing a fuel and its combustion problems. The proximate analysis simply gives the percentage breakdown of a fuel in terms of four broad components:

1. moisture
2. volatile matter
3. fixed carbon
4. ash

**1. Moisture.** Moisture may be formed from hydrogen and oxygen in the fuel, but this is not counted in the moisture analysis. The moisture is the free water or absorbed water in the fuel as delivered. To determine this moisture content, a sample of the fuel is dried at a temperature slightly above 212°F. The difference in weight between the original fuel sample and the dried weight is expressed as a percent.

The moisture in the fuel is a handicap. It is a component in the freight and handling charges. Further, when the fuel is charged into the furnace, about 1000 Btu will be lost from the useful heat of combustion for every pound of moisture to be evaporated. Indeed, it is possible for a fuel to be too wet to be self-burning. Fuels with excessive moisture content, such as hog fuel, lignite, or bagasse, require special furnaces which contain drying grates for evaporation of moisture.

**2. Volatile Matter.** If the dried sample of fuel is next heated to 1700°F *in the absence of air*, the volatile hydrocarbons in the fuel will be boiled out. The difference in sample weight as a result of this operation is expressed as percent of volatile matter. In the case of liquid fuels from wells, such volatile matter is chiefly straight-chain hydrocarbons, saturated or unsaturated. In the case of coal, which also contains large amounts of volatile matter, these hydrocarbons tend more to cyclic or aromatic hydrocarbons such as benzene.

The smoke that results from fuel burning is either black or silvery gray. The clouds of condensed steam that issue from chimneys should not be confused with smoke—all fuels contain hydrogen, therefore all chimneys transport steam. Black smoke results from inefficient firing, such as a cold furnace or insufficient air, and it can be produced from volatile matter or fixed carbon. Black smoke can be prevented by proper combustion practice. Most true chimney smoke, however, is silvery gray, like the exhaust from a smoky automobile that needs new piston rings. This smoke results from the burning of volatile matter and arises from the serious practical difficulties of getting a thorough mix of air and volatiles within the furnace or engine and the complex combustion chemistry associated with these volatiles. Burning volatile compounds at temperatures below 1400°F not only produces simple carbon dioxide and steam but also yields complex oxygenated aldehyde compounds. A typical smokeless fuel is charcoal, which is made smokeless by distillation of volatile matter from wood.

**3. Fixed Carbon.** This is the solid carbon residue remaining after the volatile matter is boiled out of the fuel. If the moisture and volatiles are driven from the fuel sample, the remainder of the sample can be burned, leaving only ash, and the percent of fixed carbon can be determined by weight loss.

Fuels with high percentages of fixed carbon, such as anthracite coal or petroleum coke, require special furnace shapes to ignite the fixed carbon.

Volatiles have ignition temperatures below 1000°F and consequently are relatively easy to burn, but fixed carbon ignites only at temperatures exceeding 1200°F. In order to burn fixed carbon, a furnace has surfaces suitably disposed to radiate heat onto the fuel bed. At the opposite extreme, a furnace that burns volatile matter simply requires adequate volume for combustion and may be merely a rectangular or cylindrical box. Volatile matter must be burned with *overfire air*, that is, air admitted to the furnace above the grate or hearth level. Fixed carbon is more efficiently burned using *underfire air*, or air admitted through the hearth or grates. Hence, an experienced person can examine a furnace design and make a shrewd guess about what fuel will be burned in it.

Improper combustion can produce fixed carbon in a furnace. Furnaces operated at too low a temperature will produce fixed carbon from volatile matter. This is familiar in automobile engine cylinders which are driven at low speeds in city driving. They are said to "carbon up." The author once had the pleasant task of removing about 150 lb of carbon (petroleum coke?) from an improperly fired boiler burning #6 oil fuel.

**4. Ash.** Ash consists chiefly of metallic oxides. The ash content of a fuel may be an advantage or a disadvantage. Ash may be used to protect the

Fig. 13-7 A hog-singeing burner to singe hog carcasses. Because this is a food-processing operation, a clean fuel such as natural gas must be used.

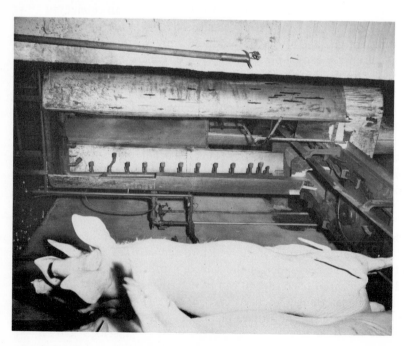

furnace grates from heat damage. But ash must be collected and removed. Some ash is collected in hoppers below the grates, but in large furnaces most ash travels toward the chimney as flyash and must be collected in flyash hoppers at the top of the furnace.

Sometimes ash imposes restraints on the operation of a furnace. In a municipal incinerator, if the furnace becomes too hot, the ash will fuse into large clinkers which are difficult to remove and which block the flow of combustion air through the grates. Other types of furnaces operate at temperatures high enough to fuse the ash so that it can flow from the furnace as a liquid.

CHARACTERISTIC PROXIMATE ANALYSES OF
SOME FUELS

| Fuel | % Moisture | % Volatile Matter | % Fixed Carbon | % Ash |
|------|-----------|-------------------|----------------|-------|
| petroleum coke | 2 | 16 | 82 | 0.05 |
| charcoal | 1 | 22 | 73 | 4 |
| lignite | 10 | 45 | 35 | 10 |
| natural gas | 0 | 100 | 0 | 0 |
| garbage | 30 | 40 | 10 | 20 |
| newspaper | 4.8 | 78.0 | 16.3 | 0.75 |
| bond paper | 6.85 | 76.0 | 13.13 | 4.15 |
| corrugated cardboard | 5.05 | 78.5 | 16.0 | 0.53 |

## 13.12 other fuel properties

Chemical analyses of fuels provide specific items of information, such as the sulfur content of the fuel. The amount of sulfur is a measure of the tendency of the combustion gases to corrode the engine or furnace passages. Gas turbine engines in particular cannot tolerate more than 0.5 percent sulfur in their fuels, since such engines use high nickel alloys which are susceptible to sulfur attack. High sulfur fuels would be equally harmful in furnaces used to heat steel products.

The fire hazard of a fuel, solvent, or other liquid is often measured by its *flash point*. The flash point is defined as the lowest temperature at which the product gives off sufficient vapor to form a flammable mixture with air. Thus if the flash point were room temperature, the fuel would present a significant fire hazard.

The *specific gravity* of a liquid fuel is referred to water as 1.0; the specific gravity of a gaseous fuel is referred to air as 1.0. The specific gravity of liquid fuels is used to convert volumes to weights and in general to evaluate petroleum products, since other properties are related to specific gravity. It is more usual to use the API gravity than specific gravity for petroleum products. The API gravity is related to specific gravity by the formula

$$°API = \frac{141.5}{sp\ gr\ at\ 60°F} - 131.5$$

As the API gravity increases,

    (a) the viscosity decreases
    (b) the heat content per pound increases
    (c) the heat content per gallon decreases
    (d) the carbon content decreases
    (e) the hydrogen content increases

The *heating value* of a fuel is expressed as the heat released by combustion per pound or per gallon. One of the chief products of combustion of hydrocarbon fuels is water vapor. If the combustion products are cooled back to room temperature, this water vapor condenses, releasing its latent heat. The addition of this latent heat of water vapor to the heating value gives the gross or higher heating value. If such latent heat is not included, the net heating value results. Most heating values are expressed as the gross heating value.

### FUEL OIL GRADES AND APPROXIMATE PROPERTIES

| Fuel Oil | Flash Point | API Gravity | Sp Gr | Btu/lb (gross) |
|----------|-------------|-------------|-------|----------------|
| 1 | 100°F | 35 | 0.85 | 19,590 |
| 2 | 100 | 26 | 0.89 | 19,270 |
| 4 | 130 | 25 | 0.90 | 19,230 |
| 5 | 150 | 12 | 0.99 | 18,640 |
| 6 | 180 | 10 | 1.0 | 18,540 |

The values in this table are only typical, and vary between geographical areas of the United States and Canada.

Municipal garbage becomes increasingly combustible because of the increasing use of plastics, rubbers, paper, and synthetic packaging materials. As a result, it is now usual to use municipal incinerators for heat and steam generation. The following table shows typical higher heating values for municipal garbage materials. The high heating value of polyethylene is explained by its composition; it is composed solely of carbon and hydrogen, two fuels of outstanding heating value.

Fig. 13-8  In contrast with Fig. 13-7, a dirty fuel. The ash in the refuse being burned in this incinerator has clinkered and the operator is attempting to remove the clinker.

| Material | Btu/lb | Lb/cu ft |
| --- | --- | --- |
| brown paper | 7250 | 7 |
| paper milk cartons | 11350 | 5 |
| corrugated cardboard | 7040 | 7 |
| coffee grounds | 10000 | 30 |
| magazines | 5250 | 35 |
| newspapers | 7975 | 7 |
| polyethylene | 20000 | 50 |
| polyurethane foam | 13000 | 15 |
| rubber tires | 10000 | 100 |
| sawdust | 8500 | 12 |

## 13.13   engine fuels

The three types of internal combustion engine, spark-ignition, diesel, and gas turbine, do not in general burn the same type of fuel. The spark-ignition or gasoline engine is used only to supply small amounts of power in mobile equipment. The fuel for this engine is a cracked gasoline containing olefins and aromatics. For ease of starting, a more volatile (i.e., lower boiling) gasoline fuel is supplied for winter operation in colder climates; this is customarily supplied by blending lighter ends such as butane into the winter gasoline. Aviation gasoline, however, has only a small olefin content and is never blended with butane. At high altitudes any low-boiling components might boil in the fuel tanks.

Gasoline requires a sufficiently high octane number to resist detonation during combustion. Octane number is expressed as follows. Isooctane is a hydrocarbon with an octane rating of 100; it is arbitrarily assigned this octane number because it has excellent combustion characteristics. Normal heptane on the other hand is a poor fuel which will knock severely if used in an engine. Normal heptane may be considered to have an octane number of zero. A blend of 80 percent isooctane and 20 percent normal heptane is assigned an octane rating of 80, that is, an octane rating is the percent isooctane in a mixture of isooctane and normal heptane. An unknown fuel is assigned the octane rating that gives the same engine performance that the unknown fuel itself does. Octane ratings must be determined in special test engines that burn mixtures of octane and heptane and also the fuels requiring octane ratings. Branched-chain hydrocarbons and aromatics have the highest octane numbers; straight-chain paraffins have the lowest. Typical octane numbers (note normal octane and isooctane):

| butane | 95 | normal octane | 0 |
| pentane | 62 | isooctane | 100 |
| hexane | 25 | cyclopentane | 100 |
| heptane | 0 | benzene | 100 |
| | | toluene | 107 |

The octane ratings of the natural gasolines pentane, hexane, heptane are so low that these compounds are unusable in modern gasoline engines.

Aircraft gas turbine engines may operate at altitudes where the temperature may be −60°F or more. They therefore require fuels which do not solidify at these temperatures. Also, since an aircraft gas turbine is expected to operate above 20,000 ft, light fractions that may boil under vacuum conditions in the fuel tanks cannot be employed. High sulfur fuels cannot be used because of their corrosiveness. The gas turbine engine, however, is basically not a fuel-sensitive engine like the spark-ignition engine, and it can be made to burn a wide range of fuels, liquid or gas. Two types of aviation turbine fuels are in general use: JP-1, an aviation kerosene, and JP-4, a wide-cut aviation gasoline. Cheaper fuels are used in stationary gas turbines.

Diesel engines rotate at low speeds and can tolerate a rather wide range of fuels, but the viscosity of the fuel must be suited to the fuel injection system. High compression diesel engines often burn fuel gases.

Knocking is possible in diesel engines as well as in gasoline engines. The cetane number of the diesel engine corresponds to the octane number of the spark-ignition engine. The two hydrocarbons which are blended to give the cetane number are cetane, rated 100, and alpha-methyl-naphthalene, rated zero. A fuel with a cetane number of 40 would have the same ignition quality as a mixture of 40 percent cetane and 60 percent AMN. Cetane numbers for good diesel fuels are not as high as the octane numbers of quality gasolines.

## 13.14 rocket propellants

These special fuels have a variety of applications, from harpoon guns, infantry bazooka weapons, and jet-assisted take-off of aircraft to large manned space vehicles. Smaller rocket engines use solid fuels, but at the present stage of rocket technology the largest rocket vehicles are confined to liquid fuels.

The most important property of a rocket fuel is its *specific impulse.* This is defined as the propulsive thrust in pounds obtainable from the burning of one pound of propellant per second. Solid fuels do not provide as high a specific impulse as liquid fuels.

A liquid propellant is pumped into a combustion chamber where it is burned; the products of combustion then leave the rear of the rocket engine by expanding to a high velocity through a nozzle. The momentum of the exhaust gases provides the reactive force that drives the rocket.

A solid propellant is cast in place in the propellant chamber. A hollow space along the axis of the chamber has the proper cross-section for the required burning rate, usually a star-shaped section so that a uniform amount of propellant is burned every second. If the propellant simply burned radially outward to the motor casing, the rate of consumption would increase with time. See Fig. 13.9. The exhaust gases expand through a nozzle as in the case of a liquid propellant. The disadvantage of the solid propel-

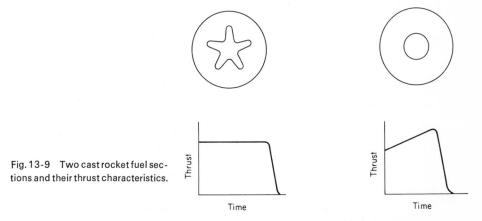

Fig. 13-9 Two cast rocket fuel sections and their thrust characteristics.

lants is their low specific impulse compared to liquid fuels. Higher specific impulse is possible only by using unstable fuels, that is, explosives which might detonate at the wrong time.

The exhaust velocity is given by the following equation:

$$c = I_{sp}g$$

where $c$ = exhaust velocity, fps

$I_{sp}$ = specific impulse

$g$ = gravitational acceleration, ft/sec$^2$.

The increase in velocity caused by consumption of the propellant is found from

$$\Delta v = c \log_e \frac{m_0}{m_f}$$

where $\Delta v$ = increase in velocity, fps

$m_0$ = launch weight of the vehicle

$m_f$ = final weight of the vehicle after propellant consumption.

The thrust in pounds = $wI_{sp}$, where $w$ = propellant consumption in pounds per second. The thrust is increased simply by increased propellant consumption.

## LIQUID PROPELLANTS

| Oxidizer | Fuel | Combustion Temperature | Specific Impulse |
|----------|------|------------------------|------------------|
| lox | ethyl alcohol | 5700°F | 276 |
| lox | hydrazine | 5660 | 300 |
| lox | hydrogen | 4800 | 388 |
| lox | JP-4 | 6100 | 285 |
| fluorine | ethyl alcohol | 7250 | 318 |
| fluorine | hydrazine | 8050 | 334 |
| fluorine | hydrogen | 6700 | 388 |
| fluorine | JP-4 | 7340 | 306 |

## SOLID PROPELLANTS

| Fuel | Specific Impulse |
|------|------------------|
| $KClO_4$ + thiokol rubber | 170–210 |
| $NH_4ClO_4$ + thiokol rubber | 170–210 |
| $NH_4ClO_4$ + polyurethane | 210–250 |

Consider the problem of obtaining 6,000,000 lb of thrust for a manned space vehicle, assuming a specific impulse of 300. This will require an enormous consumption rate:

$$\text{thrust} = wI_{sp}$$
$$6{,}000{,}000 = 300w$$
$$w = 20{,}000 \text{ lb/sec}$$

## 13.15  explosives

Explosives are materials which decompose rapidly and violently, producing large volumes of gases. Like most materials, explosives are available in a variety of properties to meet their wide range of uses, which even includes the explosive forming and explosive welding of metals.

A widely used explosive for such purposes as removing tree stumps is the homemade recipe of a mixture of Elephant brand fertilizer and diesel fuel oil. This is detonated with a blasting cap. The fertilizer mentioned is an ammonium nitrate, which is a constituent of explosives. Cellulose nitrate, the first of the plastics, is also a constituent of explosives (and the most flammable of the plastics).

Military explosives are organic nitrates such as TNT (trinitrotoluene) formulated for the purpose of violent shock. Commercial explosives include:

1. detonating explosives, which explode almost instantaneously
2. high explosives such as nitroglycerin (dynamite), which burn over a brief period
3. low explosives such as black blasting powder, which have an action between detonation and combustion

These several types present a graduated series in terms of the rapidity with which the gas pressure of the explosion products is developed. An explosion even slower than that of a low explosive would probably be termed combustion, such as occurs in a gasoline engine cylinder.

Five properties of explosives that are of chief interest here are: strength, velocity of detonation, fume production, density, and water resistance.

The *strength* of an explosive is its energy content. The energy of an explosive, as with any energy, is its ability to perform work. For an explosive, this strength is measured as the percentage by weight of nitroglycerin in dynamite. A 40 percent dynamite contains 40 percent liquid nitroglycerin (or the equivalent in some other explosive) with 60 percent of other gas-producing material such as wood pulp, ground meal, sodium nitrate, etc. Commercial explosive strengths range from 15 to 90 percent.

The *velocity of detonation*, expressed in feet per second, is the speed at which the detonation travels through a column of the explosive. Greater

Fig. 13-10    Firing of a Black Brant rocket at the Fort Churchill
rocket range.

velocity produces greater shattering effect, and when increased fragmenta-
tion is desired, higher velocity explosives are selected. Velocities may range
from 1500 to 25,000 fps.

*Density* is expressed as "cartridge count," or the number of $1\frac{1}{8} \times 8$ in.
cartridges contained in a 50-lb case.

*Fume production* is important only in underground work. The bulk of
the explosive gases are the usual products of combustion: carbon dioxide,
water vapor, and nitrogen. But a proportion of these gases, such as carbon
monoxide, are toxic. Dangerous concentrations of toxic gases must not be
used underground.

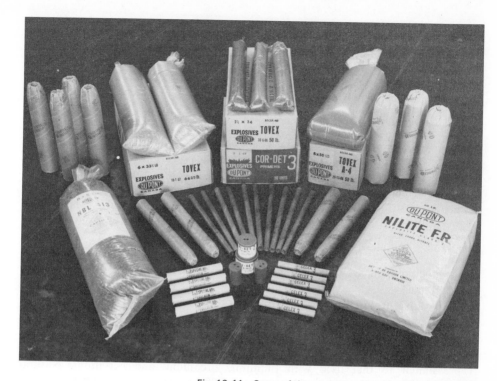

Fig. 13-11  Some of the products offered by Du Pont's explosives department to fill needs of the mining, construction, and quarrying industries. From the left, front row: NBL-413, a water-gel blasting agent for open pit work; Special Gelatin dynamite, used mainly in construction; "Trimtex," for use in controlled blasting procedures; various types of "Hi-Det" primers used in priming "Tovex" water gels; "Gelex" semigelatin dynamite, used chiefly in underground mining but also on construction projects; "Nilite" blasting agent (dyed ammonium nitrate prills and oil) widely used in underground mining, open pit mining, quarrying, and construction (when used underground it normally is loaded by pneumatic loaders); back row: a larger size of Special Gelatin; "Tovex" water gel products, largely used in open pit work; "Cor-Det" primers, a larger type of "Hi-Det" primers; and "Energex" dynamite, chiefly used on construction projects.

The high explosives include nitroglycerin dynamites, ammonia dynamites, and gelatin dynamites. The most used are the ammonias. Blasting powder, a low explosive, is a black powder containing sodium nitrate with some sulfur. It is the slowest of the explosives, giving a heaving action that does not shatter, and is valued because it does not produce excessive fines when breaking stone or coal.

The use of explosives requires such accessories as safety fuses and blasting caps. Safety fuses are used to ignite the blasting caps, which in turn detonate the explosives. A commercial safety fuse burns at a rate of about 1 yard every 2 minutes. The explosive may also be detonated electrically by means of a blasting machine wired to electric blasting caps.

## 13.16   nuclear fuels

Combustion reactions utilize the energy released from rearrangement of the electron clouds of the atoms. Such energy is of the order of a few electron volts per atom. Nuclear fuels use dissociation of the atomic nucleus for energy. The energies associated with nuclear reactions are measured in millions of electron volts per atom. Fission of an atom of U-235 releases about 200 Mev; oxidation of a carbon atom produces only 4 ev.

Nuclear fuels are spontaneously fissionable materials. Only three nuclear fuels are currently in use: Uranium-235, Uranium-233, and plutonium.

Natural uranium is distributed in the earth's crust in amounts of about 3 grams per ton, but it is mined from more concentrated ores such as pitchblende or carnotite. It is usually extracted from the ore by conversion to uranium tetrafluoride, with magnesium reduction of the fluoride, a method similar to that used with titanium, zirconium, and beryllium. Natural uranium is composed of two isotopes, 99.3 percent U-238 and 0.7 percent U-235 (neglecting 0.005 percent U-234). Plutonium does not occur in nature but is produced by neutron irradiation of U-238. U-233 is produced by similar irradiation of thorium.

For nuclear reactor use, uranium may be enriched with U-235, or used as oxide or carbide. In the United States and Russia, uranium enriched with U-235 is frequently selected. A reactor using such enriched fuel is of very compact design; an enriched fuel therefore must be used in atomic submarines or ships. Enriched fuels, however, are expensive. In Canada, natural uranium is employed. Since only the small U-235 content of the natural uranium can undergo fission, Canadian reactors have the problem of careful design for effective use of the small amount of U-235. On the other hand natural uranium is a very cheap fuel.

The reactor fuel is formed into pellets or some other shape and loaded into fuel rods of zirconium or aluminum, such as those shown in Fig. 12.7. These canning materials protect the nuclear fuel from corrosive attack, for the nuclear metals are highly reactive.

Uranium has a half life of 4.5 billion years. It has three solid phases:

$\alpha$   phase, orthorhombic, up to 660°C (1220°F)
$\beta$   phase, tetragonal, 660–770°C (1220–1420°F)
$\gamma$   phase, bcc, 770–1132°C (1420–2065°F)

Its melting point is 1132°C. The alpha phase is soft and plastic, so that many forming operations such as rolling are performed in this phase. The tetragonal beta phase is brittle, while the gamma phase is soft. Extrusion is performed in the gamma range.

Uranium is quite reactive and attacks crucibles and forming dies. It work-hardens readily, especially at temperatures below 750°F. It develops large dimensional changes from thermal cycling, and being strongly aniso-tropic, expansion effects are not completely reversible.

Uranium forms three principal oxides: $UO_2$, $UO_3$, $U_3O_8$. There are also other oxides. Because of the reactivity of the metal, the oxide is often employed as a nuclear fuel. The most stable oxide, $UO_2$, is preferred.

Hardness and strength of uranium are increased by radiation. Radia-tion also produces a decrease in thermal conductivity and dimensional increase along certain crystal axes, though such growth can be controlled by suitable heat treatment. The dioxide, however, is dimensionally stable under irradiation.

Since there are no ores of plutonium, this metal must be manufactured by irradiation in a nuclear reactor. Plutonium exists in six solid phases, the transformation temperatures between phases depending somewhat on heating and cooling rates:

$\alpha$  phase, monoclinic, up to 122°C (252°F)
$\beta$  phase, monoclinic, 122–205°C (252–400°F)
$\gamma$  phase, orthorhombic, 205–318°C (400–606°F)
$\delta$  phase, fcc, 318–450°C (606–840°F)
$\delta'$  phase, tetragonal, 450–480°C (840–896°F)
$\epsilon$  phase, bcc, to melting point

The coefficients of thermal expansion of some of these phases are remark-able. The alpha phase has the highest expansion coefficient of any metal, while the delta and delta prime phases have a negative expansion coefficient (as has the metal polonium). The electrical resistance of the phases is also remarkably high for a metal, and this resistance decreases with temperature as occurs with semiconductors, but not with other metals.

Thorium is a rather soft metal, fcc up to 1400°C (2550°F) and bcc above this temperature.

The dimensional changes and loss of thermal conductivity caused by irradiation of solid nuclear fuels present serious problems in their utiliza-tion. Other difficulties arise from the nuclear reaction itself. In the fission of an atom, two fission-product atoms are produced, both daughter atoms being foreign atoms. Thus neutron irradiation of U-235 may result in the transformations shown in Fig. 13.12. The generation of new atoms in these nuclear reactions results in a volume increase. About 10 percent of the product atoms are xenon and krypton, which can collect to form small gas pockets.

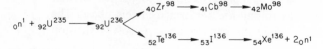

$$_0n^1 + {}_{92}U^{235} \longrightarrow {}_{92}U^{236} \Big\langle \begin{array}{l} {}_{40}Zr^{98} \longrightarrow {}_{41}Cb^{98} \longrightarrow {}_{42}Mo^{98} \\[6pt] {}_{52}Te^{136} \longrightarrow {}_{53}I^{136} \longrightarrow {}_{54}Xe^{136} + 2{}_0n^1 \end{array}$$

Fig. 13-12   Typical transformation due to neutron irradiation of
U-235.

Plutonium has a high neutron absorption for slow neutrons, so that unless the plutonium fuel elements have a very thin section, the outermost plutonium will absorb the neutrons and the core material would not participate in the nuclear reaction.

Neutron irradiation of nonfissile materials knocks atoms out of the space lattice, resulting in vacancies. Most metals are hardened, strengthened, and embrittled by irradiation with neutrons. Body-centered cubic metals develop an increase in transition temperature in Charpy tests after heavy irradiation. Polyethylene develops a degree of cross-linking and hardening under heavy doses of irradiation.

## 13.17   refrigerants

The refrigerants are used as heat transfer materials to move heat from a low-temperature region, the refrigerator, to a high-temperature region, the refrigeration unit. Refrigerants, especially the Freons, are also used to pressurize aerosol cans. It was noted earlier that the refrigerants are imperfect gases; therefore refrigerant tables are available for the determination of temperature, pressure, volume, and weight.

A few refrigerants such as ammonia are inorganic. The important group of refrigerants known as the Freons are chlorinated or fluorinated hydrocarbons, the basic hydrocarbons being methane and ethane. The American Society of Heating, Refrigeration, and Air Conditioning Engineers (ASHRAE) designates the refrigerants based on methane with numbers below 100; those based on ethane have numbers from 100 to 200. Thus Freon-12 is a methane refrigerant.

The actual system designation for refrigerants may be illustrated using F-114. The chemical formula for F-114 is $C_2Cl_2F_4$, dichlorotetrafluoroethane. The last digit gives the number of fluorine atoms in the molecule (4 fluorine atoms). The last digit but one is one more than the number of hydrogen atoms. Since there are no hydrogen atoms in F-114, this number is a 1. The last digit but two is one less than the number of carbon atoms, but if the carbon atoms number only 1, this number is omitted, as in Freon-12.

## ASHRAE STANDARD DESIGNATION OF REFRIGERANTS

| Number | Name | Formula |
|--------|------|---------|
| 12 | dichlorodifluoromethane | $CCl_2F_2$ |
| 22 | monochlorodifluoromethane | $CHClF_2$ |
| 40 | methyl chloride | $CH_3Cl$ |
| 50 | methane | $CH_4$ |
| 114 | dichlorotetrafluoroethane | $C_2Cl_2F_4$ |
| 160 | ethyl chloride | $C_2H_5Cl$ |

Refrigerants can deteriorate by three mechanisms:

1. pyrolysis —destruction by heat
2. hydrolysis—destruction by reaction with water
3. oxidation —destruction by reaction with oxygen

Both pyrolysis and hydrolysis will reduce Freons to hydrogen chloride or hydrogen fluoride. Both compounds are highly corrosive to refrigeration systems.

A large number of characteristics are desired in a refrigerant:

1. no odor, since odor may contaminate food
2. no toxicity, so that food will not be harmed
3. no explosion hazard in any concentration in air
4. no corrosive attack on the metals of the refrigeration system
5. chemical stability at both the low and high temperatures in the refrigeration system
6. insolubility in water
7. ease of leak detection
8. high electrical resistance. Many motors of refrigeration condensing units are exposed to refrigerant vapor
9. good thermal conductivity
10. low viscosity
11. ability to dissolve lubricating oil, to prevent solidification of heavy oil waxes on heat transfer surfaces
12. low cost
13. low specific volume to allow the use of equipment of small size

This is rather a demanding list to impose on any material.

Ammonia is only used in large industrial refrigeration systems. Its name is derived from Ammon, the Egyptian sun-god. If moisture is present, ammonia forms ammonium hydroxide which is corrosive to copper and copper alloys. Freon-12 is a common refrigerant in small installations. It is corrosive to natural rubber, but neoprene gaskets are resistant to it. Carbon dioxide is often selected for hospitals and ships because of safety considerations. In the form of dry ice it makes an excellent and cheap expendable refrigerant.

## PROBLEMS

1 What is the weight in pounds per cubic foot of a crude petroleum with an API gravity of 6? Of a jet engine fuel with an API gravity of 38?

2 Explain the purpose of the basic oil refinery operations of fractional distillation, cracking, and polymerization.

3 Technically, what is a grease?

4 What are the principal uses of petroleum coke?

5 In what characteristic is petroleum coke superior to coke produced from coal?

6 Can gasoline be produced from coal? How might this be done?

7 Which material, petroleum or coal, contains more paraffins and which contains more aromatics?

8 What is (a) a paraffin, (b) an olefin, (c) an aromatic?

9 Name some inorganic fuels.

10 What is the difference between a fuel and an explosive?

11 What are the differences between the liquid fuels employed in gasoline, diesel, and gas turbine engines?

12 What is the practical significance of the four components of the proximate analysis of a solid fuel?

13 Attempt to estimate the proximate analysis of wood, assuming 20 percent moisture. Consult the proximate analyses of similar materials given in the text.

14 Gas turbine engines are made of high-nickel superalloys. Why do fuel specifications for these engines limit sulfur to 0.5 percent maximum?

15 What is the objection to paraffinic fuels in automobiles?

16 (a) Why is butene blended into winter gasoline for subzero weather?
(b) Why is butane not blended into aviation gasoline for the same purpose?

17 Explain the meaning of octane number for gasoline.

18 How is the strength of an explosive measured?

19 What problems are avoided by using uranium oxide instead of metallic uranium as a nuclear fuel?

20 What is enriched uranium fuel?

21 Why is uranium fuel canned for use in reactors?

22 What is a hydrocarbon? A petrochemical?

23 Sketch the chemical formula for methane, ethane, and ethylene.

24 Sketch the chemical formula for propyl alcohol.

**25** What is a sour gas?

**26** What is the meaning of polymerization?

**27** List the following solvents in order of increasing solvent power: aromatics, chlorinated solvents, paraffins.

**28** Define the flash point of a volatile liquid.

**29** Why does a nuclear fuel increase in volume during use in a nuclear reactor?

**30** Explain why methyl chloride refrigerant is designated by the number 40.

**31** Assume that your city or town generates 4 lb of packaging materials as garbage per day from household and industrial activity. Estimate an average heating value for such packaging material. Using a combustion efficiency of 33 percent and 1000 Btu to generate 1 lb of steam, how many pounds of steam per hour could be generated from this refuse?

## 14.1 petrochemical intermediates

The petrochemical intermediates are the chemicals such as ethylene manufactured from the paraffins in petroleum and natural gas, which are then processed further into finished products. The intermediates are the raw materials for almost all rubbers and plastics. Despite their importance, the petrochemical intermediates consume only a few percent of oil and gas production. Intermediates may also be manufactured from the aromatics in coal, which is the source of about one-third of the intermediates produced on this continent.

The most important of the intermediates is ethylene. Besides such products as engine antifreeze (ethylene glycol), many plastics and rubbers are produced from ethylene. The chief refinery feed stocks used for cracking into ethylene are ethane and propane, but butane, naphthas, and gas oils may also be used. In principle, any organic chemical can be made from any fraction of petroleum or natural gas.

Other olefin intermediates are acetylene, which is also a finished product, propylene, butylene, isobutylene, and butadiene. Most of these are used in

# PLASTICS
# AND RUBBERS

## 14

the production of rubbers. Nylon may be produced from butadiene (pronounced buta-dye-een). Acetylene can be converted into such plastics as
polyvinyl chloride and Orlon.

Next to the olefins, the most important group of intermediates is the
aromatics, chiefly the BTX trio benzene, toluene, and xylene. Of the cyclic
naphthenes, the most significant intermediate is cyclohexane, used in the
production of nylon.

Insecticides, detergents, solid rocket fuels, photographic films, pharma-
ceuticals, solvents, explosives, and alcohols are a partial listing of the end
uses of the olefin, aromatic, and naphthenic intermediates mentioned.

There are several cellulose plastics. These are not produced from coal,
oil, or gas, but from wood.

Fig. 14-1   Conversion of light paraffins into unsaturated olefin intermediates.

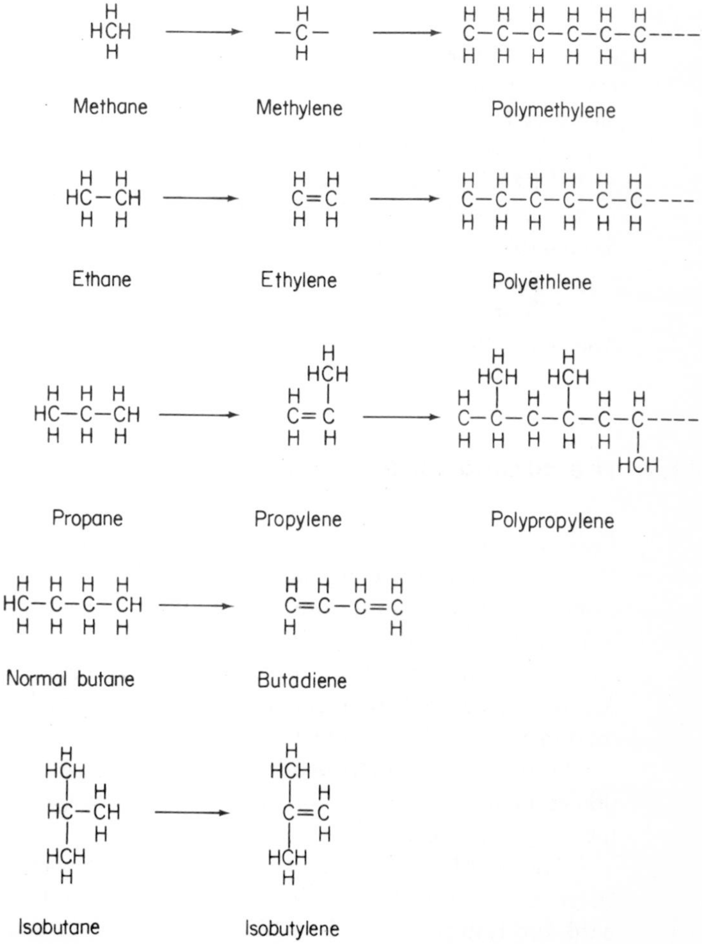

The production of polymers from intermediates will be illustrated by the simplest cases. Consider the first five paraffins:

| | |
|---|---|
| methane | $CH_4$ or $C_1$ |
| ethane | $C_2H_6$ or $C_2$ |
| propane | $C_3H_8$ or $C_3$ |
| butane | $C_4H_{10}$ or $C_4$ |
| pentane | $C_5H_{12}$ or $C_5$ |

To make each into its own polymer, each must first be converted to its corresponding olefin, methylene, ethylene, propylene, etc. See Fig. 14.1. Note that methylene has two unsatisfied bonds, so that it is an unstable and transitory compound. Butane has two common intermediates, isobutylene and butadiene.

These monomers are then polymerized by addition. Thus if 500 or more ethylene monomers polymerize, the result is polyethylene plastic.

Polymethylene has a formula similar to that of polyethylene. While polyethylene is a very flexible and soft plastic, polymethylene is brittle and therefore is not commercially available. Polypropylene is a more recent plastic and one of the lowest in price. Polypentylene is not commercially available. Polyisobutylene and polybutadiene are rubbers.

Although these polymers are diagrammed as straight chains, there is some angularity to the linkages between carbon atoms, this angle being about 110° in the case of polyethylene and similar addition polymers. The high elongation of many of these materials is explained by the straightening out of these angular linkages. Also there are always some short side chains attached to the main carbon chain.

## 14.2  the common thermoplastics

There are two basic types of polymers, *thermoplastic* and *thermosetting*. The thermoplastics are produced by addition polymerization of additional monomers to the chain and are characterized by their capacity to be repeatedly softened by heating and solidified by cooling. The thermosets are not produced by addition polymerization, and after curing they cannot be resoftened. Asphalt is a naturally occurring thermoplastic, softenable by heating. Natural thermosetting polymers are wood, cotton, wool, hair, and feathers. These can be burned, charred, or otherwise damaged by high temperatures, but they cannot be softened. For the present, attention will be directed to the thermoplastic polymers.

The word "resin" occurs frequently in association with polymers. Originally it meant a naturally occurring material used in coatings to form a hard and lustrous finish. In a technical sense a resin is an amorphous and high viscosity liquid or solid of high molecular weight which softens on heating. The term now however has been extended to include all the polymers used in the plastics industry.

Suppose that we wish to invent a plastic rain gutter for houses (this of course has already been invented). The polymers derived from the light petroleum fractions above are not suitable for this purpose. The butane polymers are rubbers; polyethylene is too soft, and all these polymers are too combustible for use as rain gutters on houses. Any chemical material made up of strings of carbon and hydrogen atoms cannot be insured against fire. If no building code or insurance code will allow polyethylene for this purpose, then the ethylene monomer must be altered to make it nonburning. The element chlorine does not burn. By replacing at least one hydrogen atom by chlorine in the monomer, a nonburning thermoplastic material results. Chlorine bonds once, like hydrogen. The modified monomer is called vinyl chloride (Fig. 14.2).

Fig. 14-2    Vinyl chloride monomer.

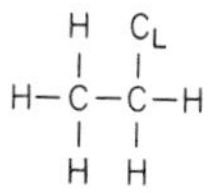

$$\begin{array}{cc} H & C_L \\ | & | \\ H-C-C-H \\ | & | \\ H & H \end{array}$$

The resulting polyvinyl chloride proves to be an excellent choice for a rain gutter. It does not burn, and unlike polyethylene, it is a stiff plastic. It is also low in price and can be made resistant to the deteriorating effects of ultraviolet radiation in sunlight. A view of a rain gutter fitting made of PVC is given in Fig. 14.3.

Fig. 14-3 An injection-molded rainwater fitting, gutter to down-spout, made of rigid polyvinyl chloride.

If we substitute two chlorine atoms for two hydrogen atoms in the ethylene atom, polyvinylidene chloride of Fig. 14.4 results. This material

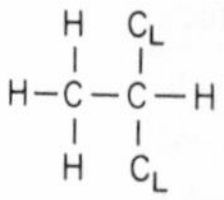

```
H   Cl
|   |
H—C—C—H
|   |
H   Cl
```
Fig. 14-4    Vinylidene    chloride
monomer.

also bears the trade name Saran. The addition of chlorine atoms can be carried to the maximum of four.

The element fluorine, with a single bond, also does not burn. One hydrogen atom in the ethylene monomer can be replaced by fluorine to give the polyvinyl fluoride of Fig. 14.5. PVF is a thermoplastic of outstanding

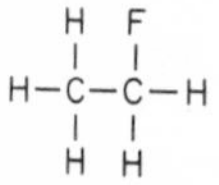

```
H   F
|   |
H—C—C—H
|   |
H   H
```
Fig. 14-5   Vinyl fluoride monomer.

properties, is nonburning, and is used as a surface film to protect other materials (such as the interior furnishings of passenger aircraft) from weathering, ultraviolet degradation, corrosion, or damage by scuffing. It is sold under the trade name Tedlar.

Two fluorine atoms give polyvinylidene fluoride, another plastic used as film like PVF, but less well known. Four fluorine atoms produces the famous Teflon, polytetrafluoroethylene, with outstanding resistance to elevated temperatures. A great many other monomer possibilities can be designed using fluorine or chlorine or both.

Clearly the chemist, or even the reader of this book, can continue to "invent" thermoplastics almost without end. Each formula so easily invented would of course have to be manufactured and tested for properties and cost, and many, such as polymethylene, would be found unsuitable for reasons of characteristics, cost, or processing limitations.

## 14.3  *four basic types*

### *of polymers*

In earlier years the polymers were separated into two broad classes, plastics and elastomers or rubbers, with the plastics divided into thermoplastic and thermosetting. Rubbers were elastic, or bounced; plastic deformed plastically. There are many inconvenient exceptions to this simple classification: Bakelite (phenol-formaldehyde) does not strain plastically even though it is called a plastic. Nor do the epoxies. Further, polyurethanes can be formulated for either rubber or plastic properties.

Based on stress-strain characteristics, four types of polymeric materials can be distinguished.

1. Flexible thermoplastics, such as polyethylene, capable of large plastic deformation.
2. Rigid thermoplastics, such as unplasticized polyvinyl chloride and polystyrene, limited in their maximum possible strain and therefore brittle.
3. Rigid thermosets, such as epoxy and phenol-formaldehyde, also brittle.
4. Elastomers or rubbers, distinguished by remarkable elastic extensibility. Most rubbers are thermosets.

In a standard stress-strain test each of the four types gives a characteristic shape of stress-strain curve. These are shown in Fig. 14.6, with typical stress and strain values.

Fig. 14-6 Typical tensile stress-strain curves for polymers. These do not disclose any tendency for creep under sustained loads, nor can behavior in compression be predicted from these curves. Values given for maximum stress and strain are typical ones. (a) Flexible thermoplastic; (b) rigid thermoplastic; (c) thermosetting plastic; (d) elastomer.

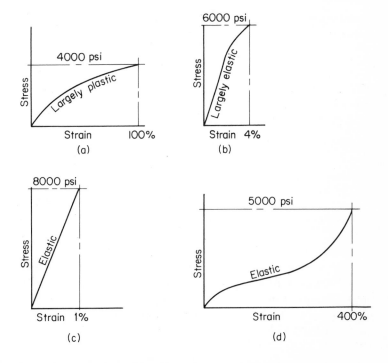

All four types of polymers exhibit considerable creep and stress relaxation under conditions of prolonged stress. If the plastic part is loaded and kept under constant strain, the stress required to maintain this strain will decay. Such decay of stress is called "stress relaxation." Stress relaxation curves for polymer materials have the exponential shape of Fig. 14.7.

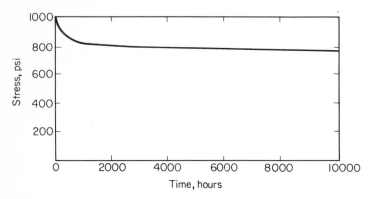

Fig. 14-7   Stress relaxation curve for a polyethylene with an initial applied stress approximating its yield stress.

The stress relaxation for the first day may possibly equal that for the next 300 days. If the stress is maintained constant, then the strain will increase in the manner of Fig. 14.8. Polymer materials therefore can sustain considerably higher loads for short times than for prolonged loading conditions.

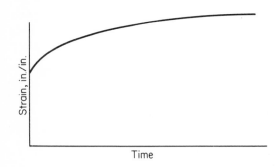

Fig. 14-8   Typical constant stress curve for plastics.

Fig. 14.9 shows the short time, long time, and allowable design stress for an unplasticized polyvinyl chloride. Using a factor of safety of 3, at room temperature or 20°C, an allowable short time stress level of about 3000 psi is possible, but for prolonged loading at this temperature a stress of 900 psi should not be exceeded.

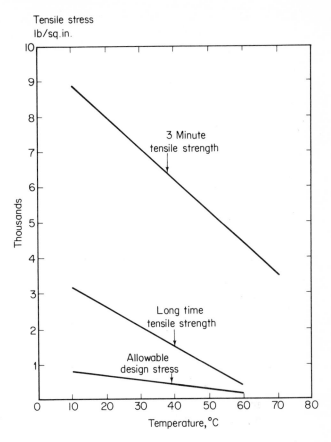

Fig. 14-9 Short time, long time, and design stresses for an unplasticized polyvinyl chloride.

## 14.4 the polymer molecule

The synthetic thermoplastics are molecules which are chains of 500 or more carbon atoms, the distance between adjacent carbon atoms being close to 1.5 A (1 A $= 10^{-8}$ cm). In discussing the length of a polymer molecule we are necessarily referring to an average length, since it is not possible, and it is not necessary, for all molecules to be composed of the same number of monomers. However some control over the length of the molecule (actually molecular weight rather than length) is possible.

Most monomers are gases. A short polymer chain of low molecular weight will be a liquid. If the molecular weight is large enough, the polymer is a solid. The entanglement and attraction between such large molecules

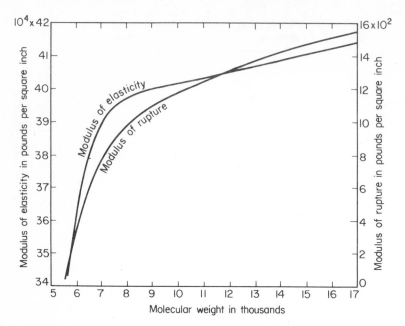

Fig. 14-10   Effect of molecular weight on modulus of elasticity and modulus of rupture of vinyl chloride-acetate copolymer.

account for their mechanical strength. Higher molecular weight results in increased strength and elastic modulus, as illustrated in Fig. 14.10. As is usual for any material, increased strength will be accompanied by decreased ductility. The increase in mechanical properties levels off as molecular weight becomes exceedingly large.

For polyethylene, ethylene is a gas, a polymer of 6 monomers of ethylene is a liquid, 36 is a grease, 140 is a wax, and 500 is the familiar plastic.

The best mechanical properties therefore are provided by highest molecular weight. However the very largest molecular weights are the most difficult to process in such plastics converting operations as extrusion, injection molding, and blow molding. Therefore the molecular weight selected is the best compromise between properties and processability. In the case of rubbers, coatings, and thermosets, a low molecular weight is used for ease of processing, and this low weight is later converted into a high molecular weight for best properties. A very high molecular weight is accompanied by a very high melt viscosity. To reduce this viscosity the processing temperature must be elevated, and the resulting excessively high temperatures degrade the plastic and increase the cost of production. At the same time the molecular weight must be adjusted to the process: an extrusion grade of thermoplastic will have a higher molecular weight than an injection-molding grade because the hot extrusion must have sufficient viscosity to support itself when it leaves the die. For a plastic foam, the melt viscosity must be low enough to permit expansion by the foaming gas, but it must be high enough so that the thin walls of the foam cells do not rupture during expansion. Hence like metals, plastics must be formulated to suit the manufacturing process.

To improve the processability of some polymers of high molecular weight, especially polyvinyl chloride, plasticizers may be added to the material. Unplasticized polyvinyl chloride (UPVC) is hard and somewhat brittle. When plasticized, it is suitable for garden hose, shower curtains, raincoats, and packaging film, all of which require a rubbery flexibility. The plasticizer is a liquid such as dioctyl phthalate that lowers the overall molecular weight.

Many thermoplastics, though not all, are deteriorated by prolonged exposure to oxygen of the air and ultraviolet radiation. Resistance to such deterioration is improved by higher molecular weight for three reasons:

1. Higher molecular weight means fewer molecules per pound or per unit volume. Fewer molecules means fewer terminal monomers in the chain, and these are necessarily the reactive monomers.
2. The longer molecule is less mobile.
3. A high molecular weight if degraded will degrade only to a medium molecular weight with acceptable properties instead of a low molecular weight.

Aging has as one of its effects the reduction of molecular weight.

## 14.5  crystallinity

The simplest of the polymer structures is polyethylene. In the molecule of polyethylene, the hydrogen atoms lie in planes perpendicular to the plane of the carbon atoms, which have a zigzag arrangement (see Fig. 14.11). Polytetrafluoroethylene (Teflon) is similar to polyethylene, with the four hydrogen atoms of polyethylene replaced by four fluorine atoms. But in PTFE the carbon chain describes a helix, with 14 carbon atoms per turn of the helix.

Fig. 14-11 Configuration of the polyethylene molecule. Each carbon atom has two hydrogen atoms attached, one above and one below the plane of the paper.

A polymer with the shape of a coil spring, and with side chains or attached methyl ($CH_3$) groups or aromatic rings, has a shape that is hardly susceptible to crystal ordering. It is not surprising that many polymer materials are amorphous or glasses. But often a degree of crystal ordering on a submicroscopic scale is possible. Such ordered regions are called *crystallites*. The whole molecule, because of its great length, may extend through several crystalline and noncrystalline regions, as illustrated in Fig. 14.12. The greater the crystallinity or the number of crystallites, the greater the specific weight of the polymer because of the closer packing of the molecules in the crystalline phase. Thus low-density polyethylene may be as much as 55–70 percent crystalline, while high-density polyethylene may be 75–95 percent crystalline.

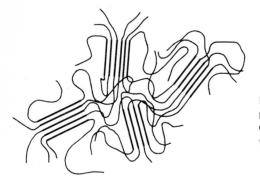

Fig. 14-12 The coiling of long polymer chains through several crystalline and noncrystalline regions.

An advantage of crystallinity is that the polymer product may be used at higher temperatures than are possible with a more amorphous product. Instead of softening gradually with increased temperature, the crystalline polymer tends to exhibit a sharp melting point characteristic of any crystalline material. A more crystalline polymer will have better resistance to water absorption and to solvent attack. The crystalline regions are closely ordered segments while the amorphous regions present an open and random arrangement which can be penetrated by water, solvents, or permeating liquids. High crystallinity makes the use of plasticizers impractical, because the plasticizer cannot penetraιe the crystallites. Plasticizers are therefore not used with crystalline polymers such as polyethylene and nylon. Similarly, the crystalline polymers tend to be impermeable to gases, a characteristic that may be useful in food packaging or in protective coatings. Such impermeability is a disadvantage in polymer fibers, which usually must receive dyes.

Ultimate tensile strength increases with crystallinity as shown for polyethylene in Fig. 14.13, but the reduced mobility of the molecules caused by crystallinity reduces ultimate elongation. The modulus of elasticity of polyethylene is greatly influenced by crystallinity, as shown in Fig. 14.14.

The relative crystallinity can be altered by heat treatment of those polymers with the capacity to crystallize. A crystallizable polyethylene can be given a crystallinity of about 80 percent by slow cooling or 65 percent by rapid cooling.

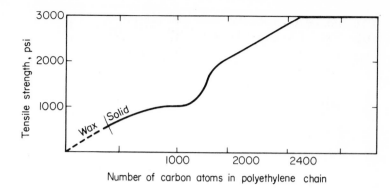

Fig. 14-13  Variation of tensile strength with length of carbon chain in a typical polyethylene.

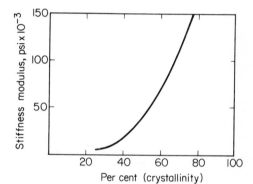

Fig. 14-14  Effect of degree of crystallinity on elastic modulus of polyethylene.

In addition to the amorphous and crystalline conditions, an intermediate condition, the *oriented* condition, can be produced. If a polymer is drawn into fibers through a die, the molecules and microcrystals of the polymer are aligned in the direction of drawing. The oriented condition gives very high tensile strength levels and impact strength in the direction of orientation. Polymer fibers, being oriented, are considerably stronger than bulk polymers. The production of polyethylene film employs extrusion with air-blowing, as seen in Fig. 14.15; such film is oriented in both the longitudinal and cross direction, with the additional advantage that the film is improved in clarity.

Crystallinity is improved by ordering in the chain of monomers. The molecule of polypropylene is shown in Fig. 14.16. If the methyl unit ($CH_3$) is attached at random locations above and below the carbon chain as in the first part of the figure, then the polymer cannot crystallize. Such an irregular orientation is called the *atactic* structure. A regular arrangement of methyl groups, as in the second part of the figure, makes crystallinity possible. The tactic and atactic structures are produced by different catalysts and processing

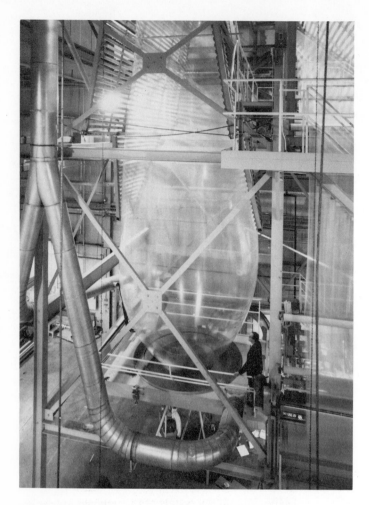

Fig. 14-15 The largest polyethylene extruder ever built in Canada. This extruder produces industrial film at Union Carbide Canada Limited's Lindsay, Ontario plant. The 8-in. extruder can make film 40 ft wide and has a capacity in excess of 1500 lb/hr. (Photo courtesy Union Carbide Canada Limited, Toronto.)

Fig. 14-16 Polypropylene, atactic, and isotactic structure.

344 / *plastics and rubbers*

conditions during polymerization. In the isotactic structure all the methyl units are located on one side of the carbon chain; they alternate on both sides in the syndiotactic configuration.

## 14.6 stress-cracking

When a polymer is subjected to a moderate stress and at the same time exposed to a weak solvent or surface-active material for a prolonged period, a brittle type of failure called *environmental stress-cracking* may occur. The chemical causing the stress-cracking may be a relatively harmless chemical such as a soap or a detergent. Stress-cracking probably is initiated at the weaker amorphous regions between crystallites, even though the condition is often more commonly met in polymers of high crystallinity. Low-molecular-weight fractions in the polymer also contribute to stress-cracking. The stress need not be an applied stress; it may be a residual (internal) stress resulting from the forming process used to shape the plastic part. A blow-molded bottle contains such locked-in stresses at its corners, and if it is filled with detergent it may then fail by stress-cracking. Polyethylene especially is susceptible to this type of failure.

## 14.7 standard abbreviations

The following discussion of the characteristics and properties of specific thermoplastics must be limited to those resins in common use or of unusual significance. All these plastic materials are made in a range of grades: an extrusion grade of thermoplastic, for example, will not be quite the same as an injection-molding grade. Therefore when values for such properties as ultimate tensile strength are reported in the text, the reader should take these values as order-of-magnitude or "ball-park" figures. A polypropylene carpet fiber will have a much higher strength than a propylene film, and the film a higher strength than an injection-molded polypropylene article.

Three plastics (polyvinyl alcohol, methyl cellulose, and polyethylene oxide) are soluble in water. They will not be further discussed. The cellulose plastics will be examined in a later chapter.

The American Society for Testing and Materials (ASTM) recommends the following abbreviations. Both thermoplastics and thermosets are included in the list, which is of course not a comprehensive listing of plastics.

| ABS, | acrylonitrile-butadiene-styrene |
|------|---------------------------------|
| CA, | cellulose acetate |
| CAB, | cellulose-acetate-butyrate |
| CN, | cellulose nitrate |
| EP, | epoxy |
| MF, | melamine-formaldehyde |
| PF, | phenol-formaldehyde |
| PAN, | polyacrylonitrile |
| PA, | nylon (polyamide) |
| PC, | polycarbonate |
| PE, | polyethylene |
| PETP, | mylar (polyethylene-terephthalate) |
| PMMA, | acrylic, plexiglas, lucite (polymethyl methacrylate) |
| PP, | polypropylene |
| PS, | polystyrene |
| PTFE, | Teflon (polytetrafluoroethylene) |
| PVAC, | polyvinyl acetate |
| PVAL, | polyvinyl alcohol |
| PVB, | polyvinyl butyral |
| PVC, | polyvinyl chloride |
| PVF, | polyvinyl fluoride |
| UF, | urea-formaldehyde |
| UP, | urethane plastic |

## 14.8 the polyolefins

The polyolefins include the plastics PE and PP and the rubbers polyisoprene and polyisobutylene.

PE is a true plastic in that it is capable of large plastic deformation. Its strength is low, with low elastic modulus, high expansion and mold shrinkage, and creep at room temperature. It is thus not suited to applications requiring dimensional stability. Despite these limitations, PE has developed remarkably large markets, because of its low cost, low weight, and pleasant feel.

PE is produced in three grades: low density, intermediate density, and high density. The specific gravities of these three types are:

$$0.912—0.925$$
$$0.925—0.940$$
$$0.940—0.965$$

The degree of crystallinity is proportional to the specific gravity, and the heat resistance improves with specific gravity. The higher densities are less waxy in appearance and to the touch.

All the polyolefins are degraded by ultraviolet radiation. When resistance to weathering and ultraviolet radiation is required, PE is blended with about 2 percent carbon black.

The largest market for polyethylene is film, most of which is consumed by the packaging industry. It is used by the construction industry as a vapor barrier for walls and for concrete which is setting. Most film is made of low-density material because of cost, although the high-density PE is a much better vapor barrier. Polyethylene film is available in clear, white, and black colors and in thicknesses of 2, 4, and 6 mils. One mil = 0.001 in.

Other uses include coatings on paper or other materials, injection-molded articles, such as kitchenware, laboratory apparatus, and toys, and blow-molded bottles, containers, and squeeze bottles. Such corrosive chemicals as hydrofluoric acid, ammonia, and rust-removing compounds are now supplied in polyethylene containers. Low-density polyethylene pipe is now familiar in cold-water service. High-density pipe is preferred for use in chemical plants because of its greater rigidity and superior chemical resistance. The superior dielectric properties of polyethylene explain its use for electrical insulation on wire and cable.

Polyethylene, like other thermoplastics, may be made to cross-link by means of gamma radiation in very heavy doses. In the course of radiation, hydrogen is liberated and carbon atoms cross-link between molecules. The polyethylene becomes infusible and withstands prolonged aging at temperatures of almost 300°F.

Propylene is a gas boiling at −48°C at atmospheric pressure. Its polymer *polypropylene* is the lightest plastic so far produced, with a specific gravity of 0.90. General physical and electrical properties of polypropylene are similar to those of high-density polyethylene. Polypropylene, however, is harder and has a higher softening point and lower shrinkage.

Polypropylene film is not as limp as polyethylene film. In addition to film, polypropylene is molded into domestic hollow ware, toys, bottles, and automobile distributor caps.

## 14.9 vinyl and related polymers

PVC is produced in two basic types: unplasticized (UPVC) and plasticized. UPVC is a hard and brittle material with poor impact resistance. The plasticized polymer is rubbery. Either can be clear or colored.

PVC is not as resistant to solvents and aggressive chemicals as PE. The plasticized resin is not expected to resist solvents or chemicals. For construction materials in outdoor use such as house siding, window frames, or rain gutters, colored UPVC must be used. When suitably compounded with ultraviolet stabilizers, UPVC weathers well, though the pure resin does not.

Plasticized PVC cannot endure long exposure to weathering, since the plasticizer migrates and leaches out. A disadvantage of PVC as a building material is its high thermal expansion, a characteristic of all plastics.

Unlike PE or PP, PVC does not burn. All three thermoplastics become soft and generally unusable at 200°F, though PP has the best heat resistance. Overheating PVC releases hydrochloric acid.

Type TW Flameseal electric wire is insulated with PVC. Metal sheeting with a colored PVC film is frequently used for building walls. Like PE, this resin is inexpensive. Its uses are very many and include phonograph records, sheeting and curtain goods, and floor tile.

Polyvinyl butyral is used as an adhesive for the laminating of safety glass for the front windows of automobiles. The familiar white household cement is another vinyl polymer, polyvinyl acetate.

*Polyvinylidene chloride* is a stiff plastic like PVC. It is copolymerized with 30 to 50 percent of vinyl chloride, the resulting copolymer being quite soft and flexible. This copolymer is called Saran by Dow Chemical Corporation. A copolymer is any polymer that contains two or more monomers in its chain; several copolymers will be found in the list in Sec. 14.7.

Saran has a strong tendency to "block" or cling to itself. It is the most impermeable of the transparent films to gases and water vapor.

PVF has outstanding weather and ultraviolet resistance. As a film covering for other materials such as wood or reinforced plastic panels, it protects the underlying material from ultraviolet degradation. Other outstanding characteristics are a very high tensile strength, resistance to high temperature, abrasion resistance, and ease of maintenance. It is virtually unstainable. PVF is available only as film, in thicknesses of 0.5, 1, 1.5, and 2 mils.

PS, like PVC and PE, is a low-cost thermoplastic, and like the other two, it is very widely used. Because of the aromatic ring in the monomer, PS does not crystallize.

PS is a crystal-clear plastic, hard, brittle, and poor in impact resistance. It has a brilliant surface, but since it is easily stress-cracked, it is not recommended for holding liquids. Polystyrene tumblers for kitchen use usually fail by stress-cracking resulting from frequent dish-washing. Solvent resistance is poor, and the resin is rapidly degraded by sunlight. Its heat resistance also is poor. Its monomer formula indicates that it is combustible, and it can be identified by its blue flame when burned (ABS also gives a blue flame).

High-impact grades of PS are produced, usually by copolymerization. ABS, a copolymer of acrylonitrile, butadiene, and styrene, is hard and rigid, but it is tough and impact resistant. Solvents do not readily attack it. In the construction industry it is chiefly used for drain, waste, and vent piping.

PVF was noted as an outstanding thermoplastic. All of the fluorocarbon thermoplastics are outstanding in their properties. There are several of these, but only the famous PTFE (Teflon) will be discussed here. The vinyl fluoride monomer contains one fluorine atom. Other fluorocarbons contain two (vinylidene fluoride) or three (chlorotrifluoroethylene), while PTFE has four.

Increasing fluorine content in these polymers is accompanied by improved electrical characteristics, higher service temperatures, lower coefficient of friction, and better corrosion resistance. The best combination of properties is therefore provided by PTFE.

PTFE is a crystalline (over 90 percent crystallinity) thermoplastic, opaque, soft, waxy, and white in color. Though its monomer boils at $-106°F$, the polymer has a service temperature for continuous exposure from absolute zero to 500°F. It has little strength and creeps readily at room temperature. Specific electrical resistance is over $10^{18}$ ohm-cm, and dielectric strength is also high. PTFE is almost immune to chemical attack, except for molten alkali metals such as sodium and a few fluorine compounds. It provides the lowest unlubricated coefficient of friction of any material: about 0.04.

Though a thermoplastic by reason of absence of cross-linking, PTFE is almost a thermosetting resin in its behavior and processing. It can be extruded or injection-molded only by sintering methods employing powders.

The electric industry uses PTFE for a wide range of specialty applications, including insulation, coaxial cable components, tube bases, plugs and sockets. The chemical and instrumentation industries value its chemical resistance for gaskets, tubing, valve packings, O-rings, valve linings, flexible couplings, and diaphragms. Its low friction and nonstick characteristics explain its numerous applications for rollers, bearings, conveyor coatings, etc.

## 14.10 monomers with aromatics
## or oxygen in the chain

The use of fluorine atom attachments to the carbon chain results in thermoplastics with outstanding properties and heat stability. Other outstanding thermoplastics (and thermosets) result from the use of oxygen or aromatic rings, especially if these are incorporated into the polymer chain. These high-performance plastics have sufficient hardness, toughness, strength, and heat resistance to make them suitable as replacements for metals in mechanical items such as gears, bearings, and fasteners.

**1. Nylons.** Nylons are polyamides, and there are several types, the usual resin being nylon 6/6. Since there is strong attraction between the polymer molecules, crystallinity is high and so therefore is the heat resistance of nylon. Because of crystallinity there is a sharp melting point.

Nylon is tough and stiff, with an intermediate hardness. It is translucent and yellowish-white. Its coefficient of friction with other materials is low and its abrasion resistance high, so that it is useful for bearings and the valves of aerosol cans. However nylon will absorb moisture, which results in a slight

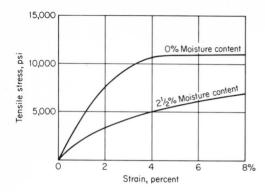

Fig. 14-17   Stress-strain curves for dry and moist nylon, tested at 70°F.

expansion and some slight loss of physical and electrical properties. See Fig. 14.17.

Nylon's largest market has been tire cord and other textile applications, including bristles for brushes. It is also used extensively for mechanical parts: bearings, gears, cams, pulleys, rollers, fasteners, automobile door latches, timing sprockets, and many others.

**2. PMMA.**   More familiarly known by its trade names such as Lucite or Plexiglas, PMMA is a crystal-clear plastic, hard and strong, with outstanding optical properties and resistance to atmospheric and ultraviolet degradation. It has a slightly higher transmission for visible light and higher transmission for ultraviolet radiation than glass, and it is used in place of glass for lighting fixtures, aircraft glazing, outdoor signs, and for window panes where van-dalism is a problem. The cast product has better optical properties and higher strength than the extruded product. The Rohm & Haas Company has sup-plied Plexiglas for almost 200,000,000 automobile tail lights.

Fractured tail lights are not unusual. The impact resistance of PMMA is better than that of glass, but this resin is nevertheless not impact resistant. Polycarbonate is far superior in this respect as a glazing material.

**3. Polyacetals (polyformaldehyde).**   This resin alternates an oxygen atom between carbon atoms in the chain. The most familiar brand name is Delrin. This is a strong and stiff thermoplastic chiefly employed in injection-molded articles such as instrument and business-machine housings, cams, and hinges.

**4. PC.**   Polycarbonate is transparent and is used as a substitute for glass in glazing applications. Though not as stiff as PMMA it is remarkably tough and impact resistant, with outstanding elongation (over 100 percent in a standard tensile test). Except in the presence of stress raisers such as sharp corners, PC is exceedingly difficult to break. It is virtually impossible to destroy a polycarbonate window by violent methods.

The creep resistance and heat stability of PC are exceptional. Molding of this material must be done at temperatures of about 600°F.

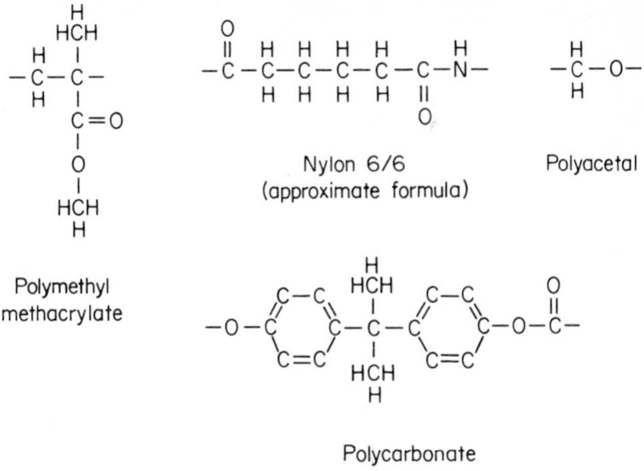

Polymethyl
methacrylate

Nylon 6/6
(approximate formula)

Polyacetal

Polycarbonate

Fig. 14-18  High-performance thermoplastics.

PC has been made into nails, screws, fasteners, kitchen utensils, hard hats, housings for portable power tools, and pump impellers. It is not bondable with the usual adhesives employed with plastics. Solvent adhesives tend to make it brittle.

**5. Special Chain Structures.**  PS has a benzene ring attached to one carbon atom in the monomer. This aromatic attachment does not contribute high strength or high heat stability. The high-performance plastics discussed in this section contain C—0—C—0 chains or incorporate aromatic rings in the chain. Sulfur atoms in the chain also result in improved strength and heat stability. The silicones, to be discussed later, have a silicon-oxygen chain, which also results in remarkable heat stability.

These special chain structures that contribute high strength and heat stability are still under aggressive development, though a few have already seen limited commercial use, such as the polyimides and ionomers. Some of these polymers are not linear chains but are ladder chains such as the polyimide of Fig. 14.19. This is partially a ladder structure because of bonding at two points on the aromatic rings at the left-hand end of the monomer. As these resins approach a completely ladder-like structure, the breakage of any single bond by heat or other attack is less likely to crack the entire molecule. Stability at temperatures exceeding 1000°F has been observed for many of these newer structures. The ladder structure also stiffens the molecule.

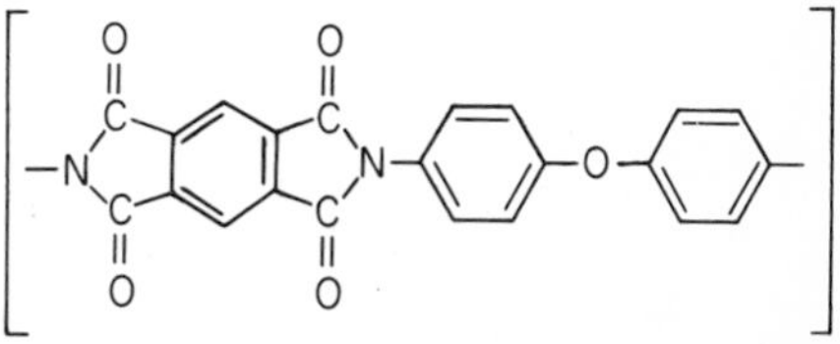

Fig. 14-19  Polyimide monomer. Note the ladder type of structure in the left half of the monomer.

Other developments directed toward heat stability use metal ions in the monomer, either in the chain, as in *chelate polymers*, or in the attachments to the chain, as in *ionomers*. The side-attached metal ion tends to bond or cross-link to an adjacent molecule, thus increasing strength and rigidity. The ladder structure and the cross-bonding of metal ions are an intermediate structure between the linear thermoplastics and the cross-linking that is the characteristic of the thermosetting plastics discussed in following sections of this chapter.

With so much chemical versatility to be exploited, it is difficult to predict what the plastics will or will not do in the decades to come.

## 14.11  characteristic properties
## of the thermoplastics

The following table is a summary of the properties of the common thermoplastic materials. Since all these basic materials are available in many formulations, the data supplied in the table are only approximate. Burning rate and impact resistance especially will vary with the formulation. A thermoplastic supplied as film is stronger than the same plastic in a solid section. Nevertheless, the table supplies order-of-magnitude information and a useful comparison between these materials.

The values for ultimate tensile stress are those obtained in the usual short-time tension test. The ultimate strength of most of these materials will decrease slightly over the years.

The thermal expansion of these materials is about ten times that of steel, and $E$-values are about one-hundredth that of steel.

The effect of sunlight (ultraviolet) on thermoplastics is sometimes difficult to summarize. The table indicates that sunlight has little effect on UPVC. The statement is not true for PVC which in not stabilized against such degradation, but if stabilized with pigments and ultraviolet absorbers such as titanium dioxide, PVC without plasticizers is one of the most weather-resistant of the thermoplastics. Carbon black is by far the best additive for ultraviolet protection, but it cannot be used if the plastic is to be colored.

## 14.12  asphalts

The asphalts are not necessarily polymers but are thermoplastics composed of long chains of carbon atoms of mixed and complex chemistry. They are mixtures of thousands of different petroleum or coal compounds, including paraffins, olefins, naphthenes, and aromatics, in straight and branched car-

## THE CONSTRUCTION THERMOPLASTICS

| | ABS | PE | UPVC | PS | Acrylic | Polycarb | PVF |
|---|---|---|---|---|---|---|---|
| 1. specific gravity | 1.04 | 0.95 | 1.4 | 1.05 | 1.2 | 1.2 | 1.5 |
| 2. tensile strength, psi | 5000 | $2\text{--}4 \times 10^3$ | 6000 | 5000 | 8000 | 9500 | 18000 |
| 3. elongation, percent | 40 | 40-400 | 10 | 1.2 | 5 | 100 | 200 |
| 4. modulus of elasticity | $250 \times 10^3$ | $25\text{--}100 \times 10^3$ | $350 \times 10^3$ | 50000 | $400 \times 10^3$ | $350 \times 10^3$ | $300 \times 10^3$ |
| 5. impact strength | good | good | poor | poor | low | high | high |
| 6. thermal expansion, per °F | 0.00005 | 0.0001 | 0.00003 | 0.00004 | 0.00005 | 0.00004 | 0.00003 |
| 7. resistance to heat, °F | 160 | 200 | 150 | 160 | 150 | 250 | 120 |
| 8. burning rate, in./min | $1\frac{1}{2}$ | 3 | none | 10 | 2 | 1 | slow |
| 9. effect of sunlight | slight | serious | slight | serious | slight | slight | none |
| 10. type of thermoplastic | flexible | flexible | rigid | rigid | rigid | flexible | flexible |

bon chains. They include in their chemical structure oxygen, nitrogen, sulfur, nickel, vanadium, and traces of other metals.

Asphalts are the residue or left-overs after the lighter constituents of petroleum are distilled off. There are also some natural rock asphalts. The words "bitumen" or "bituminous" as applied to these materials embrace both the petroleum, coal, or the natural rock asphalts, all these materials being characterized by their solubility in carbon disulfide. Coal tars, also called pitches, serve much the same general purposes as the petroleum asphalts but are extracted from coal.

All these materials, whatever their origin, are chiefly used for water-proofing: as damp-proof courses and membranes for buildings both below grade and on roofs, and for impregnating building paper, roofing felts, and building board. They are used also in floor tile, asphalt shingles, and bituminous and aluminum paints. When incorporated with fine aggregate, they are used for the surfacing of roads and parking areas. The adhesiveness of these materials to most surfaces contributes greatly to their usefulness as a moisture barrier.

Fig. 14-20 Strippable tape showing reinforcing fibers.

The viscosity and flow properties of these bituminous thermoplastic materials is significant for all their applications. A roofing asphalt applied on a steep roof must not sag under conditions of high summer temperature and direct radiation from the sun. The usual test for viscosity or consistency is a penetration test. A weight of 100 grams is applied to a penetrating needle, and the depth of penetration of the needle into the asphalt is measured after 5 seconds using a temperature of 77°F (25°C). The test is similar to a hardness test for metals, rubbers, or plastics, since it uses a penetrator loaded with a fixed weight.

Thermoplastic materials, including the coal tars and asphalts, do not have a melting point, but they soften over a range of temperature. A bitumen applied as waterproofing below grade will have a low softening point in the range of 115 to 145°F and a high penetration value. For application on roofs or vertical surfaces exposed to direct sunlight, a higher softening point and lower penetration is required.

With prolonged exposure to weather and direct sunlight, asphalts, like many thermoplastics, slowly oxidize and lose their ductility.

Bitumens are produced in three types:

1. hot asphalts, which are softened by heating and are applied hot
2. cutback asphalts, which are dissolved in solvents such as petroleum oils or naphtha
3. emulsion asphalts, which are emulsions of small droplets of asphalt dispersed in a water base

The hot asphalts bond poorly to damp surfaces. They oxidize more rapidly when exposed to ultraviolet radiation and tend to be more brittle at low temperatures than the two cold types. The emulsions give the best bond to damp surfaces.

## 14.13  thermosetting plastics

Thermosetting plastics, such as wood, wool, and Bakelite, are not softenable after polymerization. The polymerization process producing a thermoset is not the addition polymerization that gives a thermoplastic; it is *condensation polymerization*. In condensation polymerization, a chemical compound reacts with itself or another compound in a reaction that releases or "condenses" some small molecule such as water. Such polymerization may be illustrated by the reaction between phenol and formaldehyde to produce phenol-formaldehyde (Bakelite) plastic, shown in Fig. 14.21. Here water is released. A few of the more complex thermoplastics, such as nylon, are also produced by condensation polymerization.

The thermosets are generally brittle and elastic, though rubbers are not brittle even though thermosetting. The freedom of movement of the polymers under stress is greatly limited by cross-linking between the polymer chains. Thus Fig. 14.21 shows two molecules of phenol-formaldehyde cross-linked with carbon atoms. In the case of rubbers, this cross-linking is called *vulcanization*, the cross-linking agent being usually sulfur for rubbers. Linseed oil and varnish harden as oxygen from the atmosphere cross-links the molecules. The effect of cross-linking is to make a three-dimensional network polymer, reducing the plasticity of the material. The number of cross-linking bonds is always relatively small compared to the total number of possible

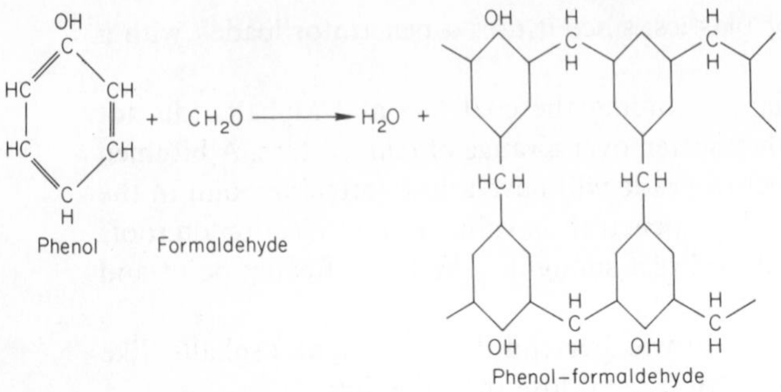

Fig. 14-21   Condensation polymerization of phenol-formaldehyde.

cross-linking sites, and the rigidity or hardness of the material is proportional to the amount of cross-linkage.

Most thermosets contain fillers as well as the resin plastic. The physical
and mechanical properties are strongly influenced by the filler material.
Carbon black is added to rubbers in high concentrations to improve the
tensile strength and resistance to abrasion and tearing, as well as to protect
against ultraviolet radiation. Aluminum powder is added for machinability,
and titanium oxide is added for whiteness and opacity. Such fillers must
have a very small particle size for large surface area, or if fibers such as fiberglass, small diameter.

The most commonly used filler is wood flour. White wood cellulose
flour is used in urea-formaldehyde for the production of white and colored
goods such as telephones. For electrical apparatus, mica or asbestos will
improve the electrical properties of the thermoset.

## 14.14   the formaldehyde resins

The several formaldehyde plastics, all of which are wholly elastic in their
stress-strain behavior, are familiar in telephones, electric switches, and melamine dinnerware. All are hard, strong, resistant to elevated temperature, and
brittle. Fig. 14.22 shows the stress-strain curve for a grade of phenol-formaldehyde: it is elastic to rupture.

Phenol-formaldehyde is the lowest-priced member of the group, familiarly known by its trade name Bakelite. It can be supplied only in dark
colors, usually browns and black. Some of its many uses include

1. wall plates for electric toggle switches and the toggle itself
2. mechanical parts of electric motor starters
3. lamp bases and receptacles
4. radio and appliance knobs
5. toilet seats

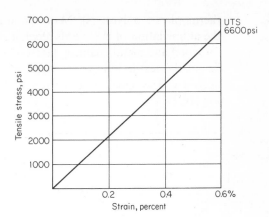

Fig. 14-22 Stress-strain curve for one grade of Bakelite. This is a brittle and elastic "plastic material".

This thermoset is used as the adhesive for exterior grade plywoods and is easily recognized in this application by its dark brown color. About 4 percent of the weight of a sheet of plywood is represented by the PF adhesive. A plywood bonded with PF will not delaminate under conditions of severe exposure to weather and water.

For the production of white or light-colored goods, urea-formaldehyde resins are used. A black telephone set will be molded from PF, a white one from UF. Decorative interior grades of plywood also require urea-formaldehyde.

Melamine-formaldehyde also may be colored. It is molded into dinnerware because it is a self-extinguishing plastic in flame tests and is the hardest of the plastic materials and therefore the most scratch-resistant.

The following are approximate properties of the several formaldehyde resins. These properties may be altered by fillers and reinforcing materials.

| | |
|---|---|
| specific gravity | 1.5 |
| ultimate tensile strength | 8000 psi |
| elongation | 0.5% |
| modulus of elasticity | $1 \times 10^6$ |
| impact resistance | poor |
| thermal expansion | 0.0002 in./in.-°F |

## 14.15 the polyesters

A polyester is any organic material produced by a condensation reaction between an alcohol and an acid. The alcohols will usually be ethylene glycol, propylene glycol, diethylene glycol, glycerol, or allyl alcohol, the acids usually either maleic, phthalic, adipic, or terephthalic. These raw materials will produce two types of polyesters: saturated or unsaturated (there are also thermoplastic polyesters). Many unsaturated polyesters are termed *alkyds*, the abbreviation of the words alcohol and acid. The alkyds are chiefly used for wood and metal enamels. Dacron and Terylene are polyester fibers familiar in the textile industries. Mylar film, often used as drafting paper, is a polyester. The remarkable strength and ductility of mylar film are shown in Fig. 14.23.

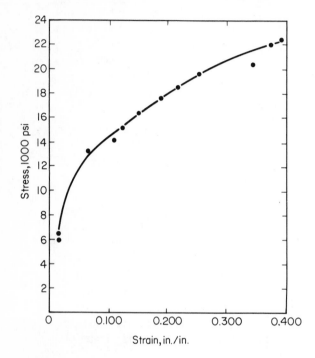

Fig. 14-23 Stress-strain curve for polyester film.

The low-molecular weight polyesters are viscous liquids or soluble solids which can be dissolved in liquid monomers such as styrene to give the viscosity needed for impregnation of glass fabric or mats. For improved weatherability, ultraviolet stability, and better color, methyl methacrylate is substituted for styrene. The resulting reinforced fiberglass plastic has high rigidity, strength, and impact resistance. Ultimate tensile strengths as high as 50,000 psi and $E$-values as high as $3 \times 10^6$ are not uncommon, and even higher strength levels are possible. The strength of the composite material is proportional to the amount of fiberglass reinforcement used. But like all plastic materials, the reinforced plastics will creep under sufficiently high long-term loading. In structural applications, a shape is preferred that will put the reinforced plastic in tension.

Other resins besides the polyesters are often reinforced with woven fabric or chopped mat. The largest market for such materials is the marine industry, for boats, decks, and superstructures. Other products include pressure tanks, septic tanks, rocket fins and nose cones, radomes, auto bodies, hard hats, dark-room developing tanks, trays, sheeting for buildings, glazing panels and skylights, lighting fixtures, appliance housings, and trailers.

## 14.16   epoxies

The epoxy resins have oxygen and aromatic rings in the polymer chain. As previously indicated, such a structure implies superior properties. The polymer molecules cross-link with oxygen. These resins are available in a wide range of formulations to provide a range of properties and curing characteristics. They are usually supplied as two components to be mixed and cured. There are also one-component epoxies which cure by the absorption of oxygen.

The epoxies are strong materials, with outstanding adhesion to most surfaces and high corrosion resistance. Their range of use is very broad, including adhesives and surface finishes. Shrinkage is very low during curing, so that they are suitable as a filler-adhesive. In civil engineering, the epoxies are used to bond concrete, repair concrete floors, and as industrial floors including epoxy terrazzo. The electrical industry uses epoxy for potting and encapsulating of electrical hardware and for insulation. Aircraft construction requires epoxy adhesive for aircraft bulkheads, floors, wing flaps, and various sandwich panels. Tooling epoxies are cast to make large and complex forming dies, such as the door-forming dies of Fig. 14.24. Tooling epoxies are

Fig. 14-24   An epoxy metal forming die. (Courtesy Rezolin, Inc., Santa Monica, Cal.)

much easier to machine than tool steels, and they are readily repaired or altered by adding additional epoxy resin.

## 14.17  the elastomers

The standard stress-strain curve displays the most important characteristic of a material: its stress-strain behavior. A number of such curves have been offered in this book, and they take various shapes. Some kinds of stress-strain behavior are undesirable, such as the totally linear and elastic performance of glass and Bakelite. No matter how strong these materials may be, they must show at least some ductility as an assurance against unexpected brittle collapse. A minimum of ductility can be accepted in strong materials which must carry severe stresses, such as punches and dies. While designers are preoccupied by stress, the necessity for elongation is often overlooked. Many components do not need significant strength, but they must have flexibility, softness, and elongation. These characteristics are required in some types of resilient floors, weather stripping, joint sealants, expansion joints, adhesives, vibration mountings, footwear, vehicle tires, and other applications of similar function.

Elastomers or rubbers are capable of extreme elastic deformation at low levels of stress. Most rubbers are capable of elongations of at least five times the original length. The strain is not proportional to stress, as may be seen in the typical rubber stress-elongation diagram of Fig. 14.25. This is a typical

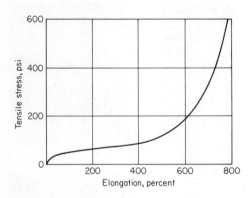

Fig. 14-25 Typical stress-strain curve for a polyisobutylene rubber.

S-curve exhibited by rubbers. Note that a tensile stress of about 100 psi corresponds to an elongation of about 500 percent.

In the case of rubbers, there is a time lag between stress and strain, sometimes many minutes long. This may be observed by indenting a rubber and watching the indentation slowly disappear. This characteristic resembles the viscosity of stiff liquids. Unlike the metals and plastics, there is no change in volume when a rubber is strained. Rubbers therefore behave remarkably like elastic viscous liquids.

Because they are not cross-linked, the thermoplastics are characterized by plastic strain and creep, while the cross-linked thermosets are elastic and brittle. The rubbers are usually addition-polymerized from unsaturated monomers such as butadiene, and they are cross-linked (vulcanized) with sulfur at approximately every 500th carbon atom. Increased cross-linking gives increased stiffness and reduced elongation.

Some of the more important rubber monomers are shown in Fig. 14.26.

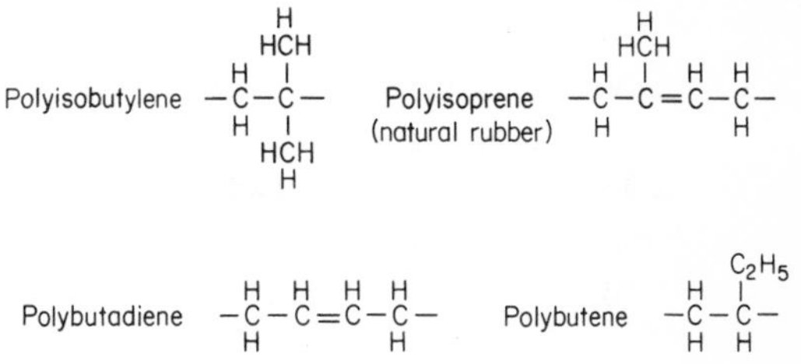

Fig. 14-26    Rubber monomers.

With few exceptions, these monomers contain double bonds. These double-bonded monomers when polymerized are susceptible to ozone attack, which results in crazing, and additives must be used to protect the elastomer. Such unsaturated types are also deteriorated by ultraviolet radiation from which they are protected by carbon black. Silicone rubber and chlorosulfonated polyethylene do not have these double bonds, and as a result they weather extremely well.

At least three-quarters of all rubber consumed in this country is synthetic rubber. Natural rubber (polyisoprene) has certain advantages which require its use in a limited range of articles, chiefly special vehicle tires. Natural rubber is stronger than the synthetic rubbers so far developed. Perhaps even more important, the heat generated in natural rubber when flexed is not so great as in synthetic rubbers (this heat generation is termed *hysteresis*). While automobile and truck tires are made of synthetic rubbers, chiefly SBR rubber, the larger tires for off-the-road heavy construction vehicles and mining trucks and aircraft tires (which must not blow out when the aircraft lands) must be made of natural rubber. Most rubber goods, tires included, are made of SBR, styrene-butadiene rubber, a copolymer.

Most rubber is molded into articles for transportation. The high friction of rubber surfaces provides traction for soles and heels of footwear, vehicle tires, conveyor belting, and rollers. Much rubber is used in the transportation of liquids through rubber hose and the transportation of energy through electric cables insulated with rubber.

## 14.18  types of rubber

**1. SBR.**  This is the general-purpose rubber used in tires, belts, hose, rubber floor tile, rubber cements, and latex paints. Tensile strength after vulcanizing and compounding with carbon black is 2500 to 3500 psi; elongation is 500 to 600 percent. It is more resistant to solvents and weathering than natural rubber. It is a copolymer of styrene-butadiene.

**2. Natural Rubber.**  This is polyisoprene, which can also be made synthetically. It is distinguished by low hysteresis and high strength, but it is readily attacked by solvents, gasolines, and ozone. When vulcanized and reinforced with carbon black, tensile strength is in the range of 3500 to 4500 psi and elongation is 550 to 650 percent.

**3. Butyl Rubber.**  This is a copolymer of isobutylene with a small amount of butadiene or isoprene. This rubber is impervious to gases and thus serves as a vapor barrier and inner tube for tires. It is used in the hoses through which polyurethane resins are pumped, since if oxygen permeates through the hose, the resin would solidify in the hose. Butyl rubber is highly resistant to the agents of outdoor weathering and as a roofing membrane it is superior to polyurethane foam.

**4. Nitrile Rubber.**  A copolymer of butadiene and acrylonitrile. These rubbers have excellent adhesion to metals and resistance to oils and solvents. Its uses include gasoline hose and hose linings, aircraft fuel tanks, printing rollers. Nitrile rubber is also the preferred rubber for O-rings. An interesting use is that of nozzles for aerosol cans to resist the fluorinated hydrocarbon refrigerant gas that pressurizes the can.

**5. Polychloroprene Rubber.**  Neoprene, as it is usually called, has good resistance to aging and weathering, is low swelling in oils and solvents, is abrasion resistant, and is unable to propagate flame because of its chlorine content. Its most important use is for electrical cable insulation, but it has a variety of other uses including gaskets, hose, engine mounts, bearing pads in bridge construction, protective clothing such as gloves and aprons, belting, and adhesives in building construction.

**6. Polysulfide Rubber (Thiokol).**  This is a rubber of low mechanical strength but with outstanding resistance to solvents, low gas permeability, and excellent weathering. Adhesion to metals is excellent, and this with its other characteristics accounts for its use as a caulking compound and sealant, especially in building construction. It has also been successfully used as a roofing membrane.

**7. Acrylic Rubbers.** Polyacrylates are cross-linked but are not vulcanized by sulfur. They have rubbery elastic properties and have outstanding resistance to oils, oxygen, ozone, and ultraviolet radiation. They are used in latex paints and roof membranes.

**8. Reclaim Rubber.** Reclaimed and reprocessed rubber is used in many rubber cements.

**9. Silicone Rubbers.** The silicone monomer is shown in Fig. 14.27. This remarkable monomer has a silicon-oxygen chain with methyl side groups. The silicone polymers have a great many uses, including water repellent treatments and mold release agents, and even high-temperature greases. The silicone rubbers retain their elastomeric properties at temperatures as low as $-200°F$ and as high as $600°F$. The silicon-oxygen chain is unaffected by ultraviolet radiation, oxygen, or ozone, but these rubbers are soft, weak (tensile strength of 1000 psi or less), and have limited elongation (400 percent or less). Electrical properties are outstanding. Tear strength is low.

Fig. 14-27  Silicone polymer chain.

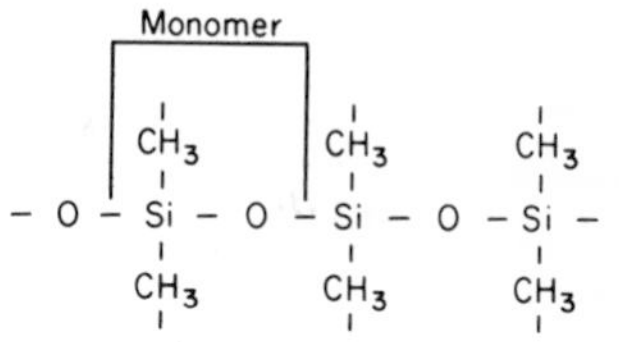

Silicone rubbers can be vulcanized at room temperature. This characteristic, together with a temperature limit of $600°F$, makes the silicone rubbers a highly useful tooling and mold material. They are especially effective as mold linings for poured polyurethane castings, since the two components of the rubber are simply mixed and then applied by spray or paintbrush. Fig. 14.28 shows an interesting use of RTV (room temperature vulcanizing) silicone rubber as an inspection tool.

**10. Rubber Hydrochloride.** This material is better known as Pliofilm. It is used in the form of transparent sheets for packaging cheese, meats, and other foods. It is easily identified by its unusual tensile and tear strength.

**11. Chlorosulfonated Polyethylene.** The monomer of this rubber is shown in Fig. 14.29. The monomer is a modified polyethylene which is vulcanized by special agents. The absence of unsaturation in the monomer indicates excellent weathering resistance, and the chlorine atom contributes flame resistance. Chlorosulfonated polyethylene (Hypalon) does not need carbon

Fig. 14-28 An ingenious use of a material. Caterpillar Tractor Co. inspects the contour of internal gear teeth with the use of Dow Corning Silastic RTV silicone rubber. The silicone rubber when cast in place, accurately reproduces the tooth contour. The rubber molding is removed and inspected (instead of the gear tooth) by optical methods. (Courtesy Dow Corning Corp.)

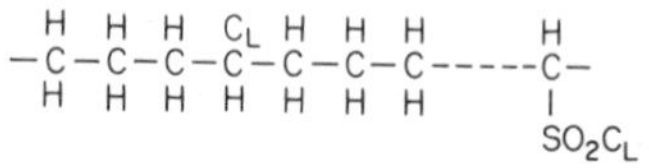

Fig. 14-29 Approximate monomer of chlorosulfonated polyethylene. There is a sulfonyl chloride group approximately every 12th carbon atom.

$$-\underset{\underset{H}{\overset{H}{|}}}{C}-\underset{\underset{H}{\overset{H}{|}}}{C}-\underset{\underset{H}{\overset{H}{|}}}{C}-\underset{\underset{H}{\overset{C_L}{|}}}{C}-\underset{\underset{H}{\overset{H}{|}}}{C}-\underset{\underset{H}{\overset{H}{|}}}{C}-\underset{\underset{H}{\overset{H}{|}}}{C}-----\underset{\underset{SO_2C_L}{\overset{H}{|}}}{C}-$$

black, and it may be pigmented to give any color, even pastel shades. It is available in sheet form or as a liquid dissolved in solvents. The liquid solution hardens quickly to a beautiful gloss. Indeed, this must be called a beautiful rubber. Although it is difficult to convey a surface appearance in a photograph, Fig. 14.30 attempts to portray the appearance of white Hypalon.

**12. Polyurethane Rubber.** Polyurethane rubbers, like silicone and Hypalon rubbers, are room-temperature curing. Hypalon is a one-component rubber, but the others require two components. A great many types of polyurethane are possible; it can be given either plastic or elastomer characteristics.

Polyurethane rubbers can be formulated to give a range of durometer hardness. A typical polyurethane rubber has a tensile strength of 5000 psi, higher than that of natural rubber, with high elongation, great tear strength, and outstanding abrasion resistance. Tensile strengths as high as 8000 psi are possible.

The high strength and abrasion resistance of polyurethane rubber suggest that this rubber would make an excellent tire material. This would be true, except for the high hysteresis (heat) developed during flexing of the tire

Fig. 14-30 A sheet of 1-mil polyvinyl fluoride (Tedlar) covers the sheet of paper over the "Tedlar" heading. Part of the PVF film has been painted with white Hypalon roofing rubber.

as it rotates. Nevertheless, polyurethane tires give outstanding life when fitted to lift trucks and other types of warehouse vehicles as cushion (noninflatable) tires.

Other uses of polyurethane include long-wearing shoe heels, printing rollers, mallet heads, oil seals, diaphragms, vibration mounts, gears, pump impellers, bowling pins, gaskets, and football helmets.

The abrasion resistance of polyurethane makes it suited to metal-forming applications. In Fig. 14.31 a block of polyurethane is used as a female braking die. This application makes use of the principle that rubbers are virtually incompressible. When the steel male die forces the workpiece into the polyurethane, the rubber yields under the die but bulges around it, forcing the workpiece up against the die to form a 90° bend without springback. The block of polyurethane can serve as the female die for a range of male forming dies, since it will conform to the shape of the male die in bending and drawing operations. Hard polyurethanes are used for these applications; Shore A hardness 90 or greater.

Fig. 14-31   A block of polyurethane used as a female bending die. The rubber block conforms to the shape of the male die.

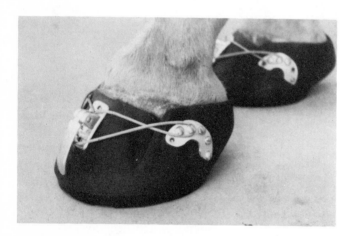

Fig. 14-32   Polyurethane horseshoes (easy boots).

Figure 14.32 shows an imaginative application for polyurethane rubber, the shoeing of a horse with polyurethane "Easy Boots." The author has used Easy Boots on one of his horses and doubts that the boots will ever wear out.

The preceding list of rubbers is a long one, yet only the common rubbers are mentioned. The variety of types and characteristics suggest that rubbers assume an increasing role in materials applications. This is especially true in building technology. The multi-storied office buildings with their curtain wall construction are critically dependent on high-performance rubbers for sealing of wall panels and large glazing sheets against rain and weather.

As an aid to understanding the application of so many types of elastometers the following summary is offered:

1. general purpose rubber, where special qualities are not required—SBR
2. low hysteresis—natural rubber
3. low gas permeability—butyl rubber
4. electrical uses—polychloroprene
5. resistance to oils, gasoline, and solvents—nitrile rubber
6. heat resistance—silicone rubbers
7. abrasion resistance—polyurethane rubber
8. resistance to weathering and ozone—butyl, polysulfide, acrylic, silicone, and chlorosulfonated polyethylene rubbers

The classification of rubbers has been set up by the SAE and ASTM societies and has been adopted by many industries. The full system is explained in ASTM Standard D735.

In this classification the following prefixes are used:

1. R    synthetic or natural compounds not oil-resistant
2. SA   synthetic compounds with very low oil swell (polysulfides or acrylonitriles)
3. SB   synthetics with low oil swell
4. SC   synthetics with medium oil swell
5. TA   synthetics with maximum resistance to high and low temperature (silicone rubbers)
6. TB   synthetics with outstanding resistance to both dry heat and oils at high temperatures

A three-digit number follows the prefix. The first digit indicates hardness (e.g., 6 indicates 60 durometer), the last two digits indicating tensile strength (e.g., 20 for 2000 psi). Finally, a letter suffix indicates special tests; for example, B indicates compression set and C weather resistance. A typical full designation might be Grade R 635G. This would indicate a rubber not oil-resistant, 60 durometer hardness, 3500 psi, the letter G indicating tear resistance.

## 14.19  a flame spread investigation

Knowledge of the relative flammability of polymer materials is critically important for their specification and use. There is no quicker method of becoming familiar with these materials and their characteristics than to make a flame spread test of them. Such a test should follow more or less closely the procedure of ASTM D635–63.

Test samples of polymer sheet are cut 5 in. long and 0.5 in. wide, and a sharp pencil mark is scribed across each sample at 1 in. and 4 in. from one end of the sample, as shown in Fig. 14.33. It is preferable to test 5 samples

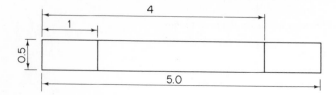

Fig. 14-33  Flame spread sample.

of each polymer type. The samples are mounted at a 45-degree angle with axis horizontal, as in Fig. 14.34. The free end of the sample is ignited with a Bunsen burner or other suitable ignition means. The time for the specimen to burn from one pencil mark to the other is found from a stopwatch or a sweep second hand, and the average time for all specimens of one type is determined. If the sample does not ignite after two attempts at ignition, it is

Fig. 14-34  Flame spread test on a sample of polystyrene.

designated "self-extinguishing." If it ignites but does not continue to burn after the igniting flame is removed, it is designated "self-extinguishing."

All pertinent data should be recorded, including thickness, and for foamed plastic the density in pounds per cubic foot must be known. In recording results, note also the character of the flame, since a flame test is the most useful method of identifying unknown plastics and rubbers. It should be borne in mind that many flammable plastics (polystyrene, for example) are available in standard and flame-retardant formulations.

The usefulness of such a test is greatly extended if it is applied to thin samples of the many woods, papers, and board stock used in construction, and if the results are compared with the polymers. If possible, obtain a sample of cellulose nitrate for this flame test; this material burns furiously and is unacceptable for most plastics applications.

## PROBLEMS

1 What is the difference in response to heat and in structure between thermoplastics and thermosets?

2 Explain the meaning of cross-linking.

3 What advantages do the epoxies offer?

4 Kitchen counter and table tops are made of laminates of formaldehyde thermosets, sold under such brand names as Arborite. Which of the formaldehyde resins cannot be used for this application, and for what reason?

5 Why is melamine-formaldehyde used for dinnerware?

6 What useful characteristics does nylon offer?

7 Why is a higher degree of crystallinity associated with greater density in a thermoplastic?

8 What are the favorable and poor characteristics of PTFE, including its frictional and electrical characteristics?

9 In what applications is natural rubber preferred over synthetic rubbers?

10 Explain the method of metal forming using rubber tooling.

11 Why is polyurethane preferred for such rubber tooling?

12 What are the characteristics of an SB 620C rubber?

13 List those rubbers which would conceivably give service as a roofing membrane. Which of these would give a colored roof?

14 Which of the rubbers is (a) an automobile tire rubber, (b) a cryogenic rubber, (c) a high-temperature rubber, (d) an O-ring rubber, (e) an electrical rubber, (f) an abrasion-resistant rubber, (g) a rubber impermeable to gases?

15 Which rubbers are vulcanized at room temperature?

16　Differentiate between PVC and PS by means of a flame test.

17　Find out whether you can differentiate between PS and ABS by means of a flame test.

18　When the point of a drafting compass is pushed into a block of thermoplastic, a small burr is raised around the hole. If the block is an elastomer, no burr is raised. Explain.

19　Classify the following as either thermosetting or thermoplastic: (a) phenol-formaldehyde, (b) silicone rubber, (c) PVC, (d) PE, (e) paper, (f) mylar drafting paper, (g) ABS, (h) foamed PS, (i) foamed polyurethane, (j) most paints and varnishes, (k) rubbers, (l) asphalts, (m) PC.

20　What characteristics of PVC make it a suitable material for a rain gutter?

21　Why does the softening point of an asphalt have to be adjusted to suit the slope of a roof?

22　Differentiate between hot, cutback, and emulsion asphalts.

23　What is the meaning of creep? Of stress relaxation?

24　Account for the resistance of chlorosulfonated polyethylene to ozone attack.

25　For what reasons is carbon black added to the rubber of vehicle tires?

26　You are offered a liquid vinyl spray that can be used as a roof coating to repair built-up roofs. The technical literature gives a tensile strength of 600 psi and an elongation of 250 percent. Is this a plasticized or an unplasticized PVC? How do you know?

27　If a paint or an adhesive is blended with a solvent that evaporates after application, is the material thermosetting or thermoplastic?

28　What influence does the number of cross-links have on the hardness and ultimate tensile strength of a thermoset?

29　What is a petrochemical intermediate?

30　What are the successive chemical operations that convert ethane into polyethylene?

31　Explain addition and condensation polymerization.

32　Which has the higher molecular weight (longer carbon chain): an extrusion or an injection-molding grade of thermoplastic? Why?

33　What is the purpose of a plasticizer?

34　Why are plasticizers not used with polyethylene?

35　What is the difference between an amorphous, an oriented, and a crystalline thermoplastic?

36　Explain stress-cracking.

37　Which thermoplastics are used for glazing purposes?

38　Which of the thermoplastics are resistant to ultraviolet degradation?

39　Name the polymer chain structures that result in outstanding properties in thermoplastics.

**40** What purpose is served by either an aluminum filler or a titanium dioxide filler in thermosets?

**41** How does a one-component epoxy cross-link?

**42** What is the usual difference in chemical structure between the monomers of a thermoplastic and of a rubber?

**43** Why is silicone rubber not degraded by exposure to sunlight?

**44** You are to suggest the possible structure of a synthetic drafting board. (a) A drafting board is a surface with certain requirements imposed on that surface. Make a careful list of the many requirements. Don't overlook the requirement that it must be and must remain flat. (b) Conceive a drafting board of synthetic materials. The surface must be nonstainable. What will you use immediately beneath the surface (refer to question 18)? What will you use for the core of the board? (Probably one of the foams discussed in the next chapter.)

**45** What plastic or rubber would you select for the blade of a plastic or rubber hockey stick?

**46** What plastic or rubber would you select for a tennis racket (not the strings)?

## 15.1 the scope
## of foamed plastics

Foamed plastics are now so diversified in their types and applications that they have virtually become a distinct technology. Developments in this material have come at a rapid pace.

Any of the organic polymers may be foamed, whether thermoplastic, thermosetting, or elastomeric. Not all foamed polymers have found receptive markets however. Foamed polyurethane is the most widely used for cushions, pads, cores, carpet underlays, and furniture, for artificial limbs, for building and container insulation, and for packaging fragile products. Foamed polystyrene is somewhat less expensive and is used in such items as paper cups, picnic baskets, packaging, and building insulation. See Fig. 15.1.

The foamed plastics can be foamed to give various densities from 1 to 60 lb/cu ft. The strength and stiffness of these foams increase with density, but the lighter foams provide best thermal insulation. The lightweight foams have lower heat resistance than their corresponding solid material, since

# FOAMS

# 15

Fig. 15-1  Foamed polystyrene protective packaging for a slide projector.

Fig. 15-2  Foamed plastics. A polystyrene beadboard on the bottom, two sections of high-density foamed PVC ceiling cove molding above it. The three materials at the top of the photograph are from left to right, foamed polyurethane, foamed polyethylene (feathery appearance) and foamed urea-formaldehyde (snowy appearance).

elevated temperatures expand and expel the gases in the foam cells, resulting in cracks and shrinkage of the foamed material.

The foamed plastic may be open-celled or closed-celled. An open-celled foam has foam cells which are interconnecting, that is, the gas phase is continuous. Closed-celled foams are preferred, since a foam of this type will absorb water only at its surface.

By a wide margin, the most outstanding cellular organic material is wood. Wood is anisotropic, showing best mechanical properties parallel to the grain. All the synthetic foams are also anisotropic, because the foam rises in a given direction. Generally, best mechanical properties are parallel to foam rise.

Three types of foamed polymer may be distinguished: insulating foam, cushioning foam, and structural foam. The sample foams of Fig. 15.2 are chiefly insulating foams. Insulating foams, which are chiefly polystyrene (Styrofoam), polyethylene (Ethafoam), rigid polyurethane, and urea-formaldehyde, are lightweight foams ranging in weight from 1 to 4 lb/cu ft. The cushioning foams must be rubbery. Flexible polyurethane is used for such service. The rigid polyurethanes may be considered as thermosetting plastics and the flexible urethane foams as rubbers.

## 15.2 structural foams

Structural foams range in weight from about 30 to 50 lb/cu ft. These foams are dense enough to sustain reasonable stresses, especially in compression. New structural foams are expected to appear soon; at the time of writing the following are in use or about to be used commercially:

1. polystyrene
2. thermoplastic polyesters
3. nylons
4. polypropylene
5. polycarbonate
6. ABS
7. high-density polyethylene
8. phenylene oxide (Noryl)
9. PVC
10. any of the above reinforced with chopped fiberglass (not woven mat)
11. polyurethane

An interesting and most useful characteristic of these structural foams is the formation of an integral skin of high density at the surface, caused by low temperature of the mold surface and the pressure of molding. This hard and dense skin increases the strength and stiffness of the foamed part. Figure 15.3 shows the skin formed on a high-density foamed PVC ceiling cove mold-

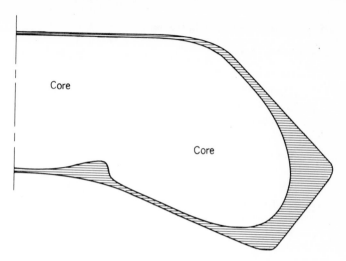

Fig. 15-3 Cross-section of the PVC foamed cove molding of Figure 14.11 showing the shape of the higher-density integral skin of this extrusion.

ing formed by extrusion. Such a part thus has a self-produced sandwich structure.

Rigid high-density foamed PVC is chiefly extruded into moldings, baseboard, window and door trim and frames, fence posts, fence palings, shelves, house siding, door sills, and pipe. Most of these applications complete with wood. In finished form such extruded shapes may be difficult to distinguish from wood when they are stained a dark brown. Even the density of these foams corresponds to that of wood—about 30 lb/cu ft. High-density foamed PVC however is inferior to wood in two properties. (1) The modulus of elasticity of such extrusions is about one-fifth that of wood, about 225,000 against $10^6$ or more for wood. Therefore for window and door frames sufficient cross-section is required for stiffness. (2) Wood has a negligible thermal expansion parallel to the grain, but these wood substitutes have the usual high thermal expansion of thermoplastics, about 0.00005 in./in.-°F.

The other structural foams listed are chiefly used in injection-molded products, such as business-machine housings, tennis racket frames, trays, and drawers. With fiberglass reinforcement, $E$-values as high as 800,000 are possible.

Polyurethane foams are best known for their use as insulation. Foamed polyurethanes in densities from 6 to 40 lb/cu ft are used for structural applications. These foams are neither extruded nor injection-molded but are mixed, poured, and foamed in inexpensive molds.

All these high-density foams will hold nails and screws if density exceeds 20 lb/cu ft. PS and ABS foamed products are not resistant to solvents. Products made from these foamed resins require a suitable barrier coating for protection.

## 15.3 urea-formaldehyde insulating foam

Urea-formaldehyde is the lightest of the insulating foams, weighing only 0.7 lb/cu ft. Its $K$-factor is 0.2 Btu/in.-°F. It has virtually no mechanical strength even in compression, is white in color, and does not support combustion. As scrap material it would delight any ecologist, since it could be converted either to fertilizer or to cattle feed.

The polystyrene and polyurethane insulating foams (to be discussed in the following sections) have closed cells (actually 95 percent closed cells and about 5 percent open) and therefore do not absorb water. About 60 percent of the cells of foamed urea-formaldehyde are open (connected) and this foam can absorb about 30 percent water by volume.

Like polyurethane foam, urea-formaldehyde is foamed on the job site. The foaming system has two components, as does polyurethane foam. One is a water solution of urea-formaldehyde and the other a water solution of the foaming agent and a catalyst for curing the resin. The material does not bond to surfaces as polyurethane foam does but flows into crevices, hollow spaces, and cavity walls before curing. Unlike polyurethane, it cures with the release of little exotherm (heat of curing). The foam dries out over a period of time to an extremely soft condition somewhat resembling wind-driven snow.

## 15.4 foamed polystyrene

The volume of solid material in the insulating foams is very small, about 2 lb/cu ft, this solid material serving to enclose cells of gas or air. Therefore the thermal conductivity of these materials is closely related to that of the gas enclosed in the cells. Air is not the best insulating gas; carbon dioxide and Freons are superior. The foaming gas for polystyrene is usually butane or pentane, giving this material a $K$-factor of 0.24 Btu/in. for insulating board.

Two methods are used to foam PS. It may be melted and mixed with a blowing agent and then extruded through an orifice. Alternately, polystyrene beads may be expanded in a steam-heated die. The molded package of Fig. 15.1 was blown in a die. Various densities are produced, the higher densities giving higher $K$-factors and greater stiffness and strength. The beaded foam is relatively isotropic in its mechanical properties, while the extruded foam is somewhat stronger in the direction of extrusion.

Foamed polystyrene can be pigmented. It is not resistant to ultraviolet radiation, has low heat resistance, and is attacked by most solvents, including many of those in paints and adhesives. Special adhesives are used for bonding this foam. The material is of course brittle and must be carefully handled. The lighter foams of 2 lb/cu ft or less can be collapsed by low temperatures when used to insulate cold storage areas. Somewhat heavier foams should be used in more extreme conditions of heat or cold. Other lightweight foams have the same weaknesses.

## 15.5 rigid polyurethane insulating foam

Polyurethane foam planks are available in various densities: 2, 3, 4, 6, 8, and 10 lb/cu ft. These are produced by foaming a very large volume of material, perhaps 4 ft $\times$ 4 ft in section and sawing this large "bun" into planks on a bandsaw. These planks have the same applications as Styrofoam planks. They are not quite so brittle, are stronger, are not attacked by most solvents, and being thermosetting will withstand higher temperatures than foamed PS. The 2-lb density is usually installed for insulation, but 4-lb foam is more resistant to collapse of the foam by heat, cold, or ice formation in roofs.

Polyurethane foams are produced by a reaction between a diisocyanate such as tolylene diisocyanate and a polyester. A small amount of water is used in the reaction in order to form carbon dioxide. This $CO_2$ is the foaming agent, though Freons may be used in foams of low density. In cold weather foaming, Freons may have to be added to obtain sufficient foam expansion. The use of Freon may provide a lower $K$-factor. A silicone oil is included in the formulation to keep the cell size small. The polymerization reaction produces linkages of urea with ethane, hence the name polyurethane. The density of the foam is determined by the type of polyester used and the amount of water, about 3 percent, which controls the amount of $CO_2$ generated.

The gas in the closed polyurethane cells is therefore either $CO_2$ or a Freon. The thermal conductivities of these gases and of air are these:

| | |
|---|---|
| Freon-11 | 0.58 Btu/in. |
| $CO_2$ | 0.102 |
| air | 0.168 |

Thus the Freon-blown foams are the best insulation. Both $CO_2$ and Freon produce a superior insulating foam to polystyrene foam. However, any comparisons must take account of an aging process that may occur in polyurethane foam. Freon-foamed polyurethane in a density of 2 lb/cu ft has an initial $K$-factor of 0.11, far superior to any other insulation. If however this insulation has access to air at its surfaces, air will permeate through the cell walls and replace the Freon. This exchange of gases raises the $K$-factor to

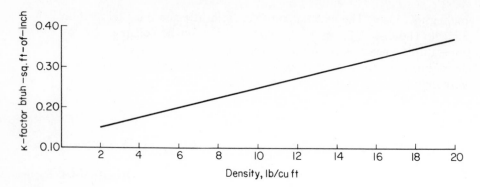

Fig. 15-4 *K*-factors for polyurethane foams.

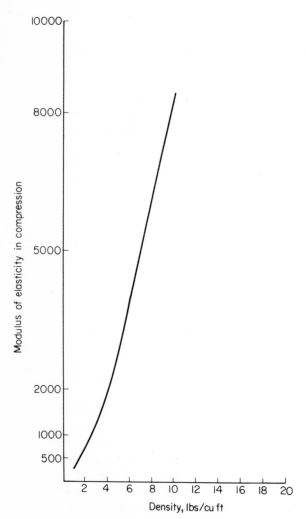

Fig. 15-5 Average values for modulus of elasticity of polyurethane foams. Specific types may vary somewhat from these values.

0.16 Btu, which however is still superior to Styrofoam. If the foam is enclosed so that air does not have access to it, the *K*-factor should remain at its initial low level.

Polyurethane foams are stronger in the direction of foam rise than in the cross-direction. Depending on which direction requires the better strength, sandwich panels may be foamed in the vertical or horizontal position.

Significant properties of polyurethane foams as a function of foam density are given in Figs. 15.4, 15.5, and 15.6. It will be noted that mechanical properties such as modulus of elasticity, tension, compression, and shear strength are not proportional to density but increase rapidly with density. All figures give properties as measured in the direction of foam rise.

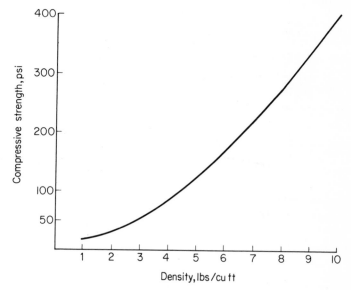

Fig. 15-6  Average values of compressive strength of polyurethane foams.

## 15.6  installation

## of polyurethane foams

For standard spray contracting, such as insulating the walls and roof of a building, a complete spray installation, truck-mounted, is necessary. The two resin components are pumped from 55-gallon drums, mixed, heated, and discharged from a special spray gun. The two components react very quickly and therefore must not mix until discharged from the gun. If they become mixed in the hose or gun, there is no solvent that can remove the cured plastic. For on-site spraying, the two components are mixed in equal volumes, though factory application of polyurethanes may use other proportions. The sprayed liquid foams immediately as it is deposited on a surface and is hard within a minute. A complete polyurethane spray installation for contracting is shown in Fig. 15.7.

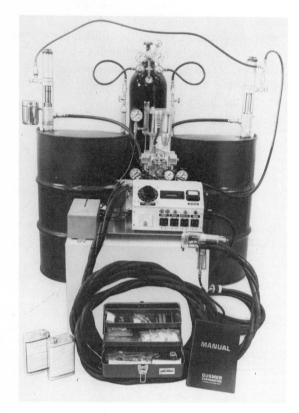

Fig. 15-7 Polyurethane spray unit. A pump is installed in each drum of resin. The nitrogen bottle between the drums provides nitrogen gas to protect the resin in the drums from atmospheric air and water vapor. The hose lines are of impermeable butyl rubber to protect the resin from air and water vapor. The orifice is on the left-hand side of the gun, the same direction as the ends of the hoses point to. (Photo courtesy Gusmer Corporation.)

For the manufacture of furniture components or other products under factory conditions, using high-density foams, equipment is more complex. Factory products are made either by spraying, pouring, or frothing the resin. In frothing, the resin is partially foamed before discharge, the foaming operation being completed after discharge from the nozzle. The small polyurethane aerosol kit shown in Fig. 15.8 uses froth.

## 15.7 flexible polyurethane foams

Flexible polyurethane foams, or foam rubber, are produced by mixing an emulsion of a polyether with tolylene diisocyanate and various catalysts. As with rigid foams, the amount of carbon dioxide generated for foaming is

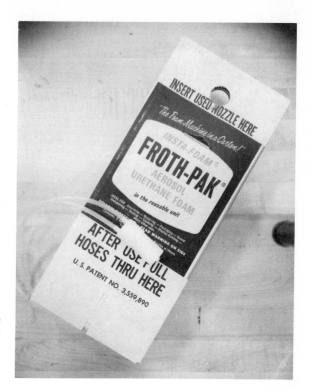

Fig. 15-8 Small aerosol kit containing two components for urethane foam complete with spray nozzle.

determined by the proportion of water in the mix, more water giving a lighter foam.

Typically such foam rubbers will have an elongation of 200 to 225 percent, which is lower than the elongations for solid rubbers. Such flexible foams if compressed 50 percent for 24 hours will fully recover this strain when the load is removed.

Such foams find their major markets in furniture, upholstery, and carpet underlays.

## PROBLEMS

1 What advantages does polyurethane offer over polystyrene as foams?

2 Explain how the $K$-factor of polyurethane foam may increase with age.

3 What are the uses of high-density polyvinyl foam?

4 What is the difference between an open-celled and a closed-cell foam?

5 Why are most foams anisotropic?

6 State the general function of an insulating, a structural, and a cushioning foam. Give an example of each type.

**7** How is a strong skin produced on a foamed plastic?

**8** What disadvantage does high-density PVC foam display as compared with wood?

**9** Fig. P15.1 shows a shipping container type MLW/MLLU 24 used in transatlantic shipping. This is a nonrefrigerated container with an allowable heat leak of 47 Btu/hour per degree temperature difference Fahrenheit over the whole area of the container. Container dimensions are 19 ft $10\frac{1}{2}$ in. long × 8 ft 0 in. high × 8 ft 0 in. wide. The container is insulated with polyurethane foam with a design $K$-factor of 0.125 Btu/in. What thickness of insulation is required in insulating the container?

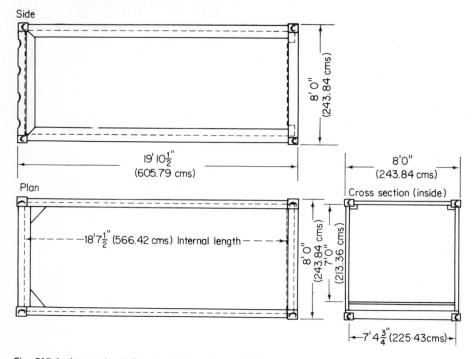

Fig. P15-1 International Standards Organization (ISO) merchandise shipping container type MLW/MLLU 24, nonrefrigerated insulated, capacity 41,260 pounds.

**10** For the above container, what insulation thickness would be required if foamed polystyrene were used. Assume a $K$-factor of 0.25 Btu/in. By how much would the weight of the insulation be increased if both polyurethane and polystyrene insulation weigh 2 lb/cu ft?

**11** Find out how beehives are constructed (honey supers). They are very simple. Select a suitable foamed plastic for a plastic hive and devise a method of manufacture. Find out how bees maintain climate control within the hive. Does your plastic hive offer advantages over a wood hive in maintaining conditions within the hive?

Adhesives and surface coatings may be related materials. Many of the organic materials are formulated for either use, such as the epoxies. In both cases a thin film must be attached to a substrate material. The adhesive must bond only. The coating must bond, but it has the additional functions of protecting the substrate and providing an attractive appearance.

The range of surface finishes is very wide and includes organic materials as well as inorganic coatings such as porcelain enamels and anodized finishes. Few manufactured parts can dispense with a finish of some kind in their manufacturing process; the durability of the part or its market acceptance may be dependent on the coating process.

## 16.1  substrates and surface films

No coating or adhesive adheres to a surface unless it can "wet" the surface. A classic example of a non-wetting material is a drop of mercury on a steel

# ADHESIVES
# AND SURFACE FINISHES

# 16

plate. The mercury has no attraction to the steel plate and draws into a tight sphere which can easily be rolled along the steel plate. Mercury, then, cannot be used to bond the steel plate to another material. A drop of a wetting material will spread over the plate until the edge of the drop forms an equilibrium angle $\theta$ with the plate, as in Fig. 16.1. Good wetting is indicated by a small angle; if the drop spreads over the whole plate to give a contact angle of zero, then the wetting is ideal.

Fig. 16-1  Welting angle.

Whether a coating material can wet a surface depends also on the condition of the surface. No paint or adhesive bond would be expected with an oil-covered or rust-covered steel plate. But there are other, more subtle surface difficulties.

Suppose the steel axle of a vehicle breaks. If the two broken parts are put together in proper alignment, the axle is not thereby mended. We take it for granted that nothing can be mended in this way. Yet it is rather remarkable that this simple repair method does not work. If the atoms of the axle attracted each other so powerfully that a large force was required to rupture the axle, then an equal force should bring the two broken parts into close enough proximity for the interatomic forces to mend the break. Yet this method of repair fails completely.

First of all, the deformation accompanying the fracture will make it impossible to bring the atoms of the two broken pieces together to the required interatomic distance of 4 Angstroms approximately. The two pieces may appear to the eye to mate well; on an atomic scale they will be a hopeless mismatch. They may be joined together only by a very large force which more or less mashes them together with considerable change of section, in the welding method called "cold welding." Such joining sometimes happens fortuitously, as when an engine shaft seizes, or when pressure and heat weld a machining chip to a cutting tool.

There is a second difficulty. All metal surfaces act as getters, or absorbers of gas molecules, particularly for water vapor. The two fractured surfaces become freshly exposed to the air and instantly absorb a layer of water vapor. This layer makes metal-to-metal contact impossible, for a film of gas intervenes which unfortunately does not possess adhesive properties.

These surface considerations help to explain the action of coatings and adhesives. Such materials must cover the surface to establish continuous contact and must in addition remove or dissolve surface impurities or gases to establish contact with the original material of the substrate. Usually a bond to the surface of metals is impossible without first cleaning the surface. Bonding to porous materials such as paper or brick may not require a cleaning operation if the adhering material can enter the pores of the substrate and thus obtain anchorage.

In outer space, repair of broken surfaces would be easier, since there are no gases to adsorb on the broken surfaces. But in space there is the possibility of a wrench seizing to a nut, a hammer to a nail, and certainly the hazard of shafts seizing to bearings.

## 16.2 metal cleaning

Metal surfaces are always contaminated with unwanted matter of various types, usually referred to as *soil*:

> mill scale and rust
> adsorbed water vapor
> oils and lubricants such as cutting oils and drawing compounds
> perspiration deposits such as fingerprints
> paints and other finishes

Before the application of paints or adhesives, such materials must be removed to produce what is called a "clean" surface. What is actually produced is a "sufficiently clean" surface, since what may be clean for one application may not be so for another.

The various metal cleaning methods may be classified as either *mechanical* or *chemical*:

mechanical: grit-blasting, barrel-finishing, scratch-brushing, ultrasonic cleaning, grinding, steam-cleaning, etc.
chemical: solvents, electrocleaning, acids, alkalis

For many applications the water-break test can be used as a test of surface cleanliness. In this simple test, if the surface will hold an unbroken film of water, it is considered clean.

## 16.3 mechanical cleaning

Small parts may be tumbled in a barrel for purposes of cleaning, degreasing, descaling, deburring, finishing, burnishing, or coloring. The method is also used to give a luster to metals and plastics. This *barrel-finishing* operation may be done wet or dry and with or without a tumbling medium.

Deburring of parts is done by adding an abrasive compound to the contents of the barrel. Without an abrasive, burrs tend to be peened over rather than removed from the part. The $pH$ of the tumbling liquid must be controlled; for most operations on steel a $p$H of about 9 is used. Typical products which use barrel-finishing methods are roller bearings, cutlery, jet engine blades, gears, and many plastic products.

*Ultrasonic cleaning* is a remarkably effective method of cleaning, even for paint stripping. The equipment requires a bath of solvent, equipped with ultrasonic transducers. The cleaning is performed by ultrasonic energy produced in a separate generator. The frequencies in use range from 20 to 40 kc, and ultrasonic power may range from 100 w in a small unit to 5 kw in a large unit. The acoustic energy produces voids in the liquid which collapse to produce cavitation pressures which are estimated at over 1000 psi. The cleaning tank must be made of hard and rigid material so that acoustic power is reflected and not absorbed by it. Typical products that are cleaned ultrasonically include printed circuit boards, transistors (for removal of soldering flux), spray nozzles, and typewriter parts.

*Abrasion methods* of cleaning may employ a wire brush, a powered scratch brush, sand, or grit-blasting. Ordinary carbon steel brushes are usually not suitable for this purpose, especially for stainless steel parts, because steel particles may embed in the work and later rust. Abrasion methods using grit or shot are the usual methods of cleaning weldments, castings, and large fabrications. Grit may be fired from a nozzle or thrown centrifugally off a rapidly rotating wheel, the latter method using considerably less power.

*Steam-cleaning* uses a steam nozzle and a portable steam generator. Low-pressure steam is not effective. High-pressure steam at about 100 psig is used, often with a cleaning compound added. Steam cleaning is often used for paint stripping and for cleaning supermarket shopping carts.

*Flame-cleaning* uses a gas flame to remove mill scale from steel or paints from metals.

## 16.4 chemical cleaning

The choice of chemical cleaner depends on the specific resistance of the metal to corrosion. Aluminum products, zinc die-castings, brass, and copper are all attacked by alkaline cleaners. Magnesium, on the other hand, will be attacked by cleaners which are not strongly alkaline, and it requires a $p$H of about 10.5.

Acid cleaners are corrosive and difficult to handle. Alkaline cleaners do not share this disadvantage and are preferred for additional reasons:

1. There is usually no chemical attack on the metal.
2. A less corrosion-resistant material may be used for tanks.
3. There is no hydrogen embrittlement of the metal to be cleaned.

Since metals, including steels, are embrittled by hydrogen, many specifications forbid the use of acid cleaning because of the presence of hydrogen. However, acid cleaners, when conditions permit their use, are cheaper and faster and are better scale removers. In addition, the etching effect of an acid cleaner will improve the adhesion of paints and adhesives.

Acid-cleaning is often termed pickling, presumably because in earlier times vinegar (acetic acid) was used for the purpose. Acid-cleaning is standard practice for the removal of mill scale from steel. The cheapest of the acids is sulfuric acid. Hydrochloric acid (muriatic acid), diluted by water three to one, is used at room temperature. If used hot, its fumes corrode nearby equipment. It is a faster cleaner than sulfuric acid, but it requires a nonmetallic tank such as rubber, fiberglass, or polyethylene. Nitric acid being more expensive is used only when other acids are not effective. Phosphoric acid is used either hot or at room temperature, in concentrations of 15 to 30 percent weight.

Alkaline cleaning is used with aluminum. Sodium hydroxide, sodium carbonate, trisodium phosphate, or other alkaline cleaners are used. A black smut may be produced on the surface of aluminum by alkaline cleaners, especially in the case of certain alloy aluminums. This smut is removed by an acid dip.

The oxide surface of aluminum must sometimes be removed, for anodizing, spot welding, or other purposes. This may be done in a sulfuric acid-sodium fluoride solution or in a phosphoric acid-chromic acid solution.

Chemical cleaning is carried out by several methods: hand wiping, dip tanks, spraying, vapor degreasing, and electrocleaning. Hand wiping is not efficient and produces imperfect results. Problems of toxicity prevent the use of hand wiping with chlorinated solvents. The latter are used in the *vapor degreasing* method. In this procedure the metal to be cleaned is suspended above the cleaning tank. The solvent in the tank is heated and its vapors condense in droplets on the cooler metal workpieces. The droplets drip back to the tank, carrying soil with them. Drips from a workpiece hung higher up must not be allowed to fall on any workpiece suspended below.

In *electrocleaning* the part to be cleaned is an electrode in a solution which carries electric current. The effect of the electric current is to electrolyze the water of the solution, hydrogen being released at the cathode and oxygen at the anode. This generation of gases provides agitation. Alkaline solutions are usual. The work may be either anode or cathode or may be switched from one to the other. Metals that form protective oxide skins, such as stainless steel or nickel, are cleaned cathodically. Brass also is cleaned cathodically to avoid loss of zinc at the anode.

After cleaning, the parts must be rinsed in hot water, since residual amounts of acids or alkalis may attack metal surfaces. The rinse water is dried from the surface as quickly as possible, often with the use of infrared lamps.

Adhesives and surface coatings are applied as soon as possible after final rinsing.

## 16.5 conversion coatings

Conversion coatings are inorganic films produced by some kind of chemical reaction with the metal surface of the workpiece. They are produced either for corrosion protection or as a base for surface finishes. Though they are extremely thin, usually much less than 0.001 in. thick, they are tightly bonded to the part since they are chemically formed from the original metallic surface. The most commonly employed conversion coatings are *phosphate*, *chromate*, and *anodic* (anodized) coatings.

Fig.16-2 Micrograph of a zinc phosphate coating on steel sheet. There is porosity between the phosphate crystals, and the coating by itself does not protect the steel.

Phosphate coatings are applied to steel or galvanized steel. The corresponding method for aluminum is usually a chromate treatment. These coatings offer the following possibilities:

1. Paint makes a strong bond to phosphate coatings.
2. While the phosphate coating by itself offers minimal corrosion protection, the addition of a finishing paint coat provides excellent protection.

3. A phosphate coat makes an excellent lubricant for the drawing of steel in presswork operations.
4. Some of the phosphating chemicals serve also as cleaners. Phosphating is applied to domestic appliances, automobile bodies, military items, and metal furniture, to name only a few applications. Such coatings are applied by brush, spray, dip, or tumbling.

For paint bonding purposes, crystalline zinc phosphate coatings ($Zn_3PO_4 \cdot 4H_2O$) are produced on metals. There is no reaction with the steel in this method, and therefore a true conversion coating is not produced. The coating is soft, and being a hydrated compound, paint baking temperatures must not be too high or the phosphated coating will be damaged. The porosity of the phosphate coat allows paint to penetrate and adhere strongly to the surface. However, too coarse a phosphate crystal structure may produce a deposit that does not adhere well to the metal surface.

An iron phosphate coating is a true conversion coating, since the phosphate is provided by a phosphoric acid solution and the iron of the part being treated. Unlike zinc phosphate, iron phosphate coatings are amorphous. The iron phosphate dip may be used both for cleaning and phosphating. Compared to zinc phosphate, such a coating has somewhat better corrosion resistance even though it is thinner, since it is less porous. After phosphating the part is usually rinsed.

Several types of conversion coatings are applied to aluminum:

1. chromate-phosphate coatings from acid solutions
2. chromate coatings
3. oxide coatings produced by electrical methods (anodizing)
4. crystalline zinc phosphate coatings, similar to those used on steels

Chromate coatings are more widely used as prepaint treatments than the other types. Like phosphate coatings on steel, these conversion coatings will not flake or chip, not even during forming of the metal sheet.

## 16.6 anodizing of aluminum

In the anodizing method, a light or heavy oxide coating is produced on aluminum by making the aluminum part the anode in an acid electrolytic bath. The anodic coating is porous and must be sealed by immersion in boiling water or linseed oil. Such coatings are applied to aluminum windows and doors, aluminum trim on highway buses, and the aluminum scales and other parts of drafting machines. Anodizing may be done for any of the following reasons:

1. improved corrosion resistance
2. paint adhesion
3. decorative and color effects—the porous oxide coat can be impregnated with dyes
4. electrical insulation
5. improved emissivity of the surface
6. abrasion resistance (hard anodizing)

The oxide coat on aluminum can be thickened by cheaper methods of chemical oxidation, but anodizing provides a more predictable and controllable quality of oxide coat and better corrosion resistance.

Many variables control the properties of the anodic film:

1. the alloy of aluminum—silicon alloys anodize to a gray color, chromium alloys tend to a yellowish color, etc.
2. prior heat treatment or welding
3. type of electrolyte
4. concentration of the electrolyte
5. voltage of the bath
6. current density in amps per square foot
7. a.c. or d.c.
8. temperature of the bath
9. anodizing time

The thickness of the oxide film is controlled by most of these variables, but chiefly by the anodizing time (Fig. 16.3). There is a practical maximum

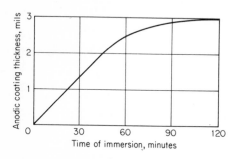

Fig. 16-3   Growth of an anodic coating.

thickness, because while the film grows from the inside out, as a conversion coating must, the electrolyte dissolves the film already formed. As a result of this chemical dissolution, the rate of film growth steadily decreases, until finally the rate of film formation and dissolution become equal, and no further increase in thickness can occur. For most applications a film of 0.5 to 2 mils is developed. By comparison, the usual thickness of the natural oxide surface of aluminum sheet is of the order of 0.000002 in. Because the outer surface of the oxide film is attacked by the electrolyte for the longest time, it is more porous than the innermost section of the film.

The electrolyte for anodizing may be sulfuric acid, chromic acid, oxalic acid, or boric acid. The latter two acids are used only for special effects.

Oxalic acid produces a yellowish tint in the oxidized surface. Boric acid is largely restricted to the production of thin and impermeable dielectric films on electrical condenser foil (the dielectric strength of an anodized film is very high). Aircraft parts may use a chromic acid electrolyte.

The sulfuric acid process is the usual one employed, in an acid concentration of 15 to 25 percent. The process is performed at room temperature, usually in a lead-lined steel tank, the lead lining acting as cathode for the circuit. Voltages of 12 to 24 v are employed, with current densities of 12 to 18 amps per sq ft and anodizing times of about 30 min. The use of higher temperatures has the effect of reducing the film thickness and increasing the porosity of the film.

"Hard anodizing" or "hard coating" is the production of an anodized coat specifically for wear resistance. The hard coat is the same oxide produced by any other anodizing process and has the same hardness, but the coating is thicker and less porous. The process is operated at a lower temperature to reduce the extent of chemical dissolution of the oxide film. Coatings as thick as 0.004 in. are produced in hard coating: such a coating will significantly increase the overall dimensions of a part produced to tolerances. Such articles as automobile pistons are hard-coated.

## 16.7 zinc coatings on steel

Zinc-coated steel, called galvanized steel, protects steels from atmospheric corrosion. If galvanized steel is to be painted, it requires a suitable primer to ensure bond of the paint to the zinc coating, though as discussed below, certain types of zinc coatings are specially prepared to receive paint.

Zinc can oxidize rapidly to white zinc oxide if the zinc surface is enclosed and if it is kept in a wet condition. This type of corrosion, called white rusting, does not occur on surfaces exposed to the weather.

Seven types of galvanized steel sheet are produced. These seven types fall into two broad classes: hot-dipped (in molten zinc) and electrocoated by electroplating methods.

### HOT-DIPPED GALVANIZED STEEL SHEET

1. Hot-dipped galvanized steel, spangled, in various coating thicknesses. This type has the familiar spangled appearance of galvanized steel. It is not suited to receive standard paints.
2. Hot-dipped and heat-treated to produce an alloyed zinc-iron coating, in various thicknesses. This type is often called Galvannealed. It has a dull gray color without spangle, and it is not intended for painting.
3. Hot-dipped and phosphatized to receive paint.

4. Hot-dipped, with a lighter zinc coating on one side than the other, also termed Differentially Coated. This type has the regular spangled appearance.

## ELECTROLYTIC ZINC-COATED STEEL SHEET

5. Electrolytic zinc-coated, with 0.1 to 0.2 oz/sq ft of sheet (total on both sides).
6. Electrolytic zinc-coated and phosphatized to receive paint.
7. Electrolytic flash zinc-coated and phosphatized for paint.

The weight of the zinc coating (total on both sides) is designated by a letter (G for galvanized and A for alloyed) and a number which is the ounces per square foot. Thus G235 has 2.35 oz/sq ft and A60 has 0.60 oz/sq ft. The useful life of a galvanized sheet is proportional to the coating thickness. The heavier coatings such as G235 are preferred for exterior applications.

Electrolytic zinc-coated steel receives a zinc coating about one-tenth as thick as the hot-dipped grades.

## 16.8 other inorganic finishes

## for metal surfaces

A range of colors is presented by *copper* and its alloys. Eleven standard colors are available, from the red of pure copper, through the golds of the brasses, to the silver of nickel silver. Until recently, these color effects in copper have been difficult to exploit because no surface treatments were available as permanent protection against tarnishing of copper. Newest developments have resulted in copper treatments with a degree of permanency against exterior exposure. These should result in more extensive decorative use of copper without the penalty of repeated cleaning treatments.

*Patterned and textured sheets* in copper, aluminum, and stainless steel are available in wide variety. Such finishes are produced by chemical etching, grinding, brushing, polishing, or by impressing a textured pattern from embossing rolls. An embossed pattern can control the reflectivity of the sheet, strengthen it by cold-working, and provide visual appeal beyond that provided by the unrelieved flat sheet. Embossed patterns are not usual with stainless steel sheet however. The most commonly used finishes for stainless sheet are #2B and #4:

#2B—cold-rolled, annealed, pickled, then given a cold-finishing pass to produce a smooth and mildly reflective surface
#4 —this is the polished finish, produced by belt-grinding with 150–180 grit

Porcelain enamelling was discussed in Chapter 6. Such enamels are capable of remarkable decorative effects, including murals, by the use of techniques such as stenciling, printing rolls, silk-screening, etc.

Today there is a strong trend toward the use of *clad steels*. These are composed of a mild steel core to provide strength and ductility, with a surface cladding to supply other service characteristics such as corrosion resistance. The cladding may be a factory-produced paint or plastic coating, a metal dip-coat such as a galvanized coat, or a metal surface bonded to the base steel. The familiar stainless steel cookware is a clad steel, actually a mild steel clad with stainless. Such a clad steel is superior to solid stainless steel for cookware, since stainless steel has low thermal conductivity.

Fig. 16-4 Anodized aluminum architectural work and anodized aluminum ashtray at the entrance to a building.

*Tin-plating* is a more expensive coating, usually reserved for food containers. *Terne sheet* is cheaper than tin plate, being an alloy of 80 percent lead and 20 percent tin. The tin is required because lead will not wet the surface of steel and hence will not bond. Most terne sheet is used for such items as fuel tanks for large and small engines and construction equipment.

*Aluminized steel* is sheet steel with an aluminum coating on both sides, produced by a hot-dip process. In appearance, this product somewhat resembles aluminum. Two grades are produced. Heat-resistant aluminized steel has an aluminum coat 0.001 in. thick on each side, while corrosion-resistant aluminized steel has a coating of 0.002 in. per side. Aluminized steel may be found in automobile mufflers and exhaust piping, ducting, and heating appliances.

*Electroplating* is the deposit of metal coatings from electrolytic baths. Plating may be done for any of several purposes:

1. appearance
2. corrosion protection
3. wear resistance
4. increasing the size of worn parts

and may be performed on any of the common metals. The part to be plated is made the cathode in the bath, and the anode is a bar of the metal to be plated. Electroplating is subject to many of the problems of anodizing, such as coating porosity and uniformity of the deposit. A heavy deposit may be plated on projections while recesses may receive only a thin deposit. Thus an electrodeposit on a screen of round wire will produce wires of elliptical section. Many electroplating processes also release hydrogen at the cathodic workpiece, a cause of embrittlement which may or may not be of concern in the service of the part.

Instead of galvanizing, zinc may be electroplated from cyanide baths. Cadmium plating is common on small hardware items; it provides corrosion protection inferior to that of zinc and may result in hydrogen embrittlement. Cadmium-plating is not permitted on food-processing equipment because its compounds are toxic.

Nickel tarnishes on exposure to atmosphere. Chromium does not tarnish, but electroplated chromium is porous. Best protection and appearance therefore is provided by first plating a thin film of nickel a few ten-thousandths of an inch thick, followed by a heavier electroplated film of chromium.

*Vacuum plating*, also called vacuum metallizing, is a method of depositing thin metal films by deposition from the vapor phase under high vacuum. Under high vacuum conditions the metal to be deposited has a high vapor pressure and a low boiling point. The vacuum deposition of cadmium avoids the hydrogen embrittlement problem of electroplated cadmium. Vacuum deposition is performed on plastics as well as metals; capacitors are made by the vacuum deposition of zinc or aluminum on paper. This is the best method for the production of reflective surfaces such as are found in sealed-beam headlights.

## 16.9 organic finishes

Despite the variety of metallic and ceramic finishes discussed in the previous sections, most surface areas are finished with various types of paints, lacquers, varnishes, and enamels. These organic finishes (which may contain inorganic pigments such as titanium oxide) contain a binder or "vehicle" which is either an oil or a synthetic resin and which usually hardens by cross-linking to produce a thermosetting finish.

There is a surprisingly large number of reasons for the painting of surfaces:

1. surface protection
2. decorative effects
3. cleanliness or "good housekeeping"
4. marketability of the product
5. advertising
6. identification (as in color coding)
7. safety and traffic control (as with traffic paints)
8. improved working conditions or morale
9. camouflage

In general, different types of finishes are required for these different objectives.

The differences between paints, varnishes, enamels, and lacquers are somewhat indeterminate, but are approximately as follows. A *paint* is a mixture of a vehicle and a pigment. A *lacquer* is a finishing material that dries quickly by evaporation of solvents and forms a film from its non-volatile constituents. *Varnish* contains no pigment. A strippable coating could therefore be termed a varnish. An *enamel* is a blend of a paint and a varnish.

The formation of the paint film occurs by either of two methods: solvent loss as in lacquers, and cross-linking as in linseed oil or epoxy finishes. In the terminology of the paint industry, the method of solvent loss characterizes a nonconvertible finish; cross-linking is the characteristic of a *convertible* finish.

The most familiar example of a nonconvertible finish is a mixture of shellac and alcohol. This is a rapid-drying lacquer. Such a finish has a number of disadvantages:

1. The loss of solvent introduces shrinkage stresses in the film, which may affect film adhesion.
2. The absence of cross-linking indicates that the solvent resistance of the film is poor.
3. A second coat will tend to redissolve the first coat.
4. Only a thin coat can be deposited from a nonconvertible finish because of solvent loss.
5. The rapid drying rate is suited to spray-painting but not to brushing.

This system may now be compared with convertible systems, which include linseed oil paints, baking enamels, and two-component paints, the latter having a synthetic resin in one can and a curing agent (also called a hardener or a catalyst) in a second can. The oil paints contain an unsaturated oil such as linseed oil which cross-links by using atmospheric oxygen as a cross-linking agent. In the case of an epoxy paint, one can contains a low-molecular weight liquid epoxy resin, the other can a curing amine perhaps, such as diethylene triamine. These two components are mixed in the proportions required for the chemical reaction. The paint must be applied before the curing reaction has advanced too far. This may be only half an hour. Material not used within this curing time must be scrapped. The *pot life* is the length of time between initial mixing and the final stiffening of the mix. Such a system will have good adhesion and chemical resistance but will tend to run or sag, since the viscosity increase must follow the curing reaction. Actually the viscosity of such a system must be adjusted with volatile solvents.

In paint testing, what is actually tested is the paint-substrate combination. Some of the more important paint tests are outlined in the following remarks.

**1. Flash Point.**   This test measures relative flammability.

**2. Hiding Power.**   The hiding power chart is illustrated in Fig. 16.5.

Fig. 16-5   Three types of hiding power charts. A paint coat has been applied to the chart on the right.

**3. Corrosion Resistance.**   A 1-mil thickness of the paint is applied to a steel panel and allowed to dry for 24 hours. Pools of acids or alkalis of the desired concentration are applied, covered with watch glasses and allowed to remain in contact for at least 6 hours. The corrodants are then removed, the surface is washed under a stream of water, and the surface examined for any change of appearance. The paint is then removed by solvent and the steel panel examined for corrosion. The line of contact of the watch glass is disregarded in the examination.

**4. Flexibility and Adhesion.**   There are many tests for these properties. The paint may be applied to a metal panel, the panel being bent 180° around a mandrel of specified diameter. Or a conical mandrel may be used, and the diameter at which cracking of the paint film first occurs is noted.

The impact test uses a steel ball with a 1-in. radius. This is dropped from varying heights and the inch-pounds of energy recorded. The direct impact test is made on the painted face of the panel; the reverse impact test drops the ball on the unpainted side of the panel. The latter test is a more severe one. Alternately, the Erichsen or Olsen cup test described in Sec. 8.15 may be used to test for flexibility and adhesion.

Still another method is used to test for adhesion. A steel rod 5/16 in. in diameter with a rounded end is used to mar the panel in two directions at right angles to each other. The rod is held perpendicular to the face of the panel and the rounded end of the rod is drawn across the finished surface using sufficient pressure to indent the film. The paint film is examined along both indentations. A less severe test requires cutting into the paint film in a cross-hatch pattern and then applying a piece of pressure-sensitive tape to the paint film. The tape is stripped, and any removal of the paint from the cross-hatched surfaces is noted.

Special apparatus is required for the testing of the abrasion, scratch, and weathering resistance of finishes.

## 16.10   the paint formulation

A paint mixture may require the following ingredients:

1. vehicle or medium
2. pigment
3. extender
4. drier
5. dyes

The finely ground solids of the paint are technically known as the pigment, and the liquid portion is called the medium or vehicle. Pigments must necessarily be white so that the coloring agents which actually pigment the paint may be added to supply the actual color. Driers are added to accelerate formation of the film. Extenders increase the "body" of the paint and improve the abrasion resistance. Dyes are used when the surface coating must be transparent but colored. Thinners are volatile solvents used to adjust the volatility of the paint. Paint which is sprayed will require more thinner than paint which is to be brushed. Volatility roughly corresponds to the evaporation rate or drying rate of the paint. One of the difficulties in formulating a latex paint is that of adjusting its volatility. The thinner in such a paint is really the water medium, and water evaporates rather slowly compared with standard solvents.

Formerly white compounds of lead were the standard white hiding pigments for paints. Because lead compounds are toxic, these are now in disfavor. Titanium dioxide is far superior as a body or pigment, and it is the material commonly used in every kind of paint except dark-colored ones and paints for special purposes such as aluminum paint. It is not toxic. Other pigment materials are zinc oxide, zinc sulfide, lithopone (a mixture of zinc sulfide and barium sulfate), and antimony oxide.

A good hiding pigment must first be white, together with a range of other requirements. For best whiteness, a high index of refraction is required, and titanium dioxide is superior to all others in this respect. The pigment must supply opaqueness or hiding power, so that any design on the substrate will not show through the paint. It must protect the paint resin against ultraviolet damage; zinc oxide is the best pigment in this characteristic and is favored for exterior paints. A variety of other requirements of some technical complexity is also needed in the pigment. All these requirements are sufficiently numerous that blends of pigments are now employed in paints. The lead pigments are generally inferior in all respects to the others mentioned, but they have characteristics useful for special paints: they form soaps which improve the wetting of undercoats and primers to steel surfaces.

Extenders control the gloss of the paint and adjust such characteristics as viscosity and brushability. Some of the common extenders are kaolin, barite (barium sulfate), diatomaceous silica, gypsum, and whiting (calcium carbonate). Whiting improves gloss, while diatomaceous silica reduces gloss.

Latex paints, which are water-thinned paints, use a different method of formulation. There is a similarity between a latex paint and the paint in an aerosol spray can. An *aerosol* is a dispersion of a liquid in a gas. An *emulsion* is a suspension of fine particles of a liquid within another liquid in which the first liquid is not soluble. Latex paints are emulsions of organic resins in water. The spherical particles of the paint are simply distributed throughout the water medium. On application to the surface being painted, the water evaporates and the resin particles coalesce to form the paint film. Such films have the advantages of low fire hazard, low odor, reduction in thinner costs, resistance to fading and peeling, and ease of cleanup simply by the use of water.

The average size of a particle of emulsion paint is about 0.1 micron, about one five-hundredth of the smallest size that can be distinguished by the naked eye. The emulsion formed must remain stable with the passage of time. Generally the particle size of an emulsion will increase with age as small particles combine into larger ones. This coalescence causes an increase in viscosity.

The resins used in latex paints are usually styrene-butadiene, vinyls, and acrylics. These may be applied to damp surfaces, including stucco and concrete. All have good permeability to moisture, and only under more severe moisture conditions in the substrate will they blister.

### 16.11 paint resins

Paints were originally based on natural unsaturated oils such as linseed oil or tung oil. These unsaturates cross-link with oxygen from the air. About 50 years ago oil-modified alkyd resins first began to replace the paints using natural oils only. Many present-day paint formulations contain no natural oils. The following list includes the more common paint resins:

PAINT RESINS

| | |
|---|---|
| Acrylic | Phenolic |
| acrylonitrile-butadiene | Polyimide |
| Alkyd | Polyester |
| amino | Rubber |
| phenolic | chlorinated |
| styrenated | neoprene |
| Amino | Silicone |
| Bituminous | Styrene |
| Cellulose | butadiene |
| nitrate | Urethane |
| ethyl | Vinyl |
| butyrate | acetate |
| Chlorinated polyether | butyral |
| Epoxy | chloride |
| amine | |
| phenolic | |
| polyimide | |

*Alkyd* resins are widely used in synthetic finishes. A number of materials can be incorporated into the alkyd molecule, including alcohols, oils, and fatty acids. Further, the alkyd resins have the capacity to improve the properties of other surface coatings, and for such purposes they are blended with styrene-butadiene, polyvinyl acetate, phenolics, and acrylics to improve water resistance, adhesion, and flexibility. Two large markets for the alkyd paints are automotive and appliance finishes.

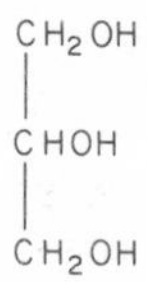

CH₂OH

CHOH

CH₂OH

Glycerol

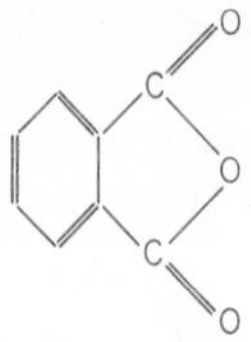

Phthalic anhydride

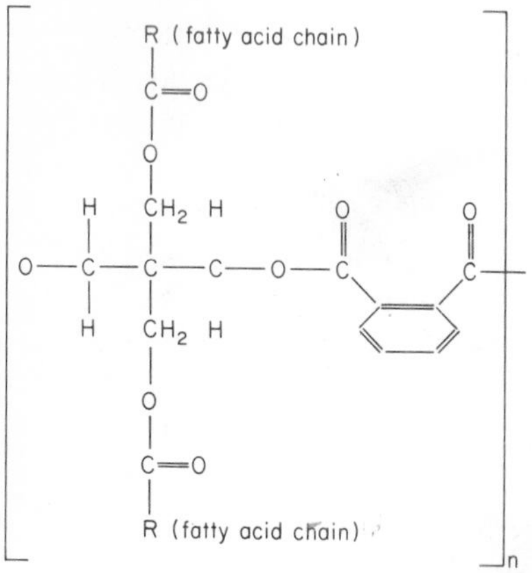

R ( fatty acid chain )

C=O

O

H   CH₂  H         O              O

O—C—C—C—O—C            C—

H   CH₂  H

O

C=O

R (fatty acid chain)

Fig. 16-6  Typical alkyd resin.

Most of the alkyd finishes are made by reacting phthalic anhydride with a polyalcohol such as glycerol (Fig. 16.6). Short-oil alkyd resins contain about 40 percent oil such as linseed oil or tung oil; long-oil resins contain about 60 percent oil. The long-oil resins are slower drying and give a softer film with better durability and flexibility. Generally speaking, long-oil alkyds are used in the maintenance, marine, and architectural markets, while short-oil alkyds are used on household appliances and furniture.

As with many of the synthetic plastics and coatings, the alkyd resins ultimately deteriorate under ultraviolet radiation. The degree of gloss failure is roughly proportional to the amount of such radiation received.

*Acrylic* resins are also used as automobile finishes, since they retain their gloss unusually well.

*Epoxy* resins offer the following outstanding properties:

1. no primer coat is required
2. they adhere well to metals, wood, concrete, and brick
3. they are resistant to solvents and cleaning fluids
4. they are flexible and resistant to abrasion
5. they may be used at temperatures as high as 300°F

For best resistance to chemical attack, as much as a 5-mil thickness of epoxy may be applied, in two or more coats. Good epoxy finishes are required to withstand 5 percent sulfuric acid or 20 percent sodium hydroxide for 72 hours without evidence of deterioration, and impacts of 30 in.-lb or more without cracking.

*Cellulose nitrate* is chiefly used in quick-drying lacquers. *Ethyl cellulose* provides highly flexible films with good acid and alkali resistance. *Cellulose acetate butyrate* has the best resistance to hydrocarbons.

Urethane enamels, like the epoxies, provide a superior surface coating, with excellent bonding, abrasion resistance, and resistance to moisture and chemical attack. They are however less resistant to ultraviolet radiation.

Chlorosulfonated polyethylene, though not a standard paint formulation, makes an excellent flexible, decorative, and protective coating for metal, wood, or masonry surfaces. It is resistant to all types of weathering and to abrasion and retains its original brilliance.

Clear and transparent finishes on exterior surfaces, especially wood, have a short life. This is explained by the absence of pigmenting materials to absorb ultraviolet radiation.

Fire-retardant paints when heated will foam into a heat-insulating mat that protects the underlying material from excessive temperature. These paints are made of noncombustible resins such as vinyls and chlorinated alkyds. They do not wear or weather well and are not suited to exterior application.

## 16.12 special finishes
## for metal surfaces

While most of the synthetic resins (most notably the epoxies) have good adhesion and suitable properties for metal surfaces, metal surfaces in general pose unique problems and have special requirements such as corrosion resistance. Unlike concrete, wood, and wallboard, metals have surfaces that are nonporous. Nevertheless, paint vehicles bond well to steel. However, paint vehicles adhere somewhat less effectively to less reactive metal surfaces. Adhesion to aluminum is still more difficult, and few paints will adhere to copper. In the protection of metal surfaces, paint films cannot exclude moisture, but they provide corrosion resistance by imposing an electrical resistance to the currents of galvanic corrosion. In severe environments the standard paints cannot give adequate protection to metals, and special formulations such as bituminous coatings must be used. Heavy coal-tar epoxy or other coal-tar base coatings are used to protect buried steel tanks and pipes. However, coal-tar coatings are far from beautiful and cannot therefore be used to protect the inside of a household refrigerator from condensation. Instead, porcelain enamels and vinyl plastics are used for refrigerator interiors.

*Aluminum paints* give excellent protection and adhesion to metals, and are used both for priming and finishing coats. Such a paint carries a dispersion of fine thin flakes of aluminum, smaller than 100 mesh size, which forms an impervious film in layers of perhaps a dozen flakes deep, bonded together with the vehicle. The arrangement is known as "leafing" (Fig. 16.7)

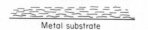

Metal substrate

Fig. 16-7 Leafing of aluminum paint.

because of the analogy with falling leaves. Such aluminum paints have remarkable hiding power and a reflectivity of about 75 percent. A film of aluminum paint 0.0005 in. thick will completely hide any colored surface. Asphalt-base varnishes are often used as the vehicle, and this makes the paint appear black in the can until the aluminum is stirred and dispersed.

*Silicone* resins are used in paints which must have great resistance to heat, moisture, oxidation, and ultraviolet radiation. They are used on ovens, boilers, and smokestacks for heat resistance and on masonry and concrete as water repellants.

*Red Lead* ($Pb_3O_4$) or lead oxide primers are used on steel surfaces. These may be red lead in linseed oil, or red lead in a vehicle of linseed oil and a long-oil alkyd resin. Because of the large linseed oil content, small amounts of tight rust or mill scale on the steel surface do not usually harm the integrity of these primer coats. Any of the primer coats used for steels may be used for aluminum except those containing red or white lead.

*Zinc chromate* is used as a corrosion-inhibiting primer on light metals such as magnesium and aluminum. Heavy coats are not employed, a coating of a third of a mil thickness giving better service than one of several mils. The chromate has a light yellow-green color, with poor hiding power, though hiding power is not the purpose of zinc chromate. Xylene or toluene thinners are used. These evaporate quickly, giving a drying time of about 5 minutes for the chromate primer.

## 16.13 adhesive bonding

Adhesive bonding, like metal finishing, requires careful surface preparation, and in general the same metal cleaning methods apply. Many of the synthetic rubbers and plastics used in metal finishes are also employed as adhesives.

Adhesive bonding has certain disadvantages when compared to other joining methods such as welding or the use of fasteners. Surface preparation is one disadvantage. Another is the relatively long time required for the adhesive to cure. The curing operation also may require heat or pressure or both. Finally, most adhesives, with the exception of sodium silicate and ceramic adhesives, have service temperature limitations of about 300°F.

Against these disadvantages must be set these significant advantages:

1. Absence of stress concentrations such as occur at rivet and bolt holes.
2. Uniform stress distribution over the large bonded area (this may not be entirely true of a lap joint).
3. Dissimilar materials may be joined, such as rubber to glass, or glass to metal.
4. Adhesive joints have excellent resistance to fatigue stresses.
5. An adhesive joint can provide electrical insulation or a moisture barrier.
6. Small particles can be bonded, such as abrasive grains to paper backing.
7. Adhesives are especially advantageous in the bonding of thin foils of paper and metal.

Structural adhesives are available for the dependable bonding of virtually any material. Their reliability is attested by their considerable use in aircraft and rocket construction, for wing flaps, bulkheads, floor panels, helicopter rotor blades, and in joining aircraft skins to fuselages. As much as 800 to 1000 lb of adhesives may be used in a single aircraft. Aerospace uses for adhesives include the bonding of sandwich structures made by joining thin metal faces to lightweight cores. Such construction offers unusual strength-to-weight ratios.

The available range of adhesives is quite wide, the following list summarizing the range:

1. animal and vegetable glues
2. inorganic adhesives and cements, such as sodium silicate and portland cement
3. solvent release thermoplastic adhesives
4. chemically curing thermosetting adhesives
5. elastomeric adhesives
6. pressure-sensitive and contact cements
7. emulsion adhesives
8. hot melt adhesives

The animal and vegetable glues include hide glue, bone glue, blood glues, starch, and casein glue. Most of these adhesives are used in the paper and packaging industries. Since paper has little tear strength or moisture and heat resistance, no especially stringent requirements are imposed on paper adhesives. Most of the sodium silicate adhesives likewise find their chief applications in paper products but are also used in the briquetting of iron ore, the flux coatings of manual welding rods, and the bonding of abrasive grinding wheels. The following list is suggestive of the multitude of uses for adhesives in paper and packaging:

1. corrugated cardboard (corn or potato starch)
2. gummed kraft paper tapes (bone glue or starch)
3. hand luggage

4. bottle labelling
5. paper laminating
6. bookbinding
7. paper milk cartons
8. sandpaper (animal glue)
9. paper tubes (starch adhesives)
10. envelopes and stamps
11. paper bags

There are two other applications of interest to cigarette smokers–the bonding of the abrasive to match-book covers and the bonding of the seam of a cigarette paper. Both joints require water resistance and a rapid-curing joint to meet the speed of manufacturing operations peculiar to these two products. Casein glues are suitable for both applications.

Asphalt falls into two of the above groups of adhesives. An emulsion asphalt is an emulsion adhesive; a cutback asphalt is a solvent release adhesive. Many resilient floor tile types employ asphalt as an emulsion, though sulfite flooring adhesives and rubber-base mastics may be used. Asphalt, like the sulfite flooring adhesives, is resistant to the alkalis in cement.

The solvent release thermoplastic adhesives cure by evaporation of the solvent, thus allowing the consistency of the adhesive to increase. The familiar white glues for household use are solvent-release polyvinyl acetate resins. Many of the elastomeric cements are of this type. The basic weakness of the solvent release method is that the volume of the adhesive will be reduced about one-third. This loss of volume may result in voids and also sets up shrinkage stresses in the bond.

The hot melt adhesives are a more recent introduction. These are mixtures of polymers which are applied hot to one surface; then the second surface is pressed on to the adhesive while it is hot. The hot adhesive wets both surfaces and cools to develop final strength. Though hot melts are relatively new, the principle is as old as the hot-mopping of asphalts in built-up roofs.

The emulsion adhesives consist of dispersed globules of adhesive in water. These bond by evaporation of the water.

## 16.14   chemically-curing thermosetting

## adhesives

The chemically-curing thermosetting adhesives include phenolformaldehyde, urea-formaldehyde, melamine-formaldehyde, resorcinolformaldehyde, and epoxies. Rigid foamed polyurethane when sprayed or poured is self-adhering, and some methods of installation use the layer of foam to bond the materials adjacent to the foam. This is often done in the construction of polyurethane sandwiches. Of all these thermosets, phenol-formaldehyde has the best weather and water resistance.

The epoxies are more expensive adhesives, but they have made a reputation for themselves in applications where reliability, high strength, hardness, weathering, abrasion resistance, water resistance, and bonding capacity to ill-fitting surfaces are required. They do not shrink on curing, and therefore they serve both as fillers and adhesives. Almost all other adhesives must be deposited in a thin film in order to develop maximum strength. The strength of an epoxy bond does not depend on the thickness of the adhesive layer.

## 16.15   elastomeric adhesives

Most of the pressure-sensitive tapes such as Scotch tape use an elastomeric adhesive. Here we are concerned with the liquid elastomeric adhesives. These are usually solvent-release rubbers, the solvent being a gasoline or alcohol, or an emulsion in water may be used. The elastomer may be SBR, acrylic elastomers, nitrile rubber, polyurethane, reclaim rubber, or sometimes other rubbers.

The nitrile rubbers are selected for resistance to oils and solvents. Neoprene adhesives are contact adhesives, that is, they stick immediately on contact with the surfaces. They are well suited to the bonding of plastics and decorative laminates to wood surfaces. SBR adhesives are also contact types used in mastics and are useful for the bonding of flooring materials to wood or concrete subfloors. Since SBR has limited tensile or shear strength, it is not used for structural and load-bearing applications.

## 16.16   characteristics
## of adhesives

Adhesives have become a science of considerable topical interest and have developed an appropriate terminology. Adhesive requirements and specifications cannot be understood without a knowledge of the following terms.

**1. Substrate.**   The material on which a coating is spread. If the coating is an adhesive, the substrate may be called the *adherend*.

**2. Adhesive.**   A substance which joins materials by attachment of their surfaces. A *cement* is the same thing as an adhesive.

**3. Pressure-sensitive Adhesive.** An adhesive that will adhere to a surface by a briefly applied pressure alone.

**4. Set.** To convert a liquid adhesive to a hardened condition. If setting occurs by polymerization or cross-linking, the operation is often referred to as a *cure*. A catalyst may be used to accelerate the cure.

**5. Tack.** Tack is the property of an adhesive which causes a surface coated with the adhesive to form a bond on immediate contact with another surface. Tack is thus the "stickiness" of an adhesive.

**6. Peel Strength.** This measures the resistance of the adhesive to stripping (Fig. 16.8) and is expressed in pounds per inch width. While many adhesives have shear strengths exceeding 5000 psi, the peel strength of any adhesive is limited, and adhesive joints must generally be protected against peeling.

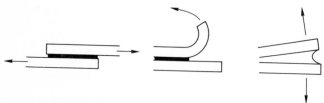

Fig. 16-8 Stressing of an adhesive joint by shear, peel, and cleavage forces. Adhesive joints are strong only in shear.

For adhesives used in lap joints there is a best film thickness, and adhesive films thinner or thicker than this optimum will be less strong. Soldered and brazed lap joints have this characteristic also. For maximum bond strength, joint thickness ranges from 2 to 5 thousandths of an inch, varying with the adhesive. Usually the thinnest possible joint is used, provided that the joint is not starved of adhesive.

Bond strengths of lap joints are proportional to the width of the joint but are not directly proportional to the joint length or overlap. As the length of overlap increases, the unit strength decreases. The first inch of overlap will carry most of the stress.

Tack is the most important characteristic of a pressure-sensitive tape, such as "Scotch tape," electrician's friction tape, drafting tape, and masking tape. These tapes have adhesives which are permanently soft, viscous, and tacky, and can be peeled from a surface without leaving any residue. Pressure-sensitive tapes do not "wet" a surface and cannot therefore be employed in heavy-duty applications or structural service. Most of the pressure-sensitive adhesives are rubber cements blended with a tackifying agent in equal amounts, such as coumarone-indene. Unmodified rubber cements have poor tack. A keying coat is used to bond the pressure-sensitive adhesive to the tape, since the two must adhere well and not strip when the tape is removed

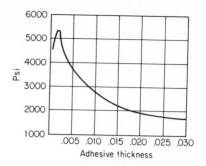

Fig. 16-9 Typical joint strength of
a synthetic adhesive.

from a surface. On the side of the tape opposite to the adhesive coat a release coat is applied to allow unwinding of the tape from its roll. These details of its formulation suggest that pressure-sensitive tape is one of the more ingenious inventions of our time.

The *mastics* used in the laying of floor and wall tile, linoleum, and countertops, also require a high degree of tack.

The sealants and caulking compounds used to close up joints in buildings must meet severe service conditions. They must remain elastic for years and in extremes of temperature, they must maintain a low elastic modulus so that they are not heavily stressed by their strains, they must bond to virtually any type of material, and they must resist water and all types of weathering attack.

The standard linseed oil glazing putty is adequate for glazing small windows only. With the passage of time it becomes hard and brittle. Present high-performance sealants may be polysulfides, silicones, polyurethanes, polymercaptans, butyl, polybutene, polyisobutylene, acrylic rubbers, polychloroprene, or Hypalon (chlorosulfonated polyethylene). The acrylics and Hypalon are one-component sealants; polyurethane sealants may be either of one or two components. Of all of these kinds, butyl has the least elastic recovery.

In selecting adhesives, the following principles are followed:

1. Load-bearing adhesive joints must be made with thermosetting adhesives
2. Joints that require flexibility must usually be made with elastomeric adhesives. Flexibility may be required because of differences of thermal expansion between materials or because the assembled joint will be flexed.

## PROBLEMS

1  What cleaning method would be used to remove (a) rust on steel, (b) paint on steel, (c) the oxide surface on aluminum, (d) fingerprints on titanium sheet?
2  What is meant by "conversion coating"?
3  What is meant by "nonconvertible finish"?

4  What is the hardening process in both a convertible and a nonconvertible finish?

5  What purposes are served by anodizing aluminum?

6  What limits the maximum thickness of an anodized coating on aluminum?

7  What is meant by wetting of a surface?

8  Why should a thermosetting surface finish be superior to a thermoplastic one?

9  What is the chief weakness of glues as adhesives?

10  What disadvantages are characteristic of a solvent release adhesive?

11  What is a contact adhesive?

12  What is a pressure-sensitive adhesive?

13  Define tack.

14  Define peel strength.

15  Explain the difference between a paint, a varnish, and a lacquer.

16  What is an emulsion paint?

17  Explain the function of the vehicle and the pigment in a paint.

18  Define volatility.

19  Why would a lead-base paint not be allowed in painting the walls of a cold storage meat freezer?

20  Why is titanium dioxide favored as a pigment in paints?

21  Why is zinc oxide used as the pigment in exterior paints?

22  Explain why lead oxide is used as a pigment in priming coats for structural steel.

23  What is a long-oil resin?

24  Define hiding power of a paint.

25  From the seven types of zinc coatings on steel, select the type which you think best suited to the following applications. State your reasons.
(a) a high-voltage electric transmission tower
(b) a sheet steel travel trailer which is to be painted
(c) a small sheet steel part which is to be formed in a die and then painted

26  Select three paint resins which you believe would be of suitable quality for painting your own automobile.

27  What is the usual vehicle in aluminum paints?

28  Select a suitable paint resin for a steel smokestack.

29  List the outstanding properties of the epoxy adhesives.

30  Obtain a quantity of white household adhesive. Make adhesive joints on carefully prepared surfaces of steel, aluminum, wood, etc., and test them in shear. Do not introduce any peeling forces into the joint. This adhesive is a typical thermoplastic adhesive. How does it perform? Compare it with a two-component epoxy adhesive.

31  If the proportion of pigment in a paint is increased, will the gloss improve or decrease?

## 17.1 *fundamental characteristics*

Biomaterials are living materials such as wood, cotton, or foods, or they may be materials processed from these materials such as linseed oil, shellac, or natural rubber. In addition, the term biomaterials may be taken to include those materials implanted into living bodies, such as artificial heart valves.

The subject of biomaterials has only recently attracted wide interest as a separate technology. It is without doubt the most interesting of the materials sciences, if only because every person is himself composed of biomaterials. It is also a technical area in which the materials practitioner rapidly acquires humility. Man-made materials and machines can appear very amateurish when compared to the incredibly sophisticated living materials and structures such as wood, blood, bone, or machine elements like the hoof of a horse or the joints of human or animal bodies. Advances in materials will in the future depend greatly upon our knowledge of biomaterials.

Animals and plants are engines, converting fuel carbohydrates to carbon dioxide with release of energy. The output of such bioengines is the life processes: growth, movement, reproduction, repair of tissues, and body heat. Animals and man obtain their fuel intakes by eating plants or other animals

# BIOMATERIALS

# 17

which have eaten plants. Only plant life can generate the life fuels directly. This they do by utilizing the energy of sunlight to produce carbohydrates in the most important of all chemical processes, photosynthesis.

The molecules or organizations of atoms of biomaterials are distinguished from those of inorganic materials by three basic characteristics:

1. they are carbon-based
2. they are highly complex
3. they have a high energy content, as is necessary if the life processes are to be carried out

The polymer materials previously discussed were more complex substances than the ceramic or metallic materials. The ultimate in chemical stability is shown by the ceramic materials. They have survived billions of years of environmental influence, and except under the stress of severely elevated temperatures, they survive also the uses to which the human race puts them. The metals are derived from the ceramic materials generally by deoxidizing or desulfurizing the ceramic ore in which the metal is chemically combined. The usual chemical activity of a metal is to revert to the ceramic by oxidizing.

A higher degree of complexity is shown by the polymer plastics and rubbers. These long chains are both linear, branched, and cross-linked, partially crystalline and amorphous. They can be oxidized or they can be depolymerized by oxygen, ozone, or ultraviolet light. The biomaterials show still higher orders of complexity, most of which are still not understood. Their most outstanding characteristic is their almost miraculous ability to repair and reproduce themselves.

The history of materials technology has followed a predictable development dictated by the characteristics of materials. This history begins in the ceramic stone ages and develops through the metals ages into the polymer age which is virtually upon us. The rapid developments in the biosciences suggest that an extensive engineering technology in biomaterials is not far distant, perhaps another decade away.

## 17.2 wood

"Wood is universally beautiful to man. It is the most humanly intimate of all materials. Man loves his association with it, likes to feel it under his hand." Thus the American architect, Frank Lloyd Wright, expresses the appeal of wood to man.

Wood is a cellular material. We have only recently been able to make foamed plastics that even approach the physical and mechanical properties of wood. Some of the more recent fiberglass-reinforced foamed high-performance plastics can attain an *E*-value of 800,000. Most woods have *E*-values twice as large as this.

Although wood has lost many applications to metals and plastics, its uses are still highly diversified. Its chief products are paper and lumber. Second to these is the use of wood as a packaging material for crating, wirebound boxes, wood boxes, skids, and pallets like the one shown in Fig. 17.1. Other uses for wood are perhaps of lesser importance: cellulose filaments, chemicals, laminated timbers, fine-art objects such as furniture, and a variety of special products such as timber piles, power poles, railroad ties, foundry patterns, cribbing, wood pipe and tanks, fisherman's floats, matches, charcoal and other luxury fuels.

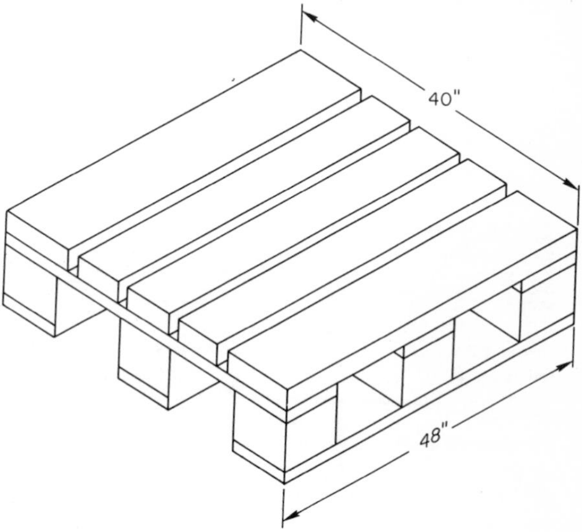

Fig. 17-1 The wooden pallet, which created a revolution in material-handling practices, material-handling equipment, and warehouse layout. The style of pallet shown is known as a four-way pallet, since it allows the forks of a lift truck to enter the pallet from any of the four sides. The 40 × 48-deck is the commonest size.

It is not possible to conceive of wood as becoming obsolete because of its unique advantages over other materials:

1. it is a self-perpetuating natural resource
2. its relative abundance in most geographical areas
3. the ease with which it can be processed
4. its light weight
5. its inherent beauty
6. it is the strongest of the cellular materials, and it is the cheapest

A remarkable property of wood is its capacity to survive fires. The fire resistance of wood is promoted simply by using large timber sections. A 12 × 12 timber might require days to burn through: no accidental fire lasts more than hours. Wood also has a low thermal conductivity and a high specific heat. Wood therefore must be considered much more fire-resistant than steel, the latter material having a high thermal conductivity and considerable loss of strength at elevated temperatures. Wood will not ignite until its temperature reaches about 550°F.

Wood has a high degree of corrosion resistance, but it will slowly "weather" if exposed continuously to the atmosphere. In weathering, its surface turns gray and roughens. Weathering apparently is a very slow corrosion process, the rate of weathering being cited as a quarter-inch per century.

The thermal expansion of wood parallel to the grain is assumed to be zero.

## 17.3 the structure of wood

A tree grows by the addition of new wood to the outer layer of the tree. This annual growth can be seen in the concentric series of growth rings (Fig. 17.2). Each year adds one more growth ring to the diameter of the tree, at least in temperate zones under normal conditions.

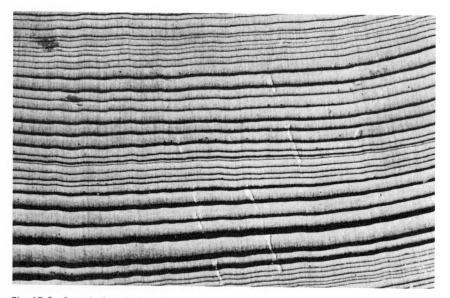

Fig. 17-2 Growth rings in Douglas Fir. At the bottom-right-hand corner, the rings indicate 10 dry years followed by 5 rainy years.

The growth rings are easily distinguished from one another because of color differences between *earlywood* and *latewood*, also known as *springwood* and *summerwood*, respectively. These two types of annual growth may also differ in hardness and density, especially in some softwoods such as Douglas fir. Latewood (summerwood) is darker in color, heavier, and harder, because later in the season the cells of the tree grow more slowly. Springwood contains cells with larger cavities and thinner walls because of the more rapid rate of growth in the earlier part of the growing season. The distinction between earlywood and latewood is more difficult in tropical hardwoods.

As the tree continues to add annual rings, the older, more central rings gradually cease to contribute to the physiological processes of the tree, such as the movement and storage of food chemicals, and provide only mechanical support to the tree. This functional change produces what is called *heartwood* in the center area of the tree. The heartwood portion is usually darker in color, drier, and harder than the surrounding layers that are still physiologically active. The outer living part of the tree is the *sapwood*. The diameter of the heartwood will decrease toward the top of the tree trunk, since the growth rings are a series of concentric cones.

Chemically, wood is a compound polymer composed of three principal polymeric constituents. The wood of a tree consists of long tubular fibers. These fibers are composed of two polymeric carbohydrates called *cellulose* and *hemicellulose*. Both are complex glucose compounds. Glucose is a sugar, which explains why fungi and insects, and even animals, can use wood as a food. These wood fibers are bound together with a third polymer called *lignin*. These fiber bundles run the length of the tree or the branch and carry food products from the roots to the leaves. Based on an oven-dry condition, the constitution of woods is the following:

|  | Hardwoods | Softwoods |
|---|---|---|
| cellulose | 40–45% | 40–45% |
| hemicellulose | 15–35 | 20 |
| lignin | 17–25 | 25–35 |

Small amounts of other substances are also present in woods.

It is cellulose that provides the strength in axial tension, toughness, and elasticity of wood. The long-chain molecules of cellulose are in bundles that run helically to form hollow needle-shaped cells or fibers. These fibers range in length from 1 to 3 millimeters. The fibers have thicker walls in hardwoods than in softwoods. The long-chain structure of cellulose makes it able to form fibers similar to other vegetable fibers such as cotton. The chemical structure of cellulose is given in Fig. 17.3 as $(C_6H_{10}O_5)$ repeated $n$ times, where $n$ is a very large number between 8000 and 10,000.

Associated with the cellulose bundles are a group of similar polymeric substances, the hemicelluloses. These are chemically similar to cellulose. The chain length is shorter, about 150 units, so that the hemicelluloses are not fibrous but are gelatins.

Lignin bonds the individual fibers together into wood, giving the wood its compressive strength. The chemical structure of lignin is not completely

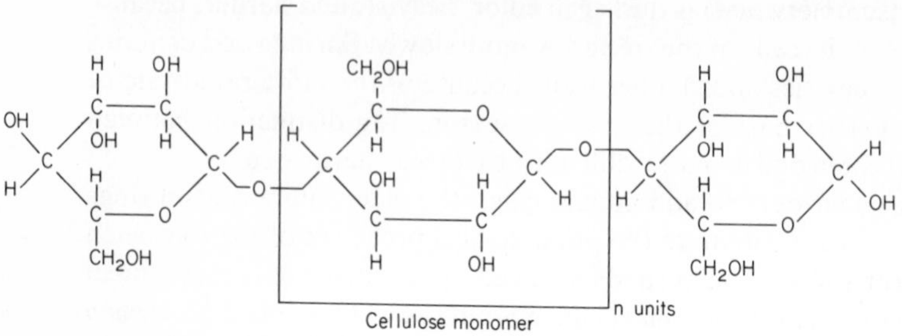

Fig. 17-3 Structure of cellulose.

known. The manufacture of paper requires the removal of most of the lignin in order to free the individual wood fibers. It is rather interesting that lignin is found only in association with cellulose, though cellulose may be found independently, as in cotton.

## 17.4 hardwoods and softwoods

The terms hardwood and softwood do not differentiate between woods that are hard or soft. The summerwood of Douglas fir, a softwood, is extremely hard, while balsa wood, a hardwood, is the softest of all woods. The hardwoods are the deciduous or broad-leaved trees which shed their leaves in the fall in cooler climates. Softwoods are the conifers that bear needles instead of leaves and produce their seeds in cones. Tamarack, which is a softwood, nevertheless sheds its needles in the fall. Softwoods generally contain more lignin and less hemicellulose than most hardwoods. Both the lightest and the heaviest, the softest and the hardest, of all woods are tropical hardwoods. The lightest is balsa, weighing from 5 to 10 lb/cu ft. The heaviest is lignum vitae or ironwood, weighing 75 to 80 lb/cu ft.

Softwoods are found in many tropical areas, but they are more characteristic of colder climates and higher altitudes, usually in large tracts of relatively few species. Hardwoods become less common as altitude increases. About 30 softwoods and 50 hardwoods are of commercial importance in the United States and Canada.

The individual character, color, and smell of the specific species is caused by extractives in wood, mainly in the heartwood. These extractives are so called because they can be removed from the wood with solvents. The extractives may also make a timber resistant to decay or to insect and fungus attack. Despite their small quantity, about 1 to 2 percent by weight, the extractives have a significant effect on the properties of the wood. These extractives explain the excellent durability of redwood, oaks, and cypress, and the durability of heartwood as compared to sapwood generally. Usually a dark color in wood indicates a durable timber. Many of these natural dyes in the wood however are destroyed or bleached by ultraviolet light or they evaporate over a period of time.

### Typical uses of some softwoods
1. pines—foundry patterns. Easily shaped, non-splitting, dimensional stability
2. Western White Pine—matches, boxes, crates
3. Southern Pine—railroad ties, boxes, tanks, millwork
4. Sitka Spruce—aircraft. Strong, few defects, light weight
5. Douglas Fir—plywood, wood tanks. Strong, tending to become brittle with age
6. Cypress—tanks, vats, and silos. Decay resistant
7. Cedar—pencils. Easily whittled, rot-resistant, but splits readily and has little strength or hardness

### Typical uses of the hardwoods
1. hickory—handles for impact tools. Hard, strong, and tough
2. ash—long tool handles (e.g., shovels), baseball bats
3. birch—butcher blocks, desks. Rots readily
4. oak—railroad ties, posts, industrial floors, pallets. A poor wood for dimensional stability
5. maple—flooring, bowling pins, boxes and crates
6. basswood—drafting boards, boxes and crates
7. aspen—pallets. A tough wood

The first trees to appear in the long history of the earth were the soft-

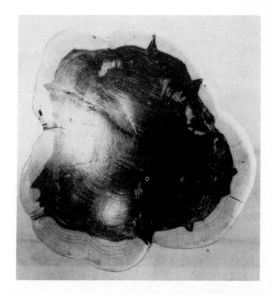

Fig. 17-4 Heartwood and Sapwood in a Red Cedar.

woods, which date from about 200,000,000 years ago. The hardwoods are about half as old as the softwoods in geological history.

## 17.5 physical properties
## of wood

The physical and mechanical properties of wood are considerably affected by the moisture content of the wood, including hardness, strength, and electrical resistance. The moisture content in the air-dried condition at equilibrium with the atmosphere is 12–15 percent.

Moisture content in percent is measured as

$$\frac{\text{weight of moist wood} - \text{oven-dried weight}}{\text{oven-dried weight}} \times 100\%$$

Generally the lighter woods can hold more moisture than heavy woods. South American balsa wood, specific gravity 0.2, may hold 400 percent moisture in the saturated condition. Very heavy hardwoods may have saturated moisture contents of only 30 percent.

Shrinkage due to moisture changes is about 0.1–0.2 percent in the longitudinal direction. Shrinkage tangential to the growth rings is almost twice the shrinkage radial to the rings for equal changes in moisture content. Woods of high lignin content, such as mahogany, show less shrinkage than woods low in lignin, such as basswood.

The specific gravity of a wood is defined as

$$\frac{\text{oven-dry weight of 1 cu ft of wood}}{62.4}$$

The specific gravity of the oven-dry cell wall material of any wood is about 1.5. The specific gravity of any species of wood must then depend on the size of the wood cells and the thickness of the cell walls. The specific gravities of most North American woods are in the range of 0.35 to 0.65.

The property of anisotropy is a design consideration in the use of wood. The tensile strength of wood on a plane perpendicular to the grain is about ten times better than the strength parallel to the grain. Shear strength parallel to the grain is poor, while shear strength perpendicular to the grain is very high. The dimensional changes in wood caused by variations of moisture content are likewise influenced by the cellular or tubular structure of wood. Changes in length parallel to the grain as a result of moisture variation are negligible. The anisotropy of wood may be reduced by laminating it as plywood, with the grain running in different directions in the several laminations. Lamination, however, reduces the apparent modulus of elasticity.

The anisotropy of wood shows in its thermal conductivity: conductivity for heat is twice as great parallel to the grain as in the other two directions. Thermal conductivity is increased with higher moisture content. Actual *K*-values per inch of thickness are about 1 Btu per hour more or less, depending on moisture content, direction of heat flow, and type of wood. The heaviest hardwoods have about twice the conductivity of the lightest softwoods.

Since the electrical resistivity of wood depends on moisture content, moisture meters usually measure the electrical resistance of the wood between probes and correlate these electrical values to moisture content.

Tensile and compressive strength of wood increases with specific gravity. The strength will be lowered by increased moisture content, however. Under long-term stress, wood will creep at room temperature. On the other hand, wood withstands impact stresses well, this feature accounting for its use in tool and implement handles.

Wood must be selected for freedom from its many possible defects. The more common defects are these:

1. *Checks.* Cracks running parallel to the grain and across the growth rings. Checks usually result from the seasoning of wood from the green condition and are especially troublesome in certain hardwoods such as oak.
2. *Cross-grain.* The grain does not run parallel to the length of the board in cross-grained lumber. Such lumber will be weak.
3. *Knots.* These occur where a branch intersects the tree trunk.
4. *Pitch pockets.* These occur between annual growth rings and are filled with wood pitch.
5. *Warp.* Any variation from two-dimensional flatness.
6. *Insect damage.*
7. *Bark pockets.* Small patches of bark embedded in the wood.
8. *Rot.* There are many forms of wood rot. Repeated drying and wetting of wood promotes one kind of rot. Another kind is promoted by the use of steel fasteners in wet wood.

### TYPICAL PHYSICAL PROPERTIES OF WOOD

| | Sp Gr | Modulus of Elasticity | Compression Parallel to Grain, psi | Tension Parallel to Grain, psi |
|---|---|---|---|---|
| Ponderosa Pine | 0.37 | $1.34 \times 10^6$ | 5200 | 9700 |
| Eastern Spruce | 0.39 | $1.2 \times 10^6$ | 5200 | |
| Douglas Fir | 0.46 | $1.6 \times 10^6$ | 7200 | 13,200 |
| Balsam Fir | 0.38 | $1.3 \times 10^6$ | 5200 | |
| Redwood | 0.38 | $1.25 \times 10^6$ | 7200 | 8800 |
| Maple | 0.60 | $1.9 \times 10^6$ | 7600 | |

## 17.6  paper products

It is customary to think of paper as newsprint or office material. Such a concept does no justice to the wide panorama of paper usage in industry, where it is even used as a structural material. The author would hazard the educated guess that the "Big Three" automotive manufacturers consume well over 100 tons of paper per day in industrial operations, not including office paper.

The paper industry defines paper as "all kinds of matted or felted sheets of fiber . . . formed on a wire screen from water suspension." Paper products are divided into two broad groups: paper and paperboard, the latter being a heavier and more rigid type of product. These two categories are produced in roughly equal amounts in the United States, about 20,000,000 tons of each. The actual consumption of paper in the United States exceeds the consumption of any metal other than steel. About 50,000,000 tons of paper is the annual consumption. In Canada the annual production of paper in tons is approximately equal to the annual production of steel. This huge paper industry consumes annually over a million tons of salt cake, half a million tons of sulfur, and a million tons of china clay.

The U.S. Bureau of the Census, Industry Division, divides statistics of paper production into 8 groups of products distributed approximately as follows:

| | |
|---|---|
| newsprint | 12% |
| book and magazine papers | 28 |
| fine and writing papers | 10 |
| wrapping papers and bags | 24 |
| tissue papers | 2 |
| sanitary papers | 11 |
| building papers | 9 |
| other papers | 4 |
| | 100% |

About one-third of this production goes into industrial and packaging uses.

## 17.7  paper-making materials

In addition to wood pulp, paper is made from waste paper and rags, and occasionally other raw stock such as bagasse (sugar cane waste). Waste paper is a difficult raw stock to use because of the presence of such foreign materials as trash, adhesive cements, and plastic film, and the problems of de-inking and bleaching the waste paper. Rag paper was exclusively used as raw material until a hundred years ago. The poorest grades of rag stock are used for roofing felt papers, though most rags are converted into specialty papers for the following purposes:

1. legal documents, insurance policies, share certificates, bonds, and bank notes
2. drafting and blueprint papers
3. high-grade letterhead bond papers and special stationery
4. cigarette and carbon papers (light-weight papers)

The two chief chemicals in wood are cellulose and lignin. Papermaking from wood pulp is based on the fact that cellulose fibers adhere to each other to form continuous sheets. Lignin is present in newsprint pulp, which is basically groundwood prepared by grinding the wood fibers out of the wood with large grinding wheels. Permanent papers must be prepared by removing the lignin with chemicals.

Wood pulp fiber is remarkably strong: about 80,000 psi tensile strength.

Neglecting special methods that produce only limited tonnages of paper, paper pulp is produced by three principal processes:

1. the kraft process, which produces the greatest tonnage
2. the sulfite process
3. the groundwood process, its most important application being the production of newsprint in Canadian paper mills.

An average kraft pulp mill produces 500 tons of paper pulp per day. For this output the daily consumption of wood will be almost 1000 cords, or a pile of wood four feet high, four feet wide, and a mile and a half long.

## PAPER PULP CHARACTERISTICS OF SOME WOODS

|  | Bark, % | Cellulose, % | Lignin, % | Fiber Length, mm |
|---|---|---|---|---|
| White Spruce | 12 | 60.5 | 29.0 | 3.5 |
| Eastern Pine | 12.5 | 60.0 | 27.5 | 3.5 |
| Hemlock | 10 | 59.5 | 30.0 | 4.0 |
| Paper Birch | 13 | 60.5 | 25.7 | 1.2 |
| Sugar Maple | 14 | 60.8 | 23.2 | 1.0 |
| Aspen | 18.5 | 65.5 | 23.4 | 1.2 |

## 17.8 wood pulp processes

Chemical pulp is produced by dissolving the lignin with chemicals, thus freeing the cellulose fiber. The *sulfite process* uses sulfurous acid and bisulfites to defiber the wood after it is cut into chips, by forming complex lignin sulfonic acids. This process was the most important pulping method until about 30 years ago, when the kraft process overtook it. The kraft method could pulp woods that were not suited to the sulfite process, such as many hardwoods and resinous woods like pine. The sulfite reaction on the wood chips is carried out in cylindrical vessels, usually of stainless steel, called *digesters* (Fig. 17.5). These are about 16 ft in diameter and about 50 ft high, with a conical bottom. Several digesters, with chip and pulp handling equipment, will be housed together in a digester building at the mill.

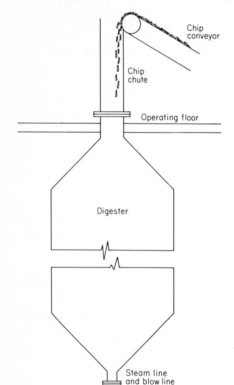

Fig. 17-5  Pulp mill digester.

The *kraft*, or sulfate, process uses alkaline caustic soda and sodium sulfide to remove the lignin. The pulp produced by this method is a stronger pulp, hence the name "kraft," the German word for "strong." Most of the kraft pulp mills of the United States are in the South, since this process can pulp both the pines and hardwoods of that area. The cooking is done in digesters of much the same construction as those used for the sulfite process.

The sulfite and sulfate processes produce chemical pulp. The groundwood process produces mechanical pulp by grinding the wood in the presence of water. Groundwood pulp is produced for newsprint, for tissue, towelling, and board paper, and for molded packaging materials such as egg containers and paper plates.

The raw pulp produced by these methods may require bleaching. In addition, the pulp must be put through a beating process, as it is called; actually it is a brushing process. This improves the strength and other properties of the finished paper. The pulp also requires various additives, such as sizing agents, dyes, and clays to improve brightness and printing characteristics. The use of fluorescent dyes is becoming common. Commonly called optical whiteners, they function by absorbing radiation in the ultraviolet and reradiating in the visible blue range, with the effect of making the paper appear whiter.

### 17.9  paper making

Paper is made on a woven wire belt in a long paper machine. The pulp stock is pumped onto the woven wire of the paper machine in the form of a thin flat ribbon from a suitable nozzle. Moisture in the stock is drained through the supporting wire mesh, after which a series of pressure rolls and a dryer remove further moisture. The paper sheet is then calendered, that is, passed through a series of rolls stacked vertically. The calendering operation controls the sheet thickness and smoothness. Finally the sheet of paper is reeled.

Figure 17.6 is a sketch of the Fourdrinier type of paper machine. The Fourdrinier wire belt is usually of bronze wire or synthetic fiber woven into a 55 to 85 mesh. Coarser mesh is used to make paperboard, and finer mesh is used for special papers such as cigarette paper. If the wire is not kept clean, the paper will be blemished. Because of abrasion and corrosion effects, the wire has a rather limited life.

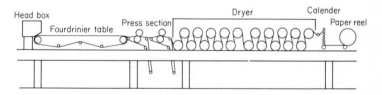

Fig. 17-6  Fourdrinier paper machine.

Paper may have to be coated with a sizing material such as starch. The sizing operation restricts penetration of printing inks into the paper and produces a hard and attractive surface. Sizing and other liquid coatings are applied by means of a flat metering blade and a puddle of the liquid coating extending across the width of the paper.

## 17.10   converting of paper

Various converting operations may be performed on the paper in order to produce the desired finished product. The word "converting," as used in manufacturing industries, refers to operations that convert semifinished products, such as paper, into manufactured articles. Converting operations on paper include laminating, coating, or forming into bags.

*Coating* is required for the many types of printing papers to enable the paper to receive the printing ink and to reproduce the printed design with sharpness, fidelity, and attractive appearance. Quality white clay (kaolin) in a starch binder is the usual coating for this purpose. Titanium dioxide, which is added to paints and porcelain enamel for whiteness, is also used in paper coatings. A waxed paper is first coated, printed, and then waxed. Building papers and roofing felts are made of paperboard coated with asphalt, this construction providing a barrier against moisture and water vapor.

*Lamination* provides a paper with a combination of properties or surfaces, or is used to produce a heavier or stronger paper or paperboard. The lamination of paper and aluminum foil is familiar in the packaging of chocolate bars and other products. Any difference in tension between the laminates during lamination will produce a curled product; this is prevented by independent tensioning controls for each sheet material.

Many papers are *creped,* especially decorative papers, sanitary papers such as towelling, and packaging papers. Creping begins by wetting the paper in a thin sizing solution and passing it through a heated roll. The paper is peeled from the roll with a blade called a creping knife. The setting of the knife, its contour, the speed of production, and the amount of size all control the amount of creping or stretch produced in the paper. The creped paper must not be tensioned before drying; otherwise the creping will be lost.

Corrugated paperboard is made of three webs of kraft paperboard, the middle one being corrugated on rolls and the outer laminations being coated to it with a suitable adhesive such as starch. The corrugated structure is highly resistant to bending or crushing. Corrugated paperboard was probably the first of the "sandwich" structures which now find applications in aerospace vehicles and highway trailers, using expanded metal or foamed plastic for the sandwich core. Larger structural paper sandwiches can be made using paper honeycomb cores bonded to a suitable outside facing material.

The chief types of paper bags are the *multiwall shipping bag* of the type used to bag portland cement, and the brown *kraft paper grocery bag*. These are made on high-speed bag machines, where they are first formed into tubes, after which a bottoming machine closes the bottom.

Cylindrical concrete forms are made of heavy paperboard. They are produced in various large diameters and lengths of 12 ft, with a wax coating, and are chiefly employed in the casting of concrete building columns. After the concrete has set, the paper is peeled away from the concrete.

The use of shipping and protective papers has increased markedly in recent years. Car liner papers are used to protect goods loaded into boxcars. Large unit loads of lumber are often shipped on flat cars wrapped in weather-proof covering paper.

## 17.11  the dimensional stability

## of paper

The dimensional changes of paper caused by absorption or loss of water are a familiar problem to draftsmen. Such lack of dimensional stability is a matter of concern in many industrial operations. It may affect register in multicolor printing operations or registering of the punched holes in punched cards. In one commonly used method of automatic oxyacetylene cutting of steels, a photocell follows a line drawn on a sheet of drafting paper while automatic controls cause the cutting torch to reproduce the shape drawn on the paper. For this type of operation, a drafting paper of suitable dimensional stability must be used.

Because of tension applied to the longitudinal dimension of paper during its manufacture, paper is anisotropic like the wood from which it is made. Longitudinal dimensional changes caused by humidity variations are minor, but in the cross-direction such changes are more troublesome, though they vary between grades of paper.

Most papers will not elongate more than 5 percent when tensioned to rupture. Special grades will elongate more than this amount, especially creped papers.

## 17.12  cellulose plastics

A considerable number of plastics are made from cellulose. Usually wood pulp is the raw material. The most important of these plastics are:

1. cellophane
2. cellulose nitrate

3. cellulose propionate
4. cellulose acetate
5. cellulose acetate-butyrate
6. methyl cellulose
7. ethyl cellulose

*Cellophane* is a regenerated cellulose and is the same polymer as cellulose. It is therefore thermosetting. It is used only as packaging film, and in this form it is usually coated with other material, such as cellulose nitrate. Cellulose film has excellent clarity and strength, but it tears readily.

*Cellulose nitrate*, better known as Celluloid, is closely similar to nitrocellulose (an explosive) in chemistry. It was the first of the synthetic polymers to be invented, being discovered about 100 years ago. It has the serious disadvantage that it burns furiously, and therefore it is not acceptable for many products, including building materials.

*Cellulose acetate* is used as a thermoplastic adhesive and as packaging and photographic film. It is transparent and permeable to water vapor and gases and is used to package fruits and vegetables that must "breathe."

# 17.13   general properties of foods

No materials are of course of more importance than foods. A considerable number of foods serve also as industrial raw materials, including potato starch, the casein in milk, peanuts, soybeans, and sunflower seeds. The processing of foods is one of the largest industries.

Most foods approximate the specific heat and latent heat of water and ice, since 80 to 90 percent of any food is water. All foods freeze between 28 and 35°F. Because of the high heat storage of water, the refrigeration and freezing of foods involves the removal of large quantities of heat and the use of refrigeration equipment of large capacity. Storage of foods also requires close control of relative humidity; this aspect also is an expensive part of food technology. All fresh fruits respire, using stored chemical constituents in the food, chiefly sugars, and oxygen from the air, to produce carbon dioxide and heat.

The specific heat of foods above their freezing points is approximated by the following formula:

$$\text{Sp ht} = 0.008\,W + 0.20$$

where $W = \%$ water content.

The following table gives the thermodynamic properties of several foods. The large amount of water even in such foods as celery and carrots should be noted.

## PROPERTIES OF PERISHABLE PRODUCTS

| Commodity | % Water | Freezing Point | Sp Ht Above Freezing | Sp Ht Below Freezing | Latent Heat, Btu/lb |
|---|---|---|---|---|---|
| apples | 84 | 29 | 0.87 | 0.45 | 121 |
| bread | 35 | – | 0.70 | 0.34 | 46–53 |
| carrots | 88 | – | – | 0.46 | 126 |
| celery | 94 | 31 | 0.95 | 0.48 | 135 |
| corn | 74 | 31 | 0.79 | 0.42 | 106 |
| butter | 16 | – | 0.33 | 0.25 | 23 |
| ice cream | 62 | – | 0.70 | 0.39 | 89 |
| milk | 87 | 31 | 0.90 | – | 125 |
| eggs | 66 | 28 | 0.74 | 0.40 | 96 |
| lettuce | 95 | 32 | 0.96 | 0.48 | 136 |
| potatoes | 78 | 31 | 0.82 | 0.43 | 111 |
| tomatoes | 94 | 31 | 0.95 | 0.48 | 134 |
| fish | 62 | 28 | 0.80 | 0.40 | 89–122 |
| beef | 62+ | 28 | 0.70–0.84 | 0.38–0.43 | 89–110 |
| pork | 32+ | 28 | 0.48–0.54 | 0.30–0.32 | 50–60 |
| poultry | 74 | 27 | 0.79 | – | 106 |

### 17.14  photosynthesis

Certain of the more significant chemical processes have been briefly described in previous chapters: the formation of hydraulic Portland cement and the production of iron and aluminum from ores. There remains to be described the most important process from any point of view, that of photosynthesis. Every year plant life produces almost 200 billion tons of carbohydrates from water taken from the soil and carbon dioxide from the atmosphere, releasing back to the atmosphere 400 billion tons of oxygen. From this basic process are derived other biomaterials such as proteins, organic fats, hormones, enzymes, and chromosomes. As the pioneer chemist Joseph Priestley said, this process reverses the effect of breathing.

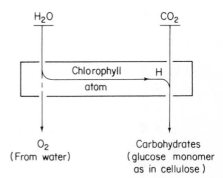

Fig. 17-7  The photosynthesis process in the chlorophyll atom.

The human eye is sensitive to a narrow range of electromagnetic radiation at the middle of the great frequency band of this radiation (Sec. 2.5). This narrow range of visible light from red to violet is the range that initiates the process of photosynthesis. The photon energy of these frequencies, about 1 electron volt, is used by the plant to split the water molecule and release its hydrogen. This hydrogen is then used to reduce carbon dioxide, with the formation of carbohydrates and release of oxygen to the air.

This chemical process of photosynthesis is carried out in the plant chemical chlorophyll. The structure of chlorophyll is shown in Fig. 17.8.

The products of photosynthesis are the glucose (sugar) structures of the celluloses (Fig. 17.3). Some of these cellulosic sugars are transformed into starches, which are similar polymers of several thousand glucose units.

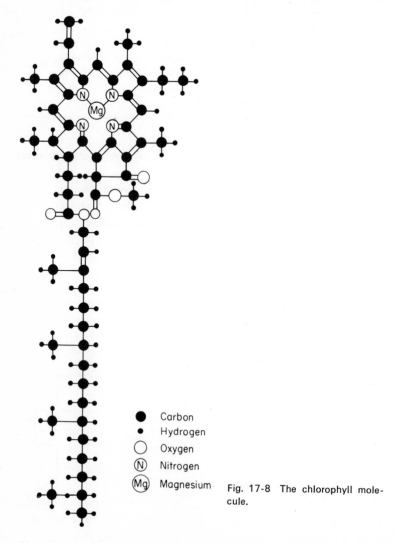

● Carbon
• Hydrogen
○ Oxygen
Ⓝ Nitrogen
㎎ Magnesium

Fig. 17-8 The chlorophyll molecule.

*Cellulose* is a polymer built from a sugar (glucose) monomer. Chemically it is called a saccharide. Cellulose is a constituent of hay and straw and of wheat bran. Other carbohydrate foods, which are various starches and sugars such as cereals, rice, corn, potatoes, and honey, are very similar chemically to cellulose. The glucose monomer of these materials is the form in which carbohydrates are transported in the bloodstream of man and animals. The carbohydrates are metabolized (burned or consumed) to carbon dioxide to supply energy to the organism. They do not supply material for tissue replacement or for growth. Rice is the staple food for more than half the world's population: it is 90 percent starch.

Saccharides, of which cellulose is the principal type, are the material of plants, associated with lignin as in wood. Radishes and parsnips as they age become woody. This is explained by the increase in lignin in these vegetables with age. Similarly, aged hay and forage crops contain more lignin than immature crops. Straw makes a very poor animal food compared with hay. Hay averages about 28 percent celluloses, hemicelluloses, and lignin, and straw 38 percent. The nondigestible lignin fraction shows the fastest increase, thus reducing the nutrient content of forage as the forage ages.

Cotton, flax, hemp, jute, and other plant fibers are celluloses. Animal fibers, including wool, alpaca, cashmere, and cattle, horse, camel, and human hair, are proteins. Wool differs from hair in being more elastic, less smooth, and smaller in diameter. Carpet wool however is coarse and resembles hair.

In contrast to plants, the solid matter in animal bodies is chiefly proteins and fats, neglecting bone. The human body analyzes

| | |
|---|---|
| water | 65% |
| proteins | 15 |
| fats | 15 |
| inorganic | 5 |
| carbohydrates | 1 |

Protein is the material of muscle or meat. The average heifer weighs in at about 1000 pounds at the stockyards; about 50 percent of this weight is meat.

Although carbohydrates provide energy, proteins are required in the food of man and animals for growth and tissue replacement and a large number of special requirements. In the absence of carbohydrate foods, proteins may also be burned in the body for energy.

There are thousands of kinds of proteins. Collagen, a protein of great importance, is a constituent of tendons, bones, and skin and provides a framework to the body of suitable mechanical properties. Hemaglobin is a protein in the red blood cells which transports oxygen for the combustion of glucose within the cells of the organism. Other proteins called enzymes are required in the digestion of food. Myosin is the protein of muscle, converting energy from carbohydrates into mechanical work of contractile movements.

Proteins are long copolymers of many amino acids. There are 22 of these amino acids, and any protein will contain several; carotene has 18 different amino acids in its chain. The amino acids are chemically related to nylon, which is a polyamide. A partial length of the simpler protein insulin is illustrated in Fig. 17.9. This figure does not show the whole molecule. Note that a protein chain is a —C—C—N— polymer with amino side

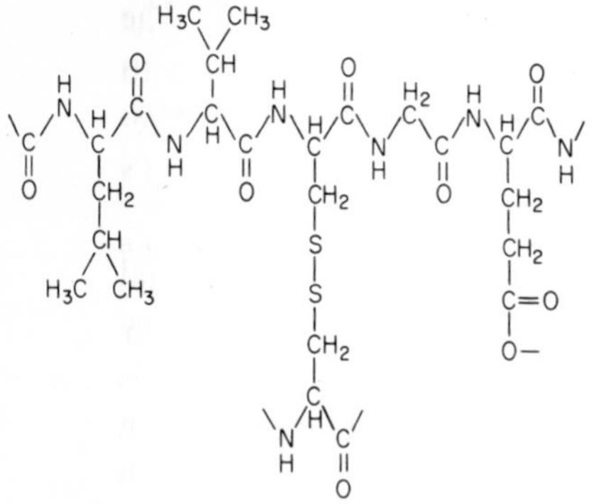

Fig. 17-9  A section of the polymer chain of the protein insulin. About one-quarter of the chain length is represented. The diversity of the amino side chains should be noted.

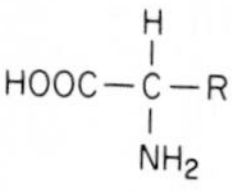

Fig. 17-10  The basic formula of the amino acids. $R$ indicates an attached group such as −H (glycine) or −$CH_3$ (alanine).

chains. The amino acids of the side chains have the formula of Fig. 17.10, where $R$ indicates different radicals. $R$ may be a hydrogen atom, or $CH_3$, or a more complex attachment.

Nitrogen is an indispensable element for the support of life, being needed to build proteins in animals and man. Cattle require up to 2.9 lb of protein feed daily in order to be finished for the market, and calves even more in proportion to their weight in order that they may grow. The nitrogenous feeds that supply protein to cattle are usually flax (linseed) meal, cottonseed meal, soybean meal, corn meal, wheat bran, clover, alfalfa, and others.

How does the organism know what kinds of proteins to build and the sequence of amino acids in each of its protein polymers? This is the same question as asking how a foal becomes a horse and how a calf becomes a cow. The information about the required proteins for the organism must somehow be coded somewhere in the organism, so that a cow will all its life remain a cow and will reproduce itself in calves and not horses. This necessary coding is contained in a protein of the organism's genes, deoxyribonucleic acid (DNA). The large DNA protein molecules are passed from one generation to the next during reproduction, carrying with them the essential protein codes to maintain the continuity of the species. A second type of large protein, ribonucleic acid (RNA), carries the coded instructions from the genes to the cells where the required proteins are assembled and polymerized. This reproductive property is the most remarkable of all material characteristics. It appears that in the future we shall be able to put this property to use. Such control in the use of materials is beyond our present concepts.

## 17.16 materials for implants

Orthopedics is the medical specialty concerned with muscle, skeleton, the joints, and their functions and movements. Stresses, movements, and joints are preoccupations of the engineering technologies, and it is not surprising that these technologies have made contributions toward the techniques of orthopedic repair of the human body.

There is considerable demand for metallic implants for the operative treatment of fractures and the replacement of diseased joints. For this purpose the following types of implants are standard:

1. wires for orthopedic fixing of fractures
2. plates and screws for the same purpose
3. pins for fixing
4. endoprostheses to replace worn bone in joints

In the mending of fractures by implants or in the replacement of joints, the implant must be strong enough to carry the stresses imposed on it, especially if the repair is to a leg. The determination of the forces carried by the bones of the leg is not a simple matter, since these forces vary greatly during the cycle of walking, nor are they collinear with the axis of the leg. In Fig. 17.11 the forces transmitted across the hip joint are shown in $X$, $Y$, and $Z$ coordinates. Since the body can adopt various configurations and the joint constantly changes shape, these components of force constantly change. Elaborate instrumentation and a computer analysis are required to analyze such stresses. The repair of a fracture by a plate screwed to the bone (Fig. 17.12) produces complex combinations of flexure, torsion, and direct stress in the plate. These stresses are quite high, and failures in such metallic implants caused by overstressing are not unknown.

Although materials are available that can withstand such stresses, the problem is one of fatigue stress. If the patient walks a mile, the maximum

Fig. 17-11 Resolution of forces at the hip joint into *X, Y, Z* components.

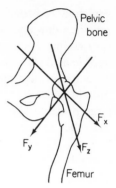

Fig. 17-12 Fracture of the femur repaired with a metal plate and screws.

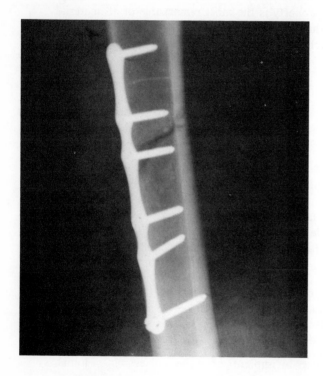

stress will be repeated 1000 times. Women and some men will walk several miles every day, hence the number of stress cycles can be millions in a year.

In addition to strength, the implant material will require a high modulus of elasticity. If the stresses are high, so will be the deflections. If the implanted pin of Fig. 17.13 in the head of the femur should bend more than a degree or so, then the break will not be able to heal. In addition, the leg will be shortened. It should be noted that the type of implant shown in Figs. 17.12 and 17.13 is normally removed after about six months.

The standard engineering concepts of stress and strain are the easier part of the problems occasioned by orthopedic implants.

In the case of a total hip prosthesis, there must be two mating parts to make the hip joint. An engineering bearing is made of two different materials, since two identical alloys will tend to seize. If this standard engineer-

Fig. 17-13  Implanted pin for repair of fracture of femur.

ing approach is used in the joints of the human body, the saline fluids of the
body will cause one of the two metals to be rapidly destroyed by galvanic
corrosion. The normal solution for a total hip prosthesis is a suitable metal
bearing against high-density polyethylene (Fig. 17.14).

Fig. 17-14  Total hip prosthesis, showing the total hip replacement of alivium alloy and the acetabular cup of high-density
polyethylene. The hip socket is called the acetabulum. (Photos
courtesy Zimmer Great Britain, London, England)

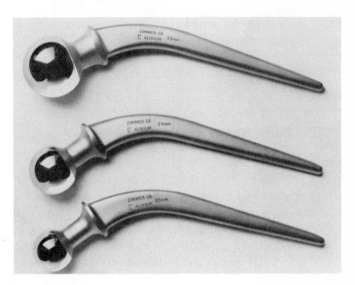

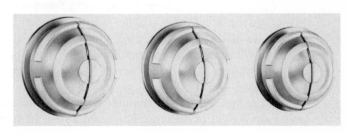

No material can be implanted which will corrode or which is toxic to the body. The material must be neither thrombogenic (causing blood clotting) nor carcinogenic (causing cancer), and it must not irritate surrounding tissues or interfere with the normal healing process. It must not shed material through wear. Early prosthetic veins, arteries, and heart valves failed because the glass and metal implants used caused blood clotting within minutes of installation. Clearly, the metallurgy of implants is no casual matter. Plastic implants are especially tricky; surface films, plasticizers, fillers, or traces of mold release agents may cause serious tissue reactions. Usually such implants must have a very smooth surface, though there are certain exceptions to this practice which will not be discussed here.

High-density polyethylene, polypropylene, silicone rubber, and PMMA bone cement have survived biological environments with success. The following metals are successful and are widely used:

1. chromium-molybdenum alloys of cobalt (Vitallium)
2. type 316L extra-low carbon stainless steel
3. titanium and certain titanium alloys

Titanium has the advantage of being lighter in weight than the other two alloys, but its $E$-value is only $17 \times 10^6$. Vitallium and titanium are preferred over 316L for long-term implants such as total joint prostheses. Lowest fatigue strength is given by Vitallium.

The success of such implants depends both on materials technology and on surgical techniques.

## 17.17   bone as an engineering

## material

Animal and human bone is an anisotropic material. In a standard stress-strain test, bone exhibits a linear range followed by plastic deformation. There is some variation in determinations of modulus of elasticity, which averages about 2.0 to $2.2 \times 10^6$ for human bone in the direction of the axis of the bone, and half as much transversely. Bovine and horse bone have a higher modulus.

Bone is a composite material, like a felted asphalt roof coating or a fiberglass-reinforced plastic. The mineral part of bone is hydroxyapatite, similar to the apatite family of minerals, with the formula $3Ca(PO_4)_2 \cdot Ca(OH)_2$. Running through the bone are strong fibers of protein collagen.

Hydroxyapatite is dense enough to hold the screws and nails of implant repair parts.

Like any other transportation device, bone must meet its stresses with a minimum weight of material. Bones are hollow in accord with the standard principle of machine design that less weight is required in carrying loads by using a tube instead of a bar. The core of the bone is not waste space: this core contains the bone marrow where blood constituents are generated.

Bone contains a mineral. Many minerals are piezoelectric, including bone. Piezoelectric signals from bone control its growth and changes of shape. If bone is bent, the region in compression is negatively charged and that in tension is positively charged. A compressive region in bone is built up with more bone material and in a region in tension bone material is removed. This process is illustrated in Fig. 17.15. The result is to reduce bending stresses to compressive stresses. Since growth of bone is governed by stress through electrical signals, an astronaut subject to prolonged periods of weightlessness can expect his bones to lose mass.

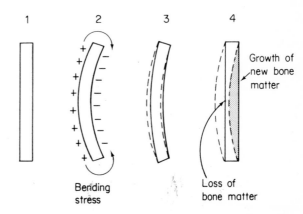

Fig. 17-15 Negative charge and bone growth. Negative charge on the concave side causes growth of new bone, and bone matter is removed from the convex side.

Such control of stress, shape, and mass would be useful in the static structures of civil engineering. An overloaded building frame receives intolerable deformation or collapses. If it functioned as bone does, it would develop material in the regions where it was required for load-bearing. As a stress method such control would be especially useful in beam-to-column connections of building frames or in the welded joints of tanks and pressure vessels.

## 17.18 artificial joints

The skeletal joints transmit stresses as high as 2000 psi, which is a respectable load for any mechanical bearing subject to movement.

For a joint to allow movement, there must be a lubricating film of liquid and a method of replenishing this lubricant as it flows out of the joint. Mechanical joints and bearings use oily liquids which are pumped through the joint by various methods. The method of physiological lubrication used in biological joints does not use inert oils but synovial fluid with a molecular weight of about $10^6$. The method of replenishing the lubricant in the joint is complex: it is a kind of weeping lubrication wherein the lubricant is squeezed from a spongy material as load is imposed on the joint. Friction coefficients for healthy non-arthritic human joints are extremely low; values as small as 0.005 have been measured. Here again biomaterials show levels of technical development unattainable in engineering.

In replacing the two mating surfaces of a movable joint, it has been re-marked that two different metals cannot be used in an animal body. The usual solution adopted for the two mating materials is shown in the total hip prosthesis of Fig. 17.14. The head of the femur is an implant metal alloy, and the acetabulum, the hollow in the hip, is given a lining of high-density polyethylene cemented in place with acrylic cement. By and large, such total hip prostheses are remarkably successful and many thousands have been installed. There are of course always some failures. Prosthetic repair of other joints of the human body is perhaps not quite so successful in restoring joint movements.

## 17.19 other internal organs

The cardiac pacemaker, an implanted heart stimulator, has probably been installed in 100,000 people by this time. This is an electronic device, as are all other stimulator implants such as those for blood pressure, bladder, and muscle functions.

Artificial heart valves are basically silicone rubber ball valves in cylin-drical metal frames or cages. The ball closes against its seat on the cage.

The artificial heart is no doubt an exciting topic for discussion and speculation, but sober assessments of developments toward such a device do not evoke optimism. An artificial heart located outside the human body is of course already in use, though only for emergency during operative pro-cedures on the heart.

The preceding sections have been a brief review of the contributions of materials technology to the support of orthopedic and physiological function of the human and animal body. So far there has been little or no technical feedback from biomaterials into engineering science. But there is much engineering to be learned from plant and animal life. Comparisons suggest that the engineering practices we employ in inert materials are crude and narrow compared with the engineering of living bodies. So far we have chosen to ignore the biomaterials as a separate, unrelated, and irrelevant area of knowledge.

Consider an example. The living wood of trees responds to stress as bone does, developing types of tissue called compression wood and tension wood in accord with applied stress. Now this is an ideal behavior for a timber structure carrying loads. Does it make sense to lose this valuable structural characteristic before building the wood into a structure?

## PROBLEMS

1   Account for the anisotropy of (a) wood, (b) paper, and (c) bone.

2   In making stress-strain tests on bone, would you expect to find differences between live and cadaveric bone? Why?

3   The wires of a cardiac pacemaker will flex with each heartbeat and therefore must be designed for fatigue. How many flexes per year will such a wire receive?

4   Why would you select high-density instead of low-density polyethylene for implants? Refer to the remarks on crystallinity in plastics.

5   What materials might be suitable for the diaphragm of a diaphragm type of artificial heart valve? What properties do you think are necessary?

6   What materials might be suitable for an artificial artery? It must be pliable. Why would you not use a plasticized PVC?

7   Suggest suitable plastics for insulating the wires of implanted pacemakers.

8   To what basic use does an organism put carbohydrates and proteins?

9   What properties of wood make it an advantageous building material?

10   What properties are disadvantages in wood?

11   Distinguish (a) springwood from summerwood, (b) heartwood from sapwood, (c) hardwood from softwood.

12   What is kraft paper?

13   Using the specific heat formula of Sec. 18.12, calculate the amount of heat that must be removed from a 500-lb beef carcass to cool it from 75 to 35°F.

14   Outline the photosynthesis process in plants.

15   What is the chemical difference between animal and vegetable fiber?

16   Explain the effect of piezoelectricity in bone.

17   What are the two chief constituents in bone?

18   Define carcinogenic material and thrombogenic material.

**19** What are the advantages of plywood over sawed lumber?

**20** Explain the meaning of the specific gravity of wood.

**21** To see what can happen when the wrong material is used for an implant, try the following investigation. It closely reproduces one effect of body fluids on sensitive metals. Heat a small piece of stainless 302 sheet in a furnace at 1200°F for two hours. Then immerse the piece for a few days in a solution of salt and water (blood is saline). Examine the edges of the piece under a low-power microscope. Can you explain what happened?

# SEMIFINISHED
# PRODUCTS

*Part* **5**

The last group of material characteristics we shall discuss are the economic ones. Although they are placed last for reasons of organization, they are ultimately the most important ones. Indeed, all material characteristics are economic characteristics. The excellent specific resistance of pure copper has only a limited economic value; if copper were to double in price, probably most electrical conductors would be made of aluminum.

It is interesting to note the high value at which the market rates some properties and the low value applied to others. Relatively little economic value is attached to a high modulus of elasticity, for example. The high *E*-value of tungsten probably has a trivial economic significance to the broad field of engineering generally; the low *E*-value of the thermoplastics is accepted. A far higher value is given, for example, to surface appearance. Stainless steel costs about ten times as much as painted steel, yet in many products where a painted steel might be acceptable, a stainless steel is substituted, as in the mail receptable of Fig. 18.1. The attractive appearance of the plastics

# ECONOMIC
# CHARACTERISTICS
# OF MATERIAL

# 18

Fig. 18-1 Mail receptacle made of stainless steel.

vastly outweighs their poor dimensional stability. The predominant market importance of appearance is best illustrated by the use of whitewall tires on cars; such tires contribute no useful function whatever to the automobile and have only two characteristics to recommend them: they are more expensive and people buy them. The latter characteristic of market acceptance cannot be safely ignored merely because it satisfies no scientific or engineering criteria. Even scientists who feel that marketing considerations are no part of their lofty calling have bought whitewall tires.

Of more immediate economic importance are such processing characteristics as machinability and weldability. Such characteristics, when properly employed, have the effect of reducing processing time, labor and overhead costs. There remain to be considered a few other economic characteristics.

## 18.2 specific weight

Specific weight means the weight of a material in pounds per cubic foot. Low specific weight is becoming an increasingly desirable characteristic, as illustrated by the following examples: the use of aluminum in transport vehicles and ship superstructures, magnesium in material-handling equipment, beryllium in rocket components, and the substitution of plastics for metals in small machine and equipment housings. The advantages of low specific weight are numerous, including low dead weight, low freight costs, lighter crating, easier material-handling, lighter supporting structures.

A number of interesting analyses have been made of the cost of extra weight in transport vehicles. Mathematical analysis suggests that the financial saving resulting from the removal of each pound of dead weight from an aircraft with a service life of 10 years is $275. This of course assumes that the one-pound weight saving becomes 1 lb of revenue freight carried under average conditions. For highway trucks, calculations indicate a saving of $5.50 per lb. The dramatic case for such weight saving lies in aerospace rockets. On the average it costs about $20,000 to put 1 lb in orbit around the earth.

Miniaturization of components, as extensively practised in the electronics industry, exploits volume rather than weight. Transistors, silicon-controlled rectifiers, and printed circuitry have contributed notably to miniaturization. The industry now talks of microminiaturization of miniaturized components. To a lesser degree, other industries also miniaturize; high strength, low alloy steels are used in the construction industry to reduce structural sections in the pursuit of smaller weight and height of buildings. In the food industry, dehydration of foods performs the same function as electronic miniaturization.

### 18.3 various economic characteristics

*Corrosion resistance* is certainly an economic factor for materials. This characteristic determines both the life and cost of the material and its maintenance costs. The final cost of a steel structure is increased because of the necessity for surface protection by paint, and allowance for regular repainting of such a structure must be made in annual budgets. Perhaps maintenance painting costs account for the use of stainless steel in the receptacle of Fig. 18.1.

*Shelf life* refers to the longest safe storage period for perishable materials. For foods, this may be only a few days. Photographic film is always supplied with a terminal date for development printed on the carton. Radioisotopes lose their activity by 50 percent every half-life.

*Cost and availability* are frequently related. Water is cheap because of its availability. So is iron ore or clay. But in the case of manufactured articles, it is volume of sales which largely determines cost. Articles made one at a time, such as replacement parts made in a maintenance shop, are very expensive owing to the high labor content. But if the techniques of mass production can be applied to large production runs, the manufacturing cost often can be made astonishingly low. Working against this low production cost is the high marketing cost and advertising cost associated with volume production, for high-volume production with low cost requires a high-volume market to absorb the products.

## 18.4  fire and explosion hazard

Fire hazard is associated with a great many materials of industry: magnesium and zirconium dust, solvents, paints, fiberglass plastics, fuels, oxygen tanks, paper, molten metals, and others. Fire protection methods often add onerously to the cost of storing, handling, and processing such materials. Explosion-proof electrical systems may be required, which are expensive. Paint booths are expensive; so are large ventilating fans, as are safety-training and enforcement of safety regulations, proper receptacles, fire equipment, fire insurance, accidents, workmen's compensation assessments, and lawsuits. The rapid development of materials technology continually opens up more possibilities of explosion and fire hazard if proper precautions are not foreseen.

## 18.5  toxicity

A vast number of materials have toxic potential. An even wider range of materials are toxic to certain individuals or produce allergies in individuals. Even seemingly innocuous materials such as firebrick mortar or machining oils may produce isolated cases of dermatitis.

The range of potentially toxic materials is surprisingly wide. Heated polyvinyl chloride may produce acid fumes. Freons may produce phosgene gas under adverse circumstances. Welders become temporarily ill if they breathe the zinc fumes when welding galvanized metal. Cadmium fumes are even more dangerous, especially since cadmium plating may be mistaken for zinc. Even copper and lead compounds present a possible hazard.

Perhaps the toxic hazards of beryllium and radioisotopes are best known. However, these materials are so dangerous that personnel will make a serious effort to observe safety regulations. In this respect less toxic materials may be more dangerous, for personnel may take chances with them. It should be no surprise that the safest of all industries (statistically) is the radiation industry: about 100,000 times more people are killed by automobiles than by radiation. Perhaps the only dangerous materials are those regarded as safe.

This chapter deals with the various shapes in which material is supplied to the fabricator for conversion into finished products. Such shapes include sheet for forming, cold-drawn bars for machining, plate for welding, and tubing or hollow bar.

## 19.1  *sheet and strip*

The most useful of all shapes for conversion into products is a sheet or a strip. Sheet too is one of the cheapest shapes to buy since it can be produced continuously on high-speed rolling mills. A sheet gives almost unlimited freedom of shape in forming operations, together with least weight, and can often be formed on simple and inexpensive equipment. Because of these advantages, a greater tonnage of metal is rolled into sheet and strip than into any other shape. Consider the variety of commodities that are formed from sheet:

| | |
|---|---|
| curtain wall panels | hinges |
| table tops | camping trailers |

# SEMIFINISHED
# PRODUCTS

# 19

| | |
|---|---|
| roof decking | trowels |
| automobile bodies | band saw blades |
| automobile frames | nuclear fuel rods |
| automobile wheels | aircraft windows |
| file cabinets | ashtrays |
| kitchen sinks | domestic can openers |
| ventilating ducts | flashlight shells |
| fan blades | electric switch boxes |
| aircraft | bottle caps |
| belt conveyors | steel strapping |
| cigarettes | |

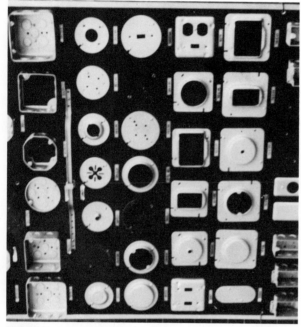

Fig. 19-1   Display of electrical hardware formed from sheet metal.

Plain carbon steel sheet containing up to 0.30 percent carbon may be formed by drawing. Both hot- and cold-rolled sheet is used for drawing and forming, though hot-rolled sheet, having somewhat better ductility, is somewhat more formable. Hot-rolled sheet is not available in the thinnest gauges. Both rimmed and killed steels are used for sheet. Killed steels have superior drawing quality, but they are expected to contain more surface defects than a rimmed steel.

Cold-rolled sheet and strip are available in several "tempers":

No. 1 or Hard Temper—very stiff, hard, and springy. This temper is used for flat work or slight deformation. $R_B$ hardness may be 90 or more.

No. 2 or Half-hard Temper—this sheet may be bent 90 degrees across the rolling direction to a radius of its own thickness. $R_B = 70$ to $85$.

No. 3 or Quarter-hard Temper—may be bent 180 degrees across the rolling direction to a radius of its own thickness. $R_B = 60$ to $75$.

No. 4 or Skin-rolled Temper—a soft and ductile sheet capable of deep drawing. It can be bent flat on itself. $R_B = 65$ max.

No. 5 or Dead-soft Temper—this temper is used for the most difficult drawing operations. $R_B = 55$ or less.

Steel decking is made by corrugating steel sheets in 24-in. wide sections which interlock. These are used for roofing and subflooring of buildings.

Expanded metal mesh (Fig. 19.2) has a wide range of uses in tote boxes, nonskid walkways and gratings, storage cabinets, partitions, ventilating screens, lockers, animal cages, etc. The usual method of designating such mesh is with two numbers, sometimes three. The first number is the size of the mesh, that is, the center-to-center distance of the diamond-shaped openings, measured the short way of the diamond. The second number designates the gauge of the metal. Thus $\frac{3}{16}$-22 is a light-weight mesh with small openings; $1\frac{1}{2}$-12 is a very heavy mesh with $1\frac{1}{2} \times 3$-in. diamond openings.

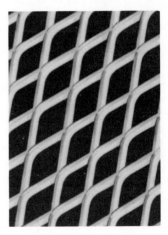

Fig. 19-2   Expanded metal mesh.

Stainless steel sheet is available in the several 300 and 400 alloys. Several finishes of stainless sheet are available, designated No. 1 to No. 8. Two are more commonly used than the others:

No. 2B—cold-rolled, annealed, pickled in acid, then given a skin pass on polished rolls to produce a smooth, mildly reflective finish.

No. 4—belt-ground with 150–180 grit abrasive to produce a polished finish.

The subject of gauge thicknesses is a matter of some confusion, since steel sheet, nonferrous sheet, and wire may use the same gauge numbers but have varying dimensions. Steel sheet uses the United States Standard

Gauge (U.S.S.) while nonferrous sheets such as aluminum use the American or Brown and Sharpe gauge system (B & S).

| Gauge No. | U.S.S. oz/sq ft | U.S.S. Thickness | B & S Thickness |
|---|---|---|---|
| 10 | 5.625 | 0.1345 | 0.102 |
| 11 | 5.000 | 0.1195 | 0.091 |
| 12 | 4.375 | 0.1046 | 0.081 |
| 14 | 3.125 | 0.0747 | 0.064 |
| 16 | 2.500 | 0.0598 | 0.051 |
| 18 | 2.000 | 0.0478 | 0.040 |
| 20 | 1.500 | 0.0359 | 0.032 |
| 22 | 1.250 | 0.0299 | 0.025 |
| 24 | 1.000 | 0.0239 | 0.020 |
| 26 | 0.750 | 0.0179 | 0.016 |

The weights in ounces per square foot in this table are those for mild steel sheet. It will noted tnat the U.S.S. system is based on weight of mild steel, not on thickness. The B & S system is based on thickness. Nonferrous sheets of any given gauge will be thinner than mild steel sheet of the same nominal gauge.

## 19.2 strapping

Nowadays materials are handled either in bulk or in unit loads. A *unit load* is a rigid arrangement of a number of items which can be handled as a single object by mechanized material-handling equipment such as lift trucks. For the unitizing of loads, strapping is often required (Fig. 19.3). For these and similar purposes various types of strapping are available in both steel and plastic. Some of the more important strapplications are these:

1. reinforcement of boxes and crates
2. closure of containers
3. baling of cloth, paper, and leather
4. securing of loads to wood pallets
5. bundling of rod, wire, and plastic pipe
6. securing of loads in railroad cars
7. tiering and stacking

The use of strapping requires a stretcher to tension the strap and a sealer to make the joint in the strap with a strap seal.

Steel strap is available in coils, in widths from $\frac{1}{4}$ to $\frac{3}{4}$ in. or wider, in a black finish. Standard gauge thicknesses are not used with material-handling

Fig. 19-3 A unit load of tires being compressed and strapped. (Machine and photo by Signode Corporation)

strap, strap thicknesses being 0.010, 0.012, 0.015, up to 0.028 in. Such strap has a tensile strength of 100,000 psi and a controlled ductility. If the strap is not sufficiently ductile, it will snap when being pulled or bent around a sharp corner. Standard steel strap has an elongation of 1 percent.

Steel strap is not suitable for unit loads which compress, such as bagged bulk material. Such material will compact under the jolting received in transportation, and a ductility of 1 percent is insufficient to accommodate this shrinkage. For such service nylon or polypropylene strapping is superior. Plastic strap is $\frac{1}{2}$ in. wide, with an elongation of about 10 percent. If the unit load should shrink 5 percent, nylon strap would still confine the load, though not with the original tension set up in the strap. Plastic strap also does not present the housekeeping problem that steel strap does, since it can be incinerated.

### 19.3  plate

How thick must a sheet become to be called a "plate"? The answer depends on a number of factors, including width of material, but roughly speaking, metal sheet which is $\frac{3}{16}$ in. or thicker over 48 in. wide, or $\frac{1}{4}$ in. thick over 6 in.

wide, is called "plate." Plate thickness may be specified either in fractions or in decimals. At the other extreme, sheet 0.005 in. or thinner is termed "foil."

For purposes of quick estimate, a convenient figure is the following. 1 sq ft of mild steel plate $\frac{1}{4}$ in. thick weighs 10.2 lb. From this information the weight of steel plate of any thickness or area can be quickly determined. The weight of aluminum plate can also be estimated from this figure, since aluminum is one-third the specific weight of steel.

Since about 1900 the standard specification for steel plate and rolled structural shapes has been ASTM Standard A7. In this specification the chemical composition of the steel was not given, except to restrict sulfur and phosphorus to 0.06 percent maximum. Carbon content was assumed to be 0.20 percent, but varied, since the steel mill adjusted the carbon content to obtain the required strength. The manganese content was similarly adjusted for ductility and ease of rolling, but it ranged from 0.3 to 0.6 percent. As a result of these variations, the Charpy 15 ft-lb transition temperature varied from $-10°$ to perhaps $+100°F$. Minimum yield strength was 33,000 psi, ultimate strength 60,000 psi or higher, and elongation in 2 in. was 24 percent. Allowable stress for structural work in this steel was 20,000 psi.

Industry has abandoned A7 steel because of brittle fracture hazard, some difficulties in welding, and its poor strength-to-weight ratio. A7 steel is now superseded by ASTM A36.

A36 steel has the following chemistry: 0.28 percent carbon maximum, 0.05 percent sulfur maximum, 0.04 percent phosphorus maximum. Manganese content is increased in heavier sections to a maximum of 1.20 percent. (While carbon promotes the brittle fracture tendency, manganese improves brittle fracture performance.) Mechanical properties of A36 steel are:

yield strength 36,000 psi minimum
ultimate tensile strength 60,000–80,000 psi
elongation in 2 in., 23%

A36 steel costs about 3 percent more than A7 steel, but it offers a yield strength and an allowable strength (22,000 psi) 10 percent greater.

## 19.4  structurals

Structural shapes are rolled in many standard sizes. In addition, certain mills roll special sizes. The usual shapes are wide flange (W) beams, channels, angles, tees, zeds, and bulb angles. These shapes are produced by rolling, if made of steel, and by extrusion if made of nonferrous metals such as aluminum. The extrusion process permits much greater latitude in the shapes that can be produced.

The usual method of designating a steel structural shape is by two numbers designating the principal dimension (depth) and the weight per lineal foot. Thus 16 W 78 indicates a W shape approximately 16 in. deep (actually $16\frac{3}{8}$ in.) and 78 lb per lineal foot. It is usual to specify angles by the length of the two legs and the thickness: thus $2 \times 2 \times \frac{1}{8}$, $6 \times 4 \times \frac{1}{2}$, etc.

The largest of the steel shapes is a W beam $36 \times 300$ lb. Larger beams are made by welding plate together into a wide flange section. The largest of the extruded aluminum shapes is a 12-in. I beam 0.437 in. thick weighing 15.44 lb per ft.

## 19.5 wire, rod, and bar

These three products are similar. In general, wire is considered to be any solid section with no cross-sectional dimension exceeding $\frac{1}{4}$ in. The term "rod" means a circular cross-section. A bar has a noncircular cross-section, such as hexagonal or square. Tube is hollow rod.

For concrete work, reinforcing bar (rebar) sizes are designated by a number which is the number of eighths of an inch in the nominal diameter. Thus a #5 bar has a diameter of $\frac{5}{8}$ in. The range of sizes is from 3 to 11. Either plain rod bars or deformed bars may be used, the latter having two longitudinal raised beads and also circumferential beads. Three strength grades are used: 55,000, 70,000, and 80,000 psi ultimate strength. The carbon content controls the strength. Structural grade rebar contains about 0.20 percent carbon, intermediate grade 0.30–0.45 percent, and hard grade 0.45–0.60 percent carbon.

Since rebar must frequently be bent to follow the stress contours of reinforced concrete, ductility is a matter of concern. Required elongation is of course much less for the hard grade of bar than for the other two grades.

While 20 ft is the standard length for most bars, pipe, and tubing, bars for machine shop work are furnished in 12-ft lengths. This shorter length is more convenient for feeding through the spindle of an automatic or turret lathe. Machining bars are usually cold-finished to rather close tolerances, the cold-finishing operation making the surface of the bar slightly harder and easier to machine. Stainless steel machining bar is commonly furnished in the HRAP condition, that is, hot-rolled, annealed, and pickled.

## 19.6 welding wire

Wire for welding filler metal is supplied either in coils for automatic welding or in short straight lengths for manual welding. Rods for the manual arc welding of mild steel have a mild steel composition of approximately 0.1

percent carbon. They are available with different flux coatings to provide different welding characteristics. Thus E6010 and E6011 rods have a flux coating that includes cellulose. The hydrogen in this coating provides an arc which penetrates deeply into the weld area. Hydrogen, being an interstitial element in steels, promotes brittleness. If this embrittling effect is to be avoided, low hydrogen electrodes are used, usually E7018 for mild steels and low alloy structural steels. Low hydrogen electrodes contain a lime or a rutile flux coating. Welding rods for the welding of stainless steels are always of the low hydrogen type.

The American Welding Society and the American Society for Testing and Materials use a four-digit number system for classifying manual arc welding rods. Taking E6011 rods for an example, the letter E signifies a rod for electric arc welding. The first two digits, 60xx, indicate the minimum ultimate tensile strength of the welding wire in thousands of psi, or 60,000 psi minimum. The third digit, xxlx, indicates the welding positions for which the rod is suited. A "1" indicates a fast-freezing rod that may be used in all positions, flat, horizontal, vertical, or overhead. A "2" indicates a rod suited only to flat and horizontal welding. The last digit does not indicate any single item of information and will not be discussed here because of its complexity.

Hence an E6011 rod will provide 60,000 psi and may be used in any welding position. A 7018 rod has a minimum tensile strength of 70,000 psi and may be used in all positions. A 6024 rod is used for flat or horizontal welding seams, such as horizontal fillet welds.

For production welding, spooled wire is automatically fed into the arc. Wire diameters of 0.030 to 0.062 in. are the usual sizes for factory welding operations, although heavier wire may be used where heavy deposition rates are required on thick plate.

## 19.7  forged rings

Probably the most reliable type of machine part for conditions of high stress is a forging. In a forging, the direction of the grain can be controlled during the forging operation to suit the stress condition, i.e., the anisotropy can be suited to the stress direction. Further, during the forging operation the material can be homogenized and strengthened by plastic working. Compared to a forging, a casting is not plastically worked, and the orientation of its grains is determined by the direction of heat flow in the material as it cools. Again, a machined part may be weak if the machined grooves cut across the direction of rolling of the raw material. Such heavily stressed articles as crane hooks, socket wrenches, and crescent wrenches are invariably forged.

Forged seamless rings are used as frames, stiffeners, and other machine parts in a wide range of equipment. For such rings the direction of grain flow must be specified. Three such directions are possible: *circumferential*, *radial*, and *axial*. See Fig. 19.4. A stiffener ring for a tank, rocket, or submarine would require a circumferential grain to resist hoop stresses. Rings with this grain direction are forged or compressed while being rotated on a supporting mandrel. On the other hand, if the ring is to be machined into a large-toothed gear ring, the grain direction should be parallel to the axis of symmetry of the gear tooth, that is, radial. Such a ring is usually termed a bored biscuit. Finally, the axial grain direction is provided by a bored and sliced multifaced round.

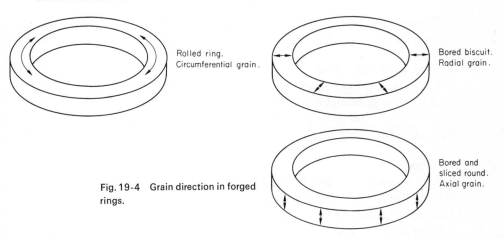

Rolled ring.
Circumferential grain.

Bored biscuit.
Radial grain.

Bored and
sliced round.
Axial grain.

Fig. 19-4   Grain direction in forged rings.

## 19.8   pipe and tube

It is not easy to differentiate between a pipe and a tube, particularly when both have the same cross-sectional dimensions. The major differences between the two may be rapidly summarized as follows:

**1. Raw Material.**   Tubing is formed from coiled strip. Pipe is manufactured from coiled skelp, which is plate rolled to less rigorous specifications than tubing strip.

**2. Tolerances on Dimensions.**   Most pipe specifications allow a variation in wall thickness of $\pm 12.5$ percent. Most tubing specifications allow a maximum variation of $\pm 10$ percent in wall thickness, and even closer tolerances may be called for.

**3. Dimensions.**   In general, pipe has only one critical dimension: the outside dimension. The outside dimension does not correspond to the nominal diameter, and neither does the inside dimension. All pipe of the same nominal diameter, despite any variations in wall thickness, has the same outside diameter; this allows the use of standard pipe threading equipment, threaded pipe fittings, and standard threaded and welded flanges. Thus all 4-in. pipe measures 4.500 in. in outside diameter. Its inside diameter may be more or less than 4 in.

Tubing on the other hand has two critical dimensions. There are actually three dimensions to a tube: O.D., I.D., and wall thickness, but only two are specified, either O.D. and wall thickness or I.D. and wall thickness. The tolerance error falls on the third and unspecified dimension.

**4. Finish.**  Mill scale is not usually removed from pipe. Tubing is usually furnished in an oiled, bright finish.

**5. Method of Manufacture.**  It is hardly possible to differentiate pipe from tubing by the method of manufacture. Most pipe, but not all pipe, is made by hot-forming, using pressure-welding methods for the longitudinal joint. Most tubing is cold-formed and welded by electric welding methods. But both pipe and tubing may be produced also by seamless methods.

Pipe is available in a wide range of materials and in several wall thicknesses. The most commonly used wall thicknesses are Schedule 40 and Schedule 80. Higher schedule numbers indicate heavier wall thickness. Schedule 40 is standard wall; Schedule 80 is about 50 percent thicker. Nominal pipe sizes range from $\frac{1}{8}$ in. to indefinitely large diameters.

### SCHEDULE 40 AND 80 PIPE

| Schedule 40 Nominal Size | O.D. | I.D. | Wall Thickness | Metal Area, sq in. | Schedule 80 I.D. | Wall Thickness |
|---|---|---|---|---|---|---|
| ½ | 0.840 | 0.622 | 0.109 | 0.25 | 0.546 | 0.147 |
| ¾ | 1.050 | 0.824 | 0.113 | 0.33 | 0.742 | 0.154 |
| 1 | 1.315 | 1.049 | 0.133 | 0.49 | 0.957 | 0.179 |
| 1¼ | 1.660 | 1.380 | 0.140 | 0.67 | 1.28 | 0.191 |
| 1½ | 1.900 | 1.610 | 0.145 | 0.80 | 1.50 | 0.200 |
| 2 | 2.375 | 2.067 | 0.154 | 1.075 | 1.94 | 0.22 |
| 3 | 3.500 | 3.068 | 0.216 | 2.23 | 2.32 | 0.276 |
| 4 | 4.500 | 4.026 | 0.237 | 3.17 | 3.83 | 0.337 |
| 6 | 6.625 | 6.065 | 0.280 | 5.6 | 5.76 | 0.43 |

Tubing is made in two principal types: mechanical and pressure tubing. Pressure tubing is used for such purposes as hydraulic cylinders. Mechanical tubing is general-purpose tubing. Both are produced to wall thicknesses ranging from 22-gauge to about 2 in. The weight of carbon steel tubing and

Fig. 19-5   Square tubing used as the frame of a sound control booth for Expo '67, Montreal.

pipe can be determined from the formula

$$W = 10.68(D - t)t \text{ pounds per lineal foot}$$

where $D$ = O.D. in inches

$t$  = wall thickness in inches

The versatility of tubing as a semifinished product is suggested by the following list of uses:

| | |
|---|---|
| bar stock for machining | grease guns |
| roller bearing races | ink reservoirs for ball |
| boiler tubing | point pens |
| conveyor rollers | soldering irons |
| warehouse racking | chrome furniture |
| hydraulic cylinders | hypodermic needles |
| bicycle frames | wrist bars to hold straps |
| drive shafts | to wrist watches |

The uses of tubing in automobiles include exhaust piping, steering column, gas tank filler tube, rear axle housings, tubular front axles, seat frames, drive shaft. This list is not complete.

Mild steel tubing is furnished in low carbon grades 1015, 1022, and 1025. Medium-carbon tubing is also available in 1030, 1035, 1040, and 1045 grades. Low alloy grades are used for higher stresses, such as 4140 for aircraft landing gear and engine mounts. The common grades of stainless tubing are used where good corrosion resistance, cleanliness, or appearance of stainless steel is desired. Nonferrous tubing is available in most metals, including the refractory metals tungsten, molybdenum, and tantalum. Finally, tubing is produced in square, hexagonal, oval, and other shapes.

When tubing is used as bar stock for machining operations, either seamless cold-finished or free-machining tubing is preferred. If the tubing is to be chucked to O.D., then O.D. and wall thickness are specified; if chucked to I.D., then I.D. must be specified. The required tube dimensions for machining stock are governed by the machining allowance and the allowances for eccentricity and other tolerances of the tube stock.

## 19.9   fasteners

Welding, brazing, soldering, and adhesive bonding produce permanent joints between components. Although some types of fasteners such as rivets are intended to produce a permanent joint (just as pressure-sensitive adhesives do not produce a permanent bond), where disassembly is to be expected, fastening devices will be used.

A *fastener* is defined in ASA B18.12-1962 (American Standards Association) as "a mechanical device for holding two or more bodies in definite positions with respect to each other." The number of types of fasteners is of course very large, mainly because many of the types are either developed for special purposes (such as aicraft bolts or track bolts) or are proprietary in their design. Most fasteners are bolts, nuts, washers, lock fasteners, staples, studs, dowels and pins, screws, or rivets.

Details on bolts, bolt heads, nuts, and washers are available from standard hardware catalogues. Special industries use special bolts. *Cap screws* are quality bolts machined all over to closely controlled dimensions. They are used in those applications where special care is required in making the joint. For example, tooling, jigs, and fixtures are made with great care and attention to detail; socket-head cap screws are ordinarily used in this work. In structural bolting, high strength bolts are employed. These develop very high clamping pressures so that any racking of the joint will be resisted by high friction forces between the mating members. Expansion bolts are used to anchor to concrete. One type contains a lead tube: as a lag screw is screwed in, it forces the lead to deform against the hole drilled in the concrete.

Fig. 19-6 An air-conditioning unit during erection. This unit uses both welded and bolted joints for its assembly.

Self-locking fasteners are shake-proof, that is, they will not release or relax in service. These use interference fits or other methods of self-locking. One simple but ingenious method is to use a two-component epoxy, one component placed on the bolt thread and the other on the nut threads.

A screw may be almost any kind of externally threaded fastener, other than a bolt. Machine screws and cap screws are used in the same manner and functions as bolts. A set screw is threaded into one part and holds the other mating part by pressure of the end of the set screw against it. A tapping screw is intended to cut its own mating thread in the part or parts to be joined. Self-tapping screws are commonly used for joining sheet metal.

A rivet is a headed fastener on which an opposite head is formed after the rivet is inserted in the aligned holes of the mating parts.

## PROBLEMS

1   For what reasons is sheet metal a preferred semifinished product for conversion into manufactured articles?

2   Why is 16-gauge aluminum sheet not the same thickness as 16-gauge steel sheet?

3  Estimate the weight of aluminum plate in a vertical cylindrical tank 30 ft in diameter and 27 ft high, using $\frac{1}{4}$-in. plate. Include bottom plates, roof plates, and wall plates.

4  Why is a forging superior to a casting or a machined part for carrying very high stresses?

5  Why do pipe and tubing of the same nominal size have different outside diameters?

6  Determine the weight of a lineal foot of 2-in. steel tubing with a 16-gauge wall.

7  Assuming that aluminum is one-third the specific weight of steel, determine the weight per lineal foot of a 3-in. O.D. aluminum tube with a $\frac{1}{4}$-in. wall.

**INDEX**

*Note:* All metals, elements, and plastics are set in **boldface** type.

Nucleus, 32
**Nylon,** 186, 349

**O** Octane number, 320
Olefins, 306, 307
Open-hearth furnace, 205
Organics, 8, 193, 300
Oriented polymers, 343
Orthopedics, 429
**Oxygen,** 6, 7, 17, 18, 92, 175, 205, 208, 209, 214, 259

**P** Paint, 395
  aluminum, 402
  fire-retardant, 401
Pair production, 22
**Palladium,** 288
Paraffins, 306
Paramagnetism, 112
Particle size, 173
Pearlite, 236, 241
Peel strength, 406
Peltier effect, 118
Periodic table, 18
Perlite, 132, 168
Permeability (magnetic), 112
Petrochemicals, 301, 302, 332
Petroleum, 301
Petroleum coke, 151, 258, 312
Phases, 47
**Phenolics,** 185, 336, 356
  **adhesives,** 404
Phosphate coatings, 388
Phosphors, 30
**Phosphorus,** 195, 202, 204, 215
Photoconductive effect, 117
Photoelectric effect, 28
Photon, 13, 16, 24
Photosynthesis, 425
Piezoelectricity, 115
Pigment, of paint, 397
Pipe, 451
Plasticity, 62
Plasticizers, 341, 342
Plastics, 9
Plate, 447

**Platinum,** 288
**Plutonium,** 273, 327
Poise, 80
Poisson's ratio, 63, 271
**Polyacetal,** 350
**Polycarbonate,** 350
**Polyesters,** 357
**Polyethylene,** 48, 182, 334, 340, 341, 342, 346
Polymers, 305, 336, 339
**Polymethylene,** 334
**Polymethyl methacrylate,** 350
**Polypropylene,** 347
**Polystyrene,** 348
  **foam,** 376
**Polyurethane enamel,** 401
  **foam,** 67, 375, 377
  **rubber,** 194, 364
**Polyvinyl chloride,** 335, 338, 347
  **foam,** 375
**Polyvinyl fluoride,** 336, 348
**Polyvinylidene chloride,** 348
Porcelain enamels, 151
Positron, 22
Pozzolans, 164
Precipitation hardening, 248
Printed circuit, 110
Propane, 99
Proportional limit, 65
Protein, 427
Proton, 13
Pyrometric cone, 126

**R** Radiation hazard, 39
**Radium,** 38
Rankine temperature, 90
Rare earths, 289
Recrystallization, 235
Red lead, 402
Reflection, of radiation, 42
Refractories, 126, 141
Refractory metals, 290
Refrigerants, 328
Reinforcing bar, 449
Relativity theory, 21
Resin (definition), 334
Rest mass, 21, 22
**Rhenium,** 293